SACRED AGRICULTURE

Also from Dennis Klocek and Lindisfarne Books

Climate

Soul of the Earth

The Seer's Handbook

A Guide to Higher Perception

SACRED AGRICULTURE

The Alchemy of Biodynamics

Dennis Klocek

Lindisfarne Books | 2013

2013
Lindisfarne Books
An imprint of Anthroposophic Press / SteinerBooks
610 Main Street, Great Barrington, MA
www.steinerbooks.org

Cover and book design by William Jens Jensen

Printed in the United States of America

Library of Congress Cataloging-in-Publication Data
Klocek, Dennis.
Sacred agriculture : the alchemy of biodynamics / Dennis Klocek.
p. cm.
ISBN 978-1-58420-141-0 (pbk.) — ISBN 978-1-58420-142-7 (ebook)
1. Agriculture—Philosophy. 2. Organic farming—Philosophy. I. Title.
II. Title: Alchemy of biodynamics.
S494.5.P485K56 2013
631.5'84—dc23

2012051422

CONTENTS

CHAPTER 1

INTRODUCTION

A Historical Survey of Agricultural Focus, from Prehistory to Contemporary Times

People sometimes ask me what I think the difference between biodynamics and regular organic method is. Sometimes I answer, "Well, it has to do with the planets and has to do with these various alchemical things." However, I think that, in the last analysis, Rudolf Steiner's contribution for a way for humanity to move into the future is that he recognized the principle of the evolution of consciousness and linked it to work on the land.

This fundamental idea is that human consciousness is not simply a straight line or a circle where we just keep recycling things again and again. The idea of simply returning again to a previous stage of consciousness because we are tired of the one we have now is actually just a default kind of thinking. Steiner's contribution to human development is that the development of consciousness is a force in the human that unites the human being with the destiny of the Earth as a spiritual being in the cosmos.

From an esoteric point of view and from Steiner's point of view, the evolution of the Earth depends on the evolution of human consciousness. They are not separate. The ancient peoples understood that. They understood that human consciousness was woven in with the destiny and life of the Earth as a spiritual being. As a result they lived in a sacred manner. Their daily round was the stuff of a priesthood. They understood the relationship between the human and the divine by seeing the Earth as the mother and the sky as the father of humanity.

It was just a given for them that nature was permeated by spiritual entities. However, that worldview had to evolve to the spot where we are today. Today the vast majority of people feel totally divorced from a real connection to the spiritual being of the Earth. The Earth primarily is a resource to be used. If you go tell your mother she is just a resource to be used, you have a lot of problems. My thesis is that the evolution of consciousness requires us to understand that our state of consciousness has an impact on the evolution of the Earth as a spiritual being.

This impact goes beyond what we do, but extends inwardly into what we carry in our hearts. What we carry in our hearts about the Earth and the spiritual nature of the Earth allows the Earth to become something that it aspires to become as a cosmic being. Rudolf Steiner's picture is that the Earth aspires to be a Sun, and we will be the spiritual beings who inhabit that Sun. That is the big picture, but that is the view of the evolution of consciousness that he strove to bring throughout his life. And the ancient peoples understood that because they knew that human beings came from the stars to live on the Earth.

They knew that all human beings are star beings who came here to learn great lessons of sharing and compassion. Moreover, they tried to remember what it was like when they were star beings; they remembered when they made rituals around that idea. Through ritual, they understood that the Earth is the mother that carried them in her womb when they came from a star to learn limitation on Earth. Most indigenous cultures have believed that we human beings are in fact omnipotent and omniscient beings who come from the stars, lacking only the spirit of limitation, which we come here to learn. The fruit of that lesson is gratitude for life.

When we learn it, we learn it not only for ourselves but also for the Earth itself, because the Earth as a living being has a biography and a destiny. The Earth will not go on forever as a physical body, just as you and I will not go on forever in our bodies. However, in our spirits, human beings and the Earth are one. The Earth spirit will go on forever and our human spirit will go on forever. That is how we are one. Nevertheless, we will both have to drop our bodies to be able to live forever in the spirit.

As a developmental stage as star beings we have learned how to deal with limitation in a clever way. We human beings call that technology. Technology allows us to say to the Earth, "Thank you for sharing your necessity for bringing darkness, but I want some light. When you turn away from the Sun, Earth, I still want to be able to read my newspaper. I want to be able to get on an airplane and fly around. I want to be able to cook my food even if the wood in the forest is wet."

So, technology is how we have learned to overcome limitation with an evolved consciousness. The way we overcame limitation in the past was we depended upon everyone else. That gave us tribal consciousness and tradition. We could depend on that because it was a living tradition. However, we moved away from that into a stage of individuality that expects to overcome limitations through the use of technology. We all have our own house and our own access to food and comfort. Technology allows us to become somewhat separate from the limitations imposed on us by having a body that must obey the rules of nature. We are able to overcome such natural limitations by becoming separate from nature, but now, being separate, we are yearning to be one again.

So, the question is this: Do we go back to the way it was for traditional societies, or is that door closed now, forcing us to go forward? In attempting to answer this unanswerable question, I would like to give you an outline of the principal of the evolution of consciousness from the esoteric view of Rudolf Steiner's work. A simple image can help us toward an answer. An evolution of consciousness takes place in human life. First, you are a kid, then you grow up a bit. In kindergarten, there is a little song that goes, "Follow, follow, follow, follow," and when the kids all hear that, they all queue up and get in line and grab one another's hands and they all follow, follow, follow.

However, imagine a teacher singing that song to the ninth grade. Not a good picture, is it? You cannot get away with that following kind of social impulse, because there has been an evolution of consciousness toward individuality and away from the collective. That evolution, in just a few short years for an individual, is an image of the evolution for all of humanity over vast stretches of time. This is not just an evolutionary process for individuals but an analog of the evolution of all humankind. Every person's individual life is lived for the whole, for the collective. That is the great esoteric mystery of existence. Thus, you contribute to the great pool of learning for the whole of humanity in what you learn from your own evolution. Whatever the lesson learned, it does not matter. What matters is that you are leaning.

You could be a single dad selling pot to support five kids. You are learning a lot and contributing what you learn to the collective consciousness. Whether you believe it or not; whether it says you contributed on your tombstone; whether they write a book about you—it does not matter. Christ never wrote a book, but we remember a great deal about what happened while he was on Earth. It is not about writing books, even though some of us try to do that. It is about how we see our lives in context of the time we spend on the Earth. My life and the life of the Earth are intimately intertwined.

The question is this: Can I honor the Earth in my life, or is it merely something that gives me a bank account? This is a particular dilemma for the agriculturalist. Therefore, I would like to give you a picture of this kind of evolution—the evolution of consciousness based on the work of Rudolf Steiner, so that the work in biodynamics comes somewhat into perspective. The biodynamic task at the present moment is to return exclusively to what Rudolf Steiner said in 1924. Was he a good gardener, planting seeds he expected to germinate in the future? Did he want us to go back and dig up the seeds he planted in 1924 and put them in a museum? That has been a question for me for a long time. Personally, I think he was a good gardener. This means there is work to be done.

Years ago, I was in Mexico City and visited the anthropological museum on the Zócalo, the Plaza de la Constitución. In that museum they have a timeline of all the civilizations of history. It was so interesting to see that because it showed how there are waves in the evolution of civilizations, whereby certain benchmarks are reached. Benchmarks are like post and lintel architecture that then form ziggurats, followed by other aspects of technology used by anthropologists to determine different developmental stages of civilizations. The curious thing about that timeline was that one could watch the various cultures move through the same developments, but they all had different rhythms for each developmental feature. They all reached the same development but at different times.

For me, the most interesting was the Japanese civilization. Most of the other cultures would go way ahead toward a particular development, but with the Japanese culture lagging behind in that development. Then, suddenly, the Japanese culture would go "wham," and leap ahead of the others, surpassing every other culture in a brilliant surge of creative activity and planting seeds for the next development. Then the other cultures would catch up and develop that seed, and the Japanese culture would again start to lag behind. It was so interesting to see these rhythmic patterns in the various cultures. Nevertheless, every culture was on the same page at some time or another. Every cultural development starts with a seed of something and evolves through a certain stage, then evolves to the next stage.

With this idea that cultures are in the process of rhythmically evolving, I would like to suggest a way to look at that. Essentially, we could say there are four levels of culture in connection with how human beings relate to the Earth as a spiritual being through the development of agriculture. Agriculture answers the most primary level of the question, "What is the most beneficial way that human beings can live on Earth?" In the human relationship to nature, we can recognize four stages. There is the child of nature, the spouse of nature, the steward of nature, and the technician. As a child of nature, the human being has had access to particular images from the art and religion that were pan-cultural. This means that no matter where in the world the culture developed, the images used to depict the relationship between human beings and the forces of the natural world were similar. When human beings were the children of nature, the pan-cultural religious impulse was known as *animism*. When you look at animistic art in religious or ritual places around the world, the images depict the same things. They were made famous by the incredible Lascaux cave paintings of bison. There is a wonderful film (*Cave of Forgotten Dreams*) that documents the discovery of a new cave in France that had never been seen before.[1] The cave is known as the Chauvet-Pont-d'Arc Cave, and the drawings look as if Picasso had painted them. The beauty of these works just blows your mind. The images were made 32,000 years ago and depict

1 The film was directed and narrated by Werner Herzog and released in 2011.

hundreds of animals in potent juxtapositions. There is only one image of a person. The religious impulse then was not "Here am I, a brilliant artist of the Italian Renaissance," but "Here am I, a child of nature. Nature provides for me my religion through the forms of the animals I resemble."

The feeling in animistic art is that the animals are so wise and so noble and powerful that the artist has to honor them by conjuring them on the wall. It is incredible how those images move the soul. Scholars theorize that this art was made using a human faculty called *eidetic imagery. Eiditic* comes from the Greek, meaning "the equivalent to." In eidetic imagery, the artists inwardly see an image that is the exact equivalent to what they have previously seen with the senses. Children often see this way. They look at the wooden knots on the wall and see faces. They can then go up and trace around the face they see.

That is the way scholars believe that these people made this art. It is a kind of projection out of such an intense inner experience of belonging and connection that the person actually sees in front of them the image in complete detail. It is then a matter of tracing around it to make it appear on the wall. Therefore, these artists were children of nature; beings who live in the womb of nature are hunters and gatherers. They follow the herd. They gather whatever is available. They do not cultivate anything. They do not need to cultivate anything, because everything is provided if they keep moving.

They hunt and they gather. They are sensitive to the nuances of the movements of the animals. They live in the consciousness of the animals. They live in the consciousness of what fruit is blooming now. What berries are available now? What roots have reached completion now? And the consciousness of awareness was provided for them from the outside just like it is for an embryo—just as it is for a young child.

Young children receive nutrition from their environment. They do not need a bank account. Therefore, children of nature possess a pan-cultural consciousness, meaning every culture at the same level has a very similar art based on images of animals. No matter where it happens, it is a universal, pan-cultural consciousness. We could say it is a state of union with nature where everyone is one being. Everyone is connected to the same stream of understanding. Everyone gets information from the same state of consciousness. The religion or dominant social consciousness is animism. "I live because things speak into me." This is the consciousness of children in preschool and kindergarten. It is called *fantasy* in psychological terms. In fantasy, the beings of the world speak into me and create an inner picture in me, and then I act out of that. The imagery, and even the rituals people find through this consciousness, are imaginations shared by everyone.

Being a hunting people, the deep indigenous people found ways to get an advantage in the pursuit of survival. They made arrowheads, spear points, and blades. Once they found the technology to flake arrowheads and to make spears and arrows, something happened. They found that there is a way to control nature. Anthropologists speak of this as a Paleolithic revolution. Around the same time that arrowheads show up (in digs from around 13,000 BCE), anthropologists describe the first evidence of a widespread tendency to make war on neighboring tribes. The theory is that, as the climate was changing, population pressure pushed people into territorial conflict. Therefore, as the evolution of consciousness moved away from the concept "we are all one with everything, follow the herd, and take down the weak ones," there was a gradual evolution away from hunting and gathering toward a more agrarian and settled kind of consciousness.

Agrarian consciousness arose in people because of a particular evolution in agriculture called "slash and burn." Instead of just gathering acorns that fell, people would burn areas, slash them, burn them, and put crops in. In this way, they did not have to move continually to follow the animals or migrate up and down the mountains as the various food crops came into ripeness. They could put things by. In slash-and-burn agriculture, people had access to a fertile area of land for a few years until it was exhausted, and then they would move on down the road and slash and burn some more, then plant again. The new soil was rich and nice and they could save seeds and start to live in small settlements. This is the beginning of what may be called the agrarian revolution.

Slash-and-burn agriculture represents a development in consciousness away from simply following migrating animals toward the practice of making a corral. This is also the time of a shift from protecting the corral from wolves to getting the wolves to guard the corral. Instead of running after the deer, why don't I just make friends with them, and I will be a Laplander.[2] This is an evolution of consciousness that moves away from simply hunting and gathering toward the beginnings of agriculture, which brought with it stone plows, the domestication of animals, and what anthropologists call "nature religion," or the Neolithic revolution. Thus, instead of having an eidetic imagery of a bison and its power, artists actually begin to set things aside as special kinds of God-type things. There is a development away from animism into independent styles of art. The arts still serve a ritual function, but the images are no longer pan-cultural. There are divisions of different kinds of art that develop as people become agrarian. Various versions of a corn god arise. One culture depends on maize, while another culture depends on barley. The local landscape may be dominated by the presence of a huge mountain on which the omnipotent god lives. This is distinct from animism. Something has evolved in the consciousness of people.

First, there were wood plows, then they put stones on the tips, then they eventually began to find they could smelt metals. As they smelted metals during the Bronze Age, there was the great flowering of the agrarian way of life. The people following the deer were children of nature. The agrarians, with their corrals and crops of grain, were wedded to nature. When people became keepers of fields and flocks; they became married to the natural world. They became spouses to nature. They became "husband men," as they used to say in the olden days.

If you have a cow, you know what that is all about. You cannot go on vacation. Forget it. You are wedded to that cow. Instead of running around and shooting it, you are wedded to it. This is an evolution of consciousness.

2 The Sami people (also called Lapp or Laplanders) are the indigenous people inhabiting the Arctic area of Sápmi, which today encompasses parts of far northern Sweden, Norway, Finland, the Kola Peninsula of Russia, and the border area between south and middle Sweden and Norway.

Did anyone see the movie *The Gods Must Be Crazy?*[3] In it, an aboriginal man is trying to return a Coke bottle to the Gods because it was wreaking havoc in his little band. Someone had thrown a coke bottle out of an airplane window and it droped into his tribal walkabout. One of the kids finds it and suddenly everyone wants it. After a bit, the kids are starting to hit each other in the head with the Coke bottle. He decides, as headman, that he is going to give this crazy thing back to the Gods. He goes on a walkabout in an attempt to return the Coke bottle to the Gods at the end of the world. He is hungry, and as he is walking through an area near a town, he sees a goat. He shoots the goat with a poison-tipped arrow, but it is someone else's goat—a fact that escapes his understanding.

As a result, the people bring the cops and they arrest the guy and put him in jail. Then he has to go before the magistrate and wonders, "Why did they put me in here? I am going to share the thing I killed with everyone, because that is what I always do." When he goes before the magistrate, he is smiling and the magistrate responds with a grim frown, and the headman thinks, "I do not get it." In his worldview, he was going to share the kill, of course. Why would he not share? But what he does not understand is what he thinks he is sharing; somebody else thinks they already own the goat. That is a shift in consciousness. So, as the agrarian consciousness develops and people become wedded to a particular piece of land, population pressure becomes an agent of social change. This pressure develops even further when people make more sophisticated tools so they can actually begin to cultivate, choose, and grow grains. They make granaries and save grain year after year after year. Now, instead of having to share the kill because I have no way to preserve it, I have a storehouse. In fact, I have got three storehouses, because I am really a good farmer. You may not have any, but I will lend you some; however, now you have to come and work for me. If you want some of this grain, you work for me. That represents a shift in consciousness. This shift of consciousness away from agrarian management that weds me to the land is that, as a landowner, I am no longer wedded to a common land, I become a steward of my own land.

3 A South African comedy by Jamie Uys, released in 1980.

This development is the beginning of so-called land tenure. A group settles somewhere and develops communal corrals with some cows; then perhaps a neighbor thinks it is okay to come in and take a few cows away. However, even though the corral is communal, those cows are your property. We no longer say, "Let's all go kill a deer and cut it up, then we will all eat it together. We will jerk out what we cannot eat, but first we will give some of it to everyone." Today, we say, "Excuse me, but those cows are mine. I built that corral with my friends, and those are my cattle because I bred them. In addition, do you see that granary over there? That is where I store feed for my cattle in the winter while you are off running around in the bushes looking for roots to dig out of the ground."

This is a great battle in consciousness, and it happened even in North America between the cowboys and the settlers at a much later time. The cowboys said, "We do not want fences," and the settlers said, "I have to fence my land so your stupid cows do not ruin my crops." That was the so-called range wars. It is the same picture of this evolution of consciousness, but now we are at the level of land stewards, not husbands of the land. This was brought about by the issue of land tenure, which becomes a big deal during the next evolution—the Iron Age. When people moved out of the Bronze Age and into the Iron Age, it saw the beginning of land ownership, as well as serfs, and slaves, who worked the land but did not own it.

If you are a hunter-gatherer—if you live off the land in an agrarian society and we are all just kind of producing things—there is no such thing as land ownership. Read the documents and the comments of Chief Seattle. Read what the native peoples thought when the white Europeans said, "Yeah, we will give you your own parcel of land." The indigenous people thought: *How can you own the land? The land is not to own. The land is the spirit of our ancestors.* However, there had been an evolution and, as that evolution unfolded, the whole issue of land tenure became a truly significant guiding force in the ways people dealt with agriculture and the role of the agriculturalists in society.

As the Iron Age began, confederacies of people were formed to pool their capital; they were the citizens with rights to land tenure. There were also non-citizens. The landed gentry were citizens; they had rights to the land. "Rights to land" means you tell others how to work your land for you. The landed gentry held the power over life. Now there were people without the rights to land, and they were forced to work for those who owned the land. Very often, that work was what we would call slavery or forced migration today. Thus, the issue of being on the land or having access to the land is a key force in human evolution; hunters and gatherers share access to the land with an agrarian society. During the Middle Ages, this went through an evolution. That was when the roots of land tenure developed, and the idea of the commons arose as a way to relieve the pressure of the land tenure issues.

In Britain today, there are still commons. You can put your cattle to graze on common land. This is great, except that the common land is often where the golf courses are located; if you play golf in England, you have to watch out for cow pies, because cows have the right-of-way on the golf course, since golf courses are on the top of the ridges where the commons are. That is where the most wind is, making it difficult to grow row crops. Instead, cattle graze on those common grounds. In the shire, there was always the lord's land along with the commons. Today, in the western states of the U.S., this is the federal rangeland. When the settlers arrived from the East to settle the West, they began fencing off the rangeland, which led to the range wars.

As people developed consciousness, the issue of the Earth *as a being* became critical, because the Earth had instead become "something to own." It represented a certain power of one group of people over another. If I owned the right to land, I could say that you are not allowed on it. One of my biggest surprises was while traveling in Europe, where every farm has passages through the fences whereby people can cross but cattle cannot get through. On the farmer's land there are pathways you can travel and that afford common access. No one will tell you to get off the land. Well, there are few places like that in California because it is all "owned." Land tenure and the idea of a commons were prevalent in the Middle Ages, but as land increased in value, there was a turning point as technology began to develop.

This is a very interesting picture. For thousands of years, land tenure was controlled by a lord, king, or emperor. However, with the rise of industrialism and the beginning of mechanization, there was a huge shift in consciousness. Mechanization represented a huge power of individuals that they did not have when they were serfs. Just consider your John Deere tractor in terms of personal power. Think of the John Deere 500—that is 500 horses. Now imagine 500 actual horses; each one is eight feet long, with a foot in front and back for the traces. That is ten feet each. Five hundred horses would amount to 5,000 feet of horses. That is nearly a mil, which seems a bit unwieldy; so we will place them four abreast, which is better. Now we have only a quarter mile of horses to control. Now imagine saying "gee-up" to that. Now, say, "haw." Now, say, "whoa." That is your John Deere 500. You have the key to 500 horses in your pocket, maybe even next to the credit card that paid for the tractor. That is the power of today's individual. You may be in hock up to your earlobes for that John Deere 500, but this a whole other thing.

As industrialism developed, an important step was the steam engine. It gave people the power to do things that they couldn't imagine doing with just a horse and a few sons. One person could deal with a huge piece of land. Nevertheless, this technological revolution began even earlier than the steam engine. During the 1700s, people manifested great ingenuity in their ability to work iron and wood into new inventions. In 1750, an invention made it possible to triple grain harvests. We do not really think of this as very earth shattering, but this remarkable technological invention was the cradle scythe. A cradle scythe was a standard scythe, but with long, thin, curved wooden fingers that grab the grain as the scythe cuts it, gathering it into a sheaf in one stroke. This invention allowed one worker to move the scythe through the grain and gather it into a sheaf with a single stroke. The cradle scythe allowed people to triple their grain harvest.

By 1850, a hundred years later, there were railroads. People went from the cradle scythe to railroads in a hundred years. Now the technological power of the steam engine could amplify people's capacity to do things in remarkable ways. The railroads made it possible to move people easily from the East Coast to the West Coast, and this began the westward migration of people in the New World. In 1850, the U.S. Homestead Act went into effect. With the Homestead Act, you might have been from Sweden or Poland and not doing so well. However, the railroad barons in the U.S. would pay your passage across the Atlantic on a steam ship, then put you on a train and send you out into the wilderness with whatever you needed to walk out onto the prairie and find a spot where no one else had staked a claim. You would walk for a day putting stakes in the ground. You were allowed to fence in the amount of land you were able to stake out in a day—generally about 160 acres—and then you had to remain and work on it for at least two years. Once you did, it was yours. This was the Homestead Act of 1850. The reason the railroads were so interested in getting people to do this was that there were so many people on the East Coast who needed food. They needed people to go out to Ohio and beyond to stake out 160-acre homesteads and send food east by railroad—a good business model! The people would go out west by themselves; they would cut sod from the prairie, make houses, and hunker down for a couple of years. They would put up some fence, maybe see if they could get some cows or whatever, and start homesteading and growing crops. That was settling the West in the U.S.—less than a hundred years after the American Revolution.

Cradle sythe

This represented a huge evolution in consciousness. Unfortunately, with this act you had a huge number of people flooding into areas where people already lived, and they had a very different picture of the way the Earth should be treated. As that clash of values began, we get the beginning of strange events around settling the New World. The native peoples were driven from their lands and forced into cycles of slavery, poverty, and depression. Moreover, there has been a tendency to think that it was only the native peoples who were dislocated and forced into depressing conditions. In fact, many other peoples, whether native or immigrant, where also put into dehumanizing situations by the tremendous forces of economic disparity brought on by technology. Within fifty years, in 1905, the majority of North Americans farmed to provide food for the cities. The majority of suicides at the time were those of farm wives. "The farmer rests with the setting sun, but a woman's work is never done."

In 1905, one typically started to work at the age of seven in a gang of children picking vegetables. Then one worked through the teens under a man in his mid-20s as one's overseer. A person would work in that labor force until one's 30s, by then perhaps having a family large enough to work a rented piece of land for oneself. Nevertheless, the food one grew was not intended for one's own family. The agricultural system had grown into an organization of truck farmers that provided food for workers in the city. The family farm had become a truck farm, and farming had become a job of growing produce for city markets.

Following the American Revolution, Thomas Jefferson became a landowner and a gardening enthusiast. He owned slaves, but he had a dream of what he called the "shirttail farmer."[4] Eventually, the land grants in the United States became that shirttail farm. The idea was that, if you give individuals enough land and let them settle it, people would be able to provide for themselves, while any surplus would provide for those unable to farm or who were engaged in service tasks. In the early years of the settling of America, ninety-five percent of the people produced things, while five percent of the population engaged in services such as education and government. That percentage is just the reverse today; ninety-five percent of the population engages in service jobs, while only five percent grow food. Five percent produce the food, because we have access to the 500-horse John Deere. It is a big shift in the evolution of consciousness.

Those were the percentages of people involved in agriculture in 1905, the height of the agricultural renaissance in the U.S. By 1930—just twenty-five years later—there was the collapse of the agricultural sector in the U.S. Mechanization had destroyed the price structure of foodstuffs. Mechanization allowed too much food to be produced so it could be sold at a profit. Because agriculture was producing too much food to sell, the

4 See Henry Wiencek, *Master of the Mountain: Thomas Jefferson and His Slaves*.

price of food went down, and because the price of food went down, farmers went out of business. The mechanization of agriculture had made food an economic commodity subject to intermediaries and price gouging. The very forces that made it possible to grow more food also produced market values that were driving people off the land.

These dubious moral implications of mechanization stimulated the beginning of the grange movement. The grange movement was founded after the Civil War in 1867 to protect people engaged in agriculture, so that those producing the food could organize and have a way to control prices. This was to prevent them from always being priced out of the markets by intermediaries. By the 1930s, the majority of those engaged in the agricultural sector were only tenants—basically serfs, kids, and criminals who worked the land for a few owners who never touched the soil. The owners were interested in commodity prices and market values and not so much in the Earth as a sacred being or in food as a divine gift.

Therefore, what began as the dislocation and dispiriting of indigenous peoples ended up dislocating and dispiriting just about everyone. Everyone was getting crushed by market values. Individuals became an abstraction—except for certain people who were non-abstractions: the owners who called the shots. This has now grown to the point where agriculture is simply a business in which soybean farmers in the Midwest hedge the commodities markets to stay in the game. Soybean farmers drive the soybean market by hedging their losses. This means they bet against their own crop coming to market. I have worked with people in the Chicago Mercantile Exchange. The biggest force in the soybean commodity market is soybean farmers who bet against their own crops. That is hedging. The reason they do this is that, if the crop fails, the commodities positions they have bought against their ability to deliver a crop will pay for the failed crops and keep them from bankruptcy. Their positions in the commodity market—that is, betting against their own crop—will support them if the crop fails. If the crop does not fail they can place that position against themselves in a weather futures market, and if they are smart they can sell that position to someone else before the delivery date. If they are savvy, maybe they will make a little cash on the side. The hedging of markets by soybean farmers, the producers, is the cause of soybean fluctuation in the market. This is what it has come to.

We are all driven by huge forces that have little to do with the idea that the Earth is a living being with a spiritual destiny linked to human evolution. Everyone is linked to this issue. Take, for instance, the transformation of corn into ethanol. There were riots in South America when people were unable to get corn to make tortillas, because corn farmers up north were using the corn to make ethanol. People could not afford food because the profits were greater when corn was transformed into fuel.

The alienation of those who work the land from the spirit of the land has been a gradual evolution of consciousness away from feeling such as, "I am a child of nature, at one with everything in the universe and all the people around me." Very gradually, as the human individuality becomes stronger, human beings will start to express that individuality. Fundamentally, to express my individuality is at odds with my need to feel that I am part of a collective.

We could ask this question: Is this an aberration or evolution? We could probably say yes to both. Yes, it is evolution, but the way it is happening is an aberration. If we could ask Rudolf Steiner why this has happened, I believe he would say something similar to what he told Ehrenfried Pfeiffer after the agriculture course in 1924.[5] Pfeiffer had asked him why it was so difficult for people in the movement to get along with one another, even when they are aware of these high and cosmologically significant ideas. Steiner replied that he thought it was a problem of nutrition. He said that, in the future, people will be sitting at groaning boards piled with food, but that the food they eat will not give them the forces to have actual spiritual perception. It will satisfy certain needs to keep the organism intact, but it will not open human consciousness to the spiritual world. I think that prophesied time is already here.

In the past, when nature provided the foods through animals, fertile soil, and primitive methods, food linked human beings to the cosmos. When

5 Published as *Spiritual Foundations for the Renewal of Agriculture,* June 7–16, 1924 (CW 327).

people went out on the prairie, people could ride on horseback into the buffalo grass and not be seen. It was said that the virgin soil in the prairie was so deep and the grasses so strong that you could hardly open the ground up with a plow and a team of oxen. People had to invent special plows to break the sod. The people who used those plows were known as sodbusters. The vitality of the land was the stuff of legend, because the natural life had sustained it for millennia. Therefore, if you are eating foods from that kind of soil, you are getting forces in your soul that can unite you to the higher consciousness behind all of evolution, especially the consciousness of the Earth as a living, spiritual being.

Rudolf Steiner's mission with the farmers was to bring practices back into the agricultural sphere to allow the plant and the Earth to intuit, once again, the action of the planets. If you read the agriculture course, many of the pictures have to do with making plants more intelligent for receiving what he called "cosmic nutrition"—that is, cosmic forces from the realms beyond the Earth. We could call these cosmic forces "condensed light." Condensed light is consciousness. Agriculture is intimately linked to the evolution of consciousness. Steiner entered a sphere in Central Europe in which the idea was still intact that what I do on the land has a spiritual root through a kind of European shamanic tradition or nature religion. You may be aware of that kind of nature religion; it is another term for alchemy. You may be aware that alchemy is a hidden language in the agriculture course. It is a pictorial, imaginative, cognitive kind of thinking that allows the soul of a person working the land to receive real understanding when actually touching the land.

You may think that this is a convenient fable from a weird guy from California, but I can tell you it is a good way to learn how to write books. When I get stuck for an understanding, I take inner images of nature and the natural world into sleep in an alchemical meditative practice. The next day, I go out and work in the soil. As soon as I touch that soil, which I have been working for fifteen years, images begin to flow, and I have to go inside to grab a pen. I love the soil in my garden, because I have seen it transform from pottery clay to black loam. I have the distinct feeling that the soil loves me when I go and touch it and suddenly know what I need to do. Then, however, my practice tells me that I have to wait until someone else corroborates what I have imagined before I can actually act on it. This kind of alchemical meditating into nature can develop a deeper spiritual soul connection to the Earth in a person. Just because you have experiences like this does not mean you belong on the funny farm.

The question is this: Can I have an experience like that and then find something in the scientific world to corroborate it? To me, that is the direction of research in biodynamics. I heard a story about a reporter who once came to interview Steiner in his study. The reporter asked, "Dr. Steiner, what is clairvoyance?" The story goes that Steiner turned and there was a table in the room stacked three feet high with books with little pieces of paper stuck in the pages as references. Steiner walked over to the table and opened a book and pointed to a page and said, "This is clairvoyance." Clairvoyance is study.

Nevertheless, study, when coupled to a meditative alchemical practice, is different from merely going to study at a university. There is an imaginative method that Rudolf Steiner brought to the work of agriculture and to the work in the natural world. That method is available to us and allows us to begin again to imagine being connected to the soul of the Earth in a direct way. This is not a return to fantasy. Human consciousness today is no longer the consciousness of Earth's ancient children; nor is it the consciousness of people who considered themselves spouses of the Earth. Human beings have been through the consciousness of technology, and one may even be a steward to a piece of land, but there is a different need today regarding human beings and the spirit of the Earth. People need to *collaborate* with the Earth in ways to which the Earth can respond. The Earth as a spiritual being interacts with its own beings. It has brothers and sisters. It has family reunions. It has issues. It has events that are significant for its destiny. The Earth has challenges, much as you do when Uncle Elmo comes to a family reunion and starts a political argument with Grandpa. You know what happens when Uncle Elmo gets going with Grandpa.

For the Earth, Elmo arguing with Grandpa is Mars conjuncting Saturn. The cosmic dimension of the agricultural work is the environment in which the Earth lives continually interacts. We are actually working in the tradition of alchemy when we study the cosmic dimensions of agriculture by basing our work on the motions of the planets. When I begin to understand some of the rhythms of how the cosmos acts and look at a plant, the plant speaks to me in the language of the great poet Goethe. When I begin to understand a little bit of the cosmic rhythms and look at an animal, the animal beings to speak to me about its ontogeny, or how it evolved from an archetype. When I can see how the ontogeny of an animal is related to the development of a particular plant, and I see how a particular organ in an animal is an image of a particular sector in which a plant grows, and if I can bring those two images together in my consciousness in a rhythmic pattern of winter and summer, then I begin to have an imagination of what are known as the *biodynamic preparations.*

These pictures that Rudolf Steiner brought are not random but very precise, elegant, alchemical imaginations that have to do with the way the soul of the Earth is trying to contact people again. The soul of the Earth is like your mom. Your mom is wondering why you are not calling home more often. The Earth is waiting for you to do that. Calling home simply means you honor the spiritual being by asking, "How's your life? How is it going with you? Have you had any interesting relationships lately? Have those relationships created anything in the place where I am living and the soil I am trying to develop? If I go and tickle you with my fingers, dear Earth, will you help me understand the ways that allow me to collaborate with you for our mutual healing?" The Earth is a living being. It is our dear mother waiting for us to understand that there has been an evolution of consciousness and that the next evolution involves the human need to collaborate consciously with the living Earth. We do this best by creating inner images from nature that we then take into sleep. Such a practice can open our hearts to imaginations where we can experience in a direct way how our own life is entwined with the life and evolution of the spiritual being of the Earth.

CHAPTER 2

TRANSFORMATION OF SUBSTANCES

I. Nature and Human Consciousness • Four Elements • Solve and Coagula • Inner Surfaces • World Mineral Plant • Colloids • Gestural Motif

In this chapter, I would like to provide some of the theory and phenomenology behind working with the biodynamic preparations. I will give the alchemical reasoning behind Rudolf Steiner's work, especially as it relates to biodynamics. As I understand it, Steiner's worldview was to bring the Spirit and matter into contact with each other through a human consciousness. That is thinking that is in line with the alchemical worldview.

The core of alchemical thinking in the ancient times was this: nature was created by Heaven, and the Fall of humankind from nature divided nature and heaven into two separate entities. Because the Fall from Paradise was seen as the source of this split, the task of human beings is to solve the problem that arises because of the Fall. The human task is to return the natural world to contact with the divine in order to heal the split. Rudolf Steiner sees this task as the struggle to give nature back to God. To do this, we first have to realize it has been separated from the divine world. That task is big. The great theological question for millennia has been that, since nature has been separated from the divine by the Fall of the human from paradise, is the divine still present in nature as we experience it as human beings.

A strong aspect of the issue surrounding this split and the Fall is the human invention of technology. The force of technology is not just cars and computers, but also includes plowing, making a hoe, or sharpening a digging stick. So let us not just talk about computers as technology, but also about a digging stick that makes it easier to plant your corn, a technology that upsets the natural order and makes human beings into creatures that have some God-like qualities of power over nature. Any time a person enters the natural world and changes something, that change is not natural. If you go out and select seeds from a particular plant and reject seeds from another, that selection is not natural; just ask Darwin. If you choose only certain plants to promote through your selection, this is not natural selection; it is human selection, and such human intrusions into the natural world can create difficulties for the archetypal links between nature and the divine world.

All of the harmonies in the divine world are reflected through the planets and down into the growth cycles of plants, the evolution of minerals, animal migrations and physiology, and all of the ordered relationships in the natural world. All such natural phenomena are reflections of the divine world. However, human beings then make bulldogs through selection. Darwin talks about this. Bulldogs can be bred so intensely that they can drown in their own mucus. They drown because they cannot breathe anymore because their skull is too short. Their skull is too short because they have been bred that way to deal with bulls. When they bite the bull on the neck they have to open their mouth so wide that they would suffocate unless their nose was pulled back and pushed up to the top of their muzzles.

So they were bred over time to have nostrils higher and higher on their muzzles. The breeding was so intense that their nostrils were on top of their muzzles and the mucus in them could not be cleared so some of them actually lost the ability to breathe. They would drown in their own mucus. Border collies were bred to go between gates and through a flock very easily. They were bred to be increasingly narrow in their frame to make moving among sheep more effective. They became so narrow in their frame and brain box that they became stupid. They were very intelligent dogs until they were bred into stupidity. These examples come from Darwin's *Origin of Species.*

Thus, when we enter nature, we bring with us forces that separate the divine plan for border collies and the natural order that produces the bodies of animals. We separate it because we have something that the natural world does not, and that is freedom in our thinking. A code word for that is *akasha,* which in Sanskrit translates as *space, ether,* or *the totality of consciousness.* Human beings have access to the consciousness of akasha. People can manipulate akashic consciousness. Human beings can manipulate their consciousness with a degree of freedom not available to other creatures in the natural order.

Alchemists understand that, if a person wishes to enter nature in a deeper way, it is necessary to work with consciousness so that what they call "the artist" can harmonize one's consciousness with a specific level of consciousness found in nature. The alchemical artist must find how nature is working so that one can choose certain times for working to amplify what is natural. This strategy is needed to transform the substances of nature into phenomena such as medicines, metal smelting, crop rotation, weeding, and cultivating out plants that would otherwise exist only in a certain area. These processes are not natural. Eventually, selecting only certain crops for a parcel of land leads to mono-cropping. To alchemists, the consciousness of a person entering nature is a tool that needs to be enhanced by observing phenomena in an exact way. One can then take what has been observed of the natural world and, performing certain meditative practices, can actually touch the phenomena of nature. Then, when one touches the organs of a cow, the individual's consciousness is heightened and one is not just messing around with a pile of cow guts. We will be able to see the whole web of forces behind the cow's stomach or intestines so that we understand what Rudolf Steiner calls *biodynamics*. If I just take the recipes, this is okay. If I just buy biodynamic preparations and put them in the water and stir, this is okay, too. However, if I wish to take the gift that Rudolf Steiner has given and take it further into transforming substance, I need to change my consciousness in a whole other way. Our consciousness is the greatest tool on our land. Our consciousness is the most potent force for change.

The alchemy of the forces and processes that operate behind the consciousness needs to be harmonized to the operation of the natural world with which the alchemist wishes to engage. This means consciously linking personal consciousness with the rhythms and patterns in the natural world—we could say in the natural order—to understand first how the natural order operates. Once that has been accomplished, the second task is to understand how any action I would do could be violating an operating principle

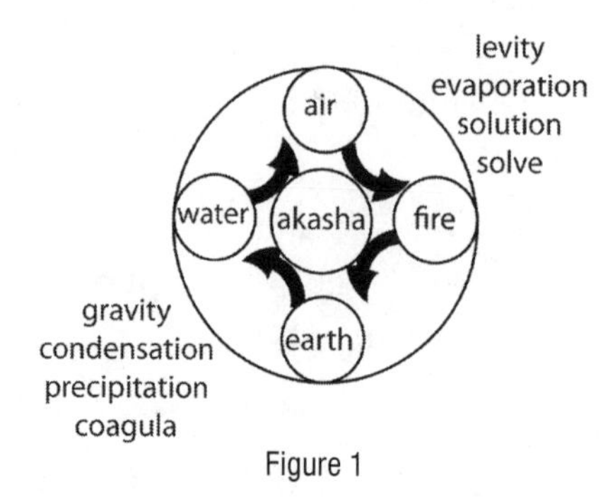

Figure 1

of the natural order. Third, how can I get my actions in harmony with that operating principle? It is one thing to recognize the natural order and to recognize whether what I am doing is working with the natural order, and it is another issue to correct my thinking and bring it in harmony with the natural order out of freedom. I need to do this work in freedom if I am to have the creativity to imagine future biodynamic preparations.

Rudolf Steiner gave the motifs for this kind of work through his own alchemical insights. My task is to explain, to the best of my knowledge, how those motifs work so that there can be memories of these ideas available to you as you manipulate things or observe a plant, a tree, or an organ of a cow. The consciousness you bring to the thing you are doing is the agent of change. That is what change is; it is the elemental world in which your work is linked to you through various levels of your consciousness. Your consciousness changes the relationships to the elemental beings and alters the ability of the elemental beings to harmonize their own relationships. Your consciousness dictates this. It is a scary thing, especially today with bulldozers, nuclear energy, and whatever. We have machines to amplify the will, but not machines to amplify our awareness of the motives in our will. We have only machines that can amplify information. It is up to us to change our awareness.

This preamble is intended to provide tools for understanding what Rudolf Steiner was trying to accomplish through his course on new agriculture, or what we may call "sacred agriculture"—that is, a spiritual act, not just the manipulation of resources.

Look at *figure 1,* with *akasha* in the center. Then look at *figure 2.* Alchemists work with what they call the four-element theory, which holds that every change in nature, every transformation in nature, goes through certain rhythmic sequences in which what physicists call a "phase shift" happens, for example when something changes from a solid to a liquid or a liquid to a gas or a gas to pure warmth. These are called phase shifts. Alchemists understand that everything is not only chemistry, but everything also goes through phase shifts in certain sequences. They recognize that even consciousness goes through phase shifts. In this chapter we will consider various elements and show on the mandala of earth, water, air, and fire how the phase shifts of the elements work in the different realms of mineral, plant, animal, and human consciousness. Human consciousness has its own phase shift called learning. Making an error and correcting it, getting an insight, understanding a process, going to sleep and dreaming, and waking from a dream and realizing you are here. These are all phase shifts of earth, water, air, and fire in the consciousness. These states of consciousness are the equivalent of water precipitating out of the air and forming ice, then melting and turning into a gas, and processes in the reverse.

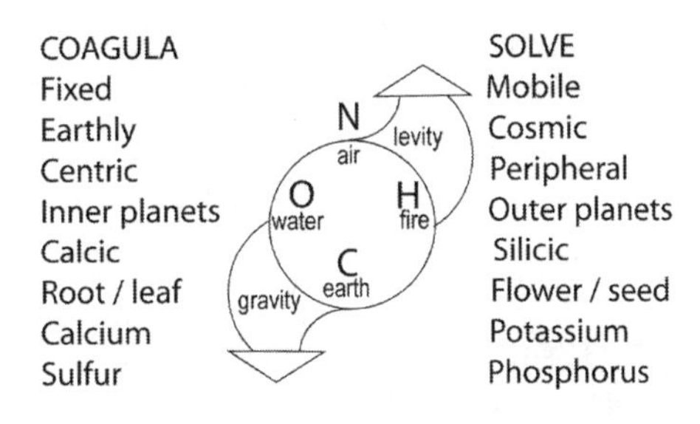

Figure 2

Alchemists would use the mandala of the four stages of earth, water, air, and fire as a mapping device for finding resonances between the various kingdoms in nature and the various human soul states. With a mandala of the four elements, we can determine where a particular reaction is and where it is moving. We can ask these questions: What is happening with the organ of the animal I wish to use? Is it an earth process, a water process, or an air process? What kind of plant is this? Is this a plant that is more earth or more air? What part is more air of a plant than earth? What minerals represent fire? What minerals represent air? What minerals are active in the flow of soil sap from the soil into the *mycorrhiza*,[1] then into the root of a plant? What are those phase shifts in the plant sap? What do they represent? What does calcium represent? Does calcium support water, fire, or air?

1 In a mycorrhizal association, fungus colonizes the host plant's roots, either intracellularly as in arbuscular mycorrhizal fungi (AMF or AM) or extracellularly as in ectomycorrhizal fungi. They are an important component of soil life and soil chemistry.

It is possible to answer these questions through exact imaginative science based on the alchemical mandala of the four elements. Alchemists understand that calcium has a certain elemental quality and that it is different from phosphorous. When we have a soil solution problem, one can think about the soil solution also with the mandala of earth, water, air, and fire. The exchange of gasses in the root zone, the exchange of gasses with the leaf in the atmosphere, the formation of phenolics and oils in the oil-forming protein process in a seed—these factors can also be thought of in terms of earth, water, air, or fire. The beauty of using the alchemical mandala as a tool for agriculture is that we can find a harmony between a mineral that is in a particular spot on the mandala and a plant that needs that elemental quality and an animal organ that is active in processing that elemental quality. That is the beauty of what the alchemists brought and why Rudolf Steiner used alchemical reasoning in forming the ideas behind biodynamics.

At a recent workshop, there was a woman in the front row who is a consultant for vineyards and who has a degree in soil science from a large agricultural university. After the presentations, in which I had lectured on the chemistry, soil chemistry, and mineral movement in plants, she came to me and said, "You know, I have my degree in soil science, but what you brought this weekend has made it make sense." She's a bright person. She's an agricultural consultant but when I gave the background of the alchemy of the earth, water, air, and fire for calcium, potassium, sodium, and phosphorous, the penny suddenly dropped for her because it was given in picture form. She could certainly think circles around me in the specifics of chemistry, but the alchemical mandala as a picture of the information helped her see relationships that her abstract university learning could not provide. She said to me, "Oh boy, I cannot wait to go back through my notes." The pictorial dynamics of the alchemical mandala had helped her to see how she could make a connection between the plants and the soil they are growing in, as well as the composts made from animal residues. This kind of pictorial thinking is not about getting more information but about making sense of the information you already have.

For alchemists, the mandala of the four elements is a meditation device for solving problems presented by some of the more difficult mysteries in nature. For example, the ability of water to go up from the bottom of a 300-foot redwood to the growing tip at the top is a mystery. Just consider the hydraulics of this. The pressures needed to push a head of water three hundred feet against gravity would explode the root where it joins the base of the stem. How this is done in nature is a mystery. Alchemically, that force that allows the water to move that far up is a counter-force to gravity. Alchemists called that force *levity.* In the alchemical mandala of the elements, the earth and the water elements represent the force of gravity.

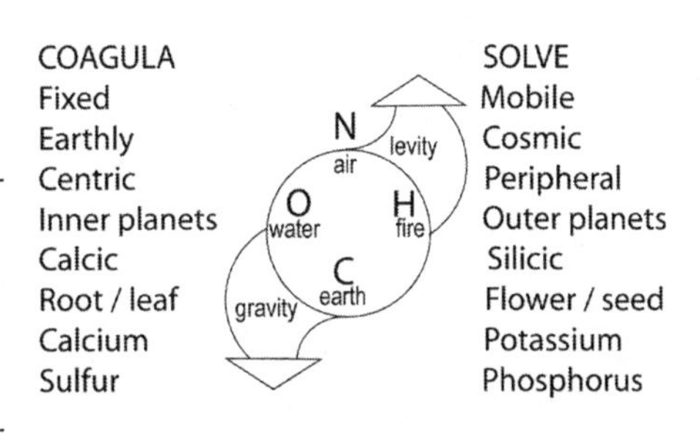

Figure 2

Under the influence of gravity, things go down because they are "cold." Cold in this sense could be 200° if the polar quality "hot" is 500°. Gravity, or what they call cold, just means things go down and become denser. Water is denser than air but it is a compound made of two gasses or "airs." When air gets cold, it moves into condition of gravity and falls through gravity into water. Water and earth come under the influence of gravity, whereas air and fire come under the influence of levity. Levity and gravity give rise to two great divisions of the mandala of the four elements. Earth and water on the left in *figure* 2 represent gravity. Gravity has the fundamental quality that alchemists call *coagula;* gravity is coagulating and supports the coming together and dropping of substances into denser conditions. Through the action of gravity, systems of forces that are in flux undergo a process of coming together and falling, or slowing down enough so that substances fall into manifestation from an invisible realm. In coagula, the molecules do not vibrate so much. This condition is coagula.

Figure 2 represents a selection of terms that Rudolf Steiner uses to characterize these forces of coagula or gravity. He said they are *centric,* that they are earthly. He means that they are on the side of coagula. On the other side of *figure 2,* the polar opposite forces to gravity are those of levity. Steiner uses various synonyms for those forces. In one context he calls the levity forces cosmic. In another they are described as peripheral, or planar. These terms all refer to the "levity forces." Thus, when he uses this kind of language in the agriculture course, we can use this diagram to follow his thinking, even though the terminology shifts. Coagula is the gravity pole and the pole opposite to levity is called solve alchemically.

Solve has hidden in it *sol,* which is the sun. The Sun is out there on the periphery of the earth. Solar forces characterize the action of the light pole, the light, and the warmth. This pole of the mandala is *solve* instead of *gravitas.* So out of solve you get conditions in which minerals suddenly come under the influence of levity in a solution, becoming so fine that they no longer fall out of the water. They become soluble through the action of a solvent. They become "sunned" and are taken to the cosmic dimension.

Earth is gravity, because our center of gravity and the earth force is coagula. That is what makes embryos able to form out of juice. Alchemically, an embryo is coagula. Coagula is what makes rain that precipitates out of clouds. This is a kind of meta-language that alchemists used, and if you begin to understand it, much becomes available to you, especially in the work of Rudolf Steiner regarding new ways of working with agriculture.

Let's return to the solve side in *figure 2.* We have mobility, cosmic forces, peripheral forces of the outer planets, silica, flower and seed process, potassium and phosphorous. This won't make any sense at all except for the two big divisions; but as we go through and I keep referring to certain things, when you have a question I will say, "Look at *figure 2.*" That is a list of many words that mean the same thing in the alchemical language Rudolf Steiner uses in the agriculture course. He interchanges them, and it is helpful to know that there are just two big divisions of qualities and that each new word does not describe a completely new quality. His choice of terms is not all over the board. The different terms refer to the two great polarities of solve and coagula.

In *figure 1,* we have a mandala, and in the center is *akasha.* On the left we have *gravity, condensation, precipitation,* and *coagula.* That is what we have just transposed from *figure 2* over to *figure 1: gravity, condensation, precipitation,* and *coagula*. On the upper right, we have *levity, evaporation, solutions,* and *solve*. However, *figure 1* shows you that if I wish to go from earth to water, the arrow shows that I have to go and touch *akasha* to do it. This is the great secret of the alchemical work—that the phase shifts you work with have to do with your consciousness and the ability to see differences in things: Oh, this is different from that, but this is changing into that. Oh, this seems to be connected to that, and here I can find a transition. The transition—the place where this shift happens between earth and water—is a kind of consciousness between water and air. It is a kind of consciousness of what Steiner calls "beings of the elemental realm."

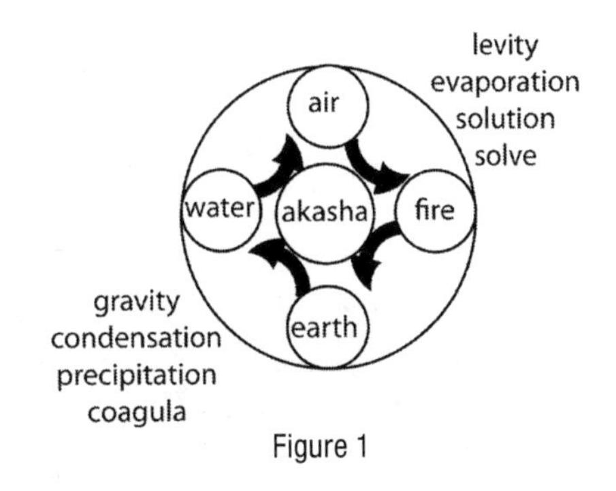

Figure 1

Later, we will discuss consciousness and the meditative exercises that help you to take motifs from nature into sleep. This is a great help when you are trying to solve problems in your gardens. If you understand the way the pictures work with regard to the elemental worlds, you can go through consciousness and give precise pictures to the elemental world as you pass from waking consciousness into sleep. Then, when you go out and touch the soil the next day, you start to get imaginative pictures of possible ways of working. The key is that you have to put a viable seed picture into the soil of the elemental world before it can germinate into an imagination. If you keep your vegetable seeds in a jar on a shelf, they will not germinate in the spring. You have to put the seeds into the soil. Where is the soil? It is your sleep. Where do you form the seed? You form it in your mind through good thinking. Clear thinking makes the seed pictures viable. When you can put clear seed pictures in the soil of sleep, that is called "imaginative thinking," or as Steiner puts it "living-picture thinking." If living-picture thinking becomes a meditative practice, then some morning there will be a little green thing coming up in your consciousness. That is called an inspired dream.

When I want to enhance my ability to see into nature, I can use these kinds of activities to help myself see more deeply. If I consciously take pictures of transformations in nature into sleep, then the elemental beings who live on the other side start to take an interest in my work on the land rather than running away from me all the time because I am doing crazy things.

What we work with is called a "gesture motif." Our goal is to develop an organ of perception in the heart so that everything that changes or patterns in nature can make an impression on me that I can lift into waking consciousness. *Figure 3* shows a diagram of two molecular planes in a mineral. You can see little pyramids from above and from below. Each of the pyramids is a molecular structure of a typical mineral silicate. In the silicates, the fundamental chemical form of a molecule is a tetrahedron, which is a pyramid made of four equal triangles arranged to have a common center when placed within a sphere. In the figure, you can count four triangular planes. There is one in the back, one on the left in the front, one on the right in the front and one on the bottom. Four equal triangles within a sphere compose a tetrahedron, the plural of which is *tetrahedra*. The tetrahedral form is the most fundamental mineral form on the planet, whether a cinderblock, a quartz crystal, or whatever. If it comes from a siliceous material, which makes up eighty percent of rocks, it has a molecular tetrahedral form according to science. In crystallography the arrangements of the tetrahedral is a very interesting subject for study. Tetrahedra have various arrangements that can provide insights into latent forces in the minerals that may be useful in agriculture.

We have thousands and thousands of different silicates, but they are all silicates since they all go back to this tetrahedral form. Later we will visualize the tetrahedron as a meditation. It is really an interesting form, because it is very potent as an *archetype*. Archetypes are the sources of what we call gestural motifs. This means simply that, wherever you find silica, it will be in the form of some type of tetrahedral molecular structure. Whether it is in an amethyst or in a tourmaline, any silicate mineral has a tetrahedral molecular structure as its basis. The arrangement of granites, feldspars, and micas—even basalt—is a silicate with a tetrahedral molecular structure.

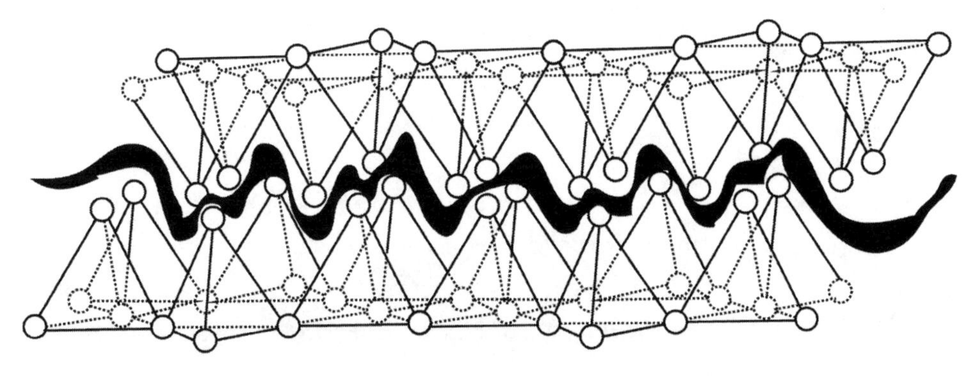

Figure 3

In *figure 3* you have one of the most common arrangements. You have planes composed of tetrahedra that are linked by their points. There is one plane above and one below. The points of the tetrahedral in the planes are directed toward each other like little dogteeth. You can see a space between them, which represents the spaces between molecules through which rainwater enters the rock and extracts other minerals. This process is called *weathering,* and it makes minerals available to your plants. Without those molecular spaces, there would be no way for rocks to weather minerals into soil and no way for the root hairs to interact with the soil. Without the weathering process, silica would not be available. Plants need potassium, calcium, phosphorous, and nitrogen to grow. The fact that these silicates have this form—a tetrahedral form that allows for inner surfaces—is a huge force in nature. The inner surfaces are where energy transfer happens; thus, when I wish to understand growth, how minerals are transformed, how organs are built, or how water works, I have to understand the action of surfaces. In the mineral silica, we have an infinite inner surface that constitutes the action of the alchemical element earth. Silica is arranged into very fine inner surfaces that provide a scaffold for life.

In the third lecture of his agriculture course, Rudolf Steiner talks about carbon. He says, "Carbon is a scaffold to which oxygen keeps dragging life." This describes the carbohydrates as scaffolding. We may wonder what carbon has to do with silica as a scaffold, but carbon and silica are octaves of each other in the periodic table. They are sisters. One is organic and one is

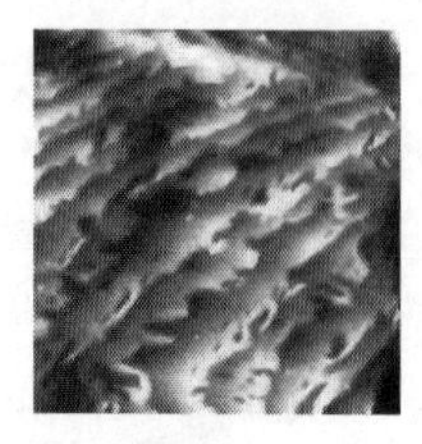

Figure 4

inorganic, but they both serve nature by providing similar archetypal gestures as scaffolds for other developments. They both sacrifice themselves so that other forms can arise.

Alchemically, it is useful to think of how the earth sacrifices. It is generally a sacrifice so that a spiritual being can incarnate. It is a sacrifice to go from a condition of mobility to suddenly becoming fixed. It is a sacrifice to go from a condition of potential to manifestation. To experience this, go out to a parking lot and look at a rock embedded in the asphalt, then think of the rock in terms of sacrifice. Look at the power of a mountain that endures for centuries and think in terms of sacrifice. This is an interesting exercise. To fall into a body is the ultimate sacrifice of a spirit. It is coagulation instead of solve; the prison of what has become instead of the freedom to become. Whatever has manifested has lost the freedom to become. Alchemically, these pictures can be used to contemplate the mineral earth, especially silica. If I look at *figure 4,* it looks like a spanakopita or a flaky croissant; it is a microphotograph of the silicate mineral mica.

When silicates go through the weathering process, certain minerals are released through "ionic exchange," or cation exchange, which stream out of the mineral body and into the soil solutions. As the liberated minerals flow through the soil, they make their own forms and variations of the tetrahedral archetype. Thus, the fluidic nature of earth is a process of flowing as solve and depositing as coagula. If you look at the structure of the mica, you can see a gestural motif that resembles leaves packed together. Alchemically, in the imagination we can develop from this, the mica represents a leaf gesture among the minerals in the earth body. Mica represents the leaf of what Rudolf Steiner has called the "world mineral plant." If you open up a piece of mica, it breaks into little simitransparent leaves. There was a time in Steiner's alchemical imagination when all the plant archetypes as spiritual templates were embedded within the mineral body. The mineral and the plant had not yet separated. However, the potential for leaf formation in the mineral created leaf after leaf in the mineral body itself. We see the activity of this global leaf-forming process in the formal gesture of mica.

At first, this sounds a bit farfetched from a natural scientific view, but these images can be found even in geological terminology. The geological process of mica formation in a rock mass is known as "exfoliation." The formation of exfoliates occurs in huge granite masses such as those in Yosemite, California. The large seams where the granite peels off of the rock faces in that area make the granite domes look like onions. The peeling process of those layers is called exfoliation. In Latin, *ex* means "from," and *folia* means "leaf." The granite faces resemble peeled-off leaves as glaciers moved over them. This geological process arises because of the particular structure of the mineral. There are weak areas that arise along the seams, where water percolates in, then freezes and expands, and pushes a huge leaf off of the face. In spring, when the ice melts, the big leaves fall off. That is exfoliation, the plant nature in the granite domes of Yosemite.

The world mineral plant is manifested in the molecular structure of the granite because of the formation of planes within the rock. Taken to the smallest level, these processes occur because the silicates have an inner tetrahedral planar formation that allows water to enter. Thus, from the micro level to the macro level of the exfoliation of a whole chunk of rock in Yosemite, similar formal processes are at work. The formal archetype is the division of the mineral substance mass into infinite inner surfaces. In a dynamic picture, the inner surfaces of the mineral support the weathering process that eventually becomes soil, which begins through a kind of inner formation of planes in the minerals. Those planes unite, and the mica nature allows the interaction of water. Potassium, sodium, and calcium, which contribute minerals to the feldspar formations, come from the mica nature. The key formative motif is the planar layers that present an abundance of surfaces. In nature, surfaces are the sites of transformation.

Figure 5 (next page) is a microphotograph of a structure called *grana*, which is the arrangement of the structures in the chloroplast of a leaf that contribute to photosynthesis. These little plates of chlorophyll are arranged to receive the light and amplify it. This arrangement is similar to plates of

Figure 5

metal immersed in the electrolyte of a car battery. Going in a micro direction, we start with the leaf; inside the leaf is a cell; inside that is the protoplast, in which we find the grana; and inside the grana are little plates that resemble the micaceous nature of the leaf of the world mineral plant. This formal motif is an analog of the silicious formation of planes through tetrahedral molecular patterning in the granite, in which the mica exhibits a proto plant motif.

If we go toward the macro, we have a leaf that is a plane, which is folia we can then take to Yosemite and see that the rock face has a formal motif resembling a plant. This kind of reasoning is known as the *method of analogy* in alchemy. The formal or gestural motif, beginning with the molecular form of the silicate, is carried all the way up into the plant in one direction, or out into the massive rock faces of mountain ranges in another evolution. Behind these formal motifs and analogs is the great granddaddy of all gestural motifs known as surface. Maximizing surface is the goal of agriculture. Maximizing surface is the central player in the formation of plant life. Maximizing inner surface is the role of humus. Humus has an infinite number of inner surfaces that allow the root hairs to find purchase in it and to find water when there is none available in the soil solution. However, the structure of it is analogous to the mineral forms just described.

Surfaces in nature provide the sites of maximum interaction between various life forms or levels of organization. In the plant, the inner surfaces of the leaf layers are used to draw in light. The grana formation serves the leaf as a kind of light antennae; it is an array of solar cells within the chlorophyll that draw in light of different wavelengths to energize the metals within the plant's sap. This is the function according to science, but the formal or gestural motif, beginning with the earth in the silica, is taken up through the surfaces in the minerals to the surfaces of the humus to the surfaces of the structures in the plant. The gestural motif of ever-more developed inner surfaces serves as a key to understanding how the life, how the energies in the organism are developing.

There is a useful exercise one can practice when working with gestural motif. We have the polarity of looking *at* a leaf and looking *with* a leaf. When you look at a leaf you usually see just the surface and a few details—enough to identify it as a leaf. Most of the time, we really do not see it; one's akasha just glosses over it. However, when we begin to form an inner picture and harmonize our akasha, or consciousness, with the consciousness of the elemental beings behind the formative processes, the leaf will begin to interact with your consciousness. This is because our consciousness determines the quality of the transformation. We could say that those beings are enthusiastic about interacting with a human being who is aware of the reality of the spirit.

The question is this: How can I actually work with that level of seeing through time as a meditative tool? When I develop a practice of looking *with* natural phenomena instead of looking *at* them, my consciousness becomes coherent like a laser, because I am merging rhythmically with the phenomenon. Thus, the first level of this work is harmonizing the mind with the phenomenon, and the second level involves the rhythmic timing of one's inner imaginations. The link is the fact that a gestural motif has a kind of rhythmic structure, which allows me to harmonize my consciousness with what Goethe called *the becoming of the phenomenon*. If I see the form just as finished, it is difficult to harmonize my mind with it. However, if I try to hold in my inner eye the process of how the form has become what it is now or how it went through a phase shift, the movement that my mind makes, or my consciousness makes, begins to harmonize with the archetype behind the phenomenon I am observing. When this happens, the deeper levels of implications of the form start to organize within my mind when I go and touch the soil, do a transplant, make a compost heap, or even smell a heap that is working. I will smell a certain something that triggers actual movements in my soul, in my limbic structure—actual chemical movements. If I train my senses, I harmonize my consciousness with those movements, and suddenly I can become aware that something is not quite right there.

The people in ancient times did not have to train for this, because their minds were not otherwise engaged with the IRS or the stock market. They just lived in harmony with these imaginations of the natural world. Today, we have lost the depth of that involvement, but this does not mean it is impossible to engage nature at the level of spirit. We can reacquire capacities to see into nature, but we must regain them consciously. In the past, people knew where the power spots were situated on the land and when all the elementals were upset because someone took the capstone off of the spring. It is practical to have this kind of imagination, but those kinds of imaginations do not help me to form concepts of what I am seeing inwardly. To do that I need to understand the laws behind what I receive in my imagination. I need to train myself to observe in a truly clear way and then rhythmically form inner pictures of the motifs. It is the rhythm that allows me to harmonize with the phenomena. This is especially true when I train my inner eye to visualize the seasonal processes.

Seasonal processes actually help me discover meditative methods for understanding phenomena. Esoterically, this is true because the seasons are actually spiritual beings who guide the transitions of natural phenomena. Rudolf Steiner calls them the *spirits of time rotation*, the highest level of elemental being. These spiritual elemental beings are what you track as you work with your planting calendar and things like that. When I meditate with pictures from nature through time, in harmony with the seasons, it is possible to form an accurate sense of when and what to do when I go out into the garden. More will be said about that later.

These methods of visualization are called the method of analogy in alchemy. In the analogical method, a mineral such as mica can be an analog of the plant. Those in turn can be analogs for the layers in the digestive organ of the intestines, which can then be an analog of the evolution of a human being from a physical body to life forces to a soul and to a spirit. We could call this the evolution sandwich of layers of similar motifs. Moreover, the motifs of the gestures are the same—with variations, of course, depending on what layer or kingdom I am visualizing. This kind of scientific process is based on what we call anecdotal evidence. Empirical scientists say that two anecdotes do not make one data bit. The great need here is to develop my inner eye in such a way that I can determine whether what I am seeing is a true analog or just something I am making up. This comes down to the fact that anecdotal evidence needs to be verified by very clear thinking, otherwise any discoveries will be discounted by those who matter in science. We need to make bridges to researchers willing to accept some of the more intuitive processes in biodynamics. If we can actually show some basis for this method of analogy, the work would be furthered through cooperation with others.

Figure 6: Phantom crystal

With *figure 6* we will try to do this kind of visualization process with analog. For instance, consider what is flowing in the mineral matrix as potassium. Potassium is the thing that builds the ion channels in the formation of the protoplasm in a leaf. It does exactly the same thing in the leaf that it is doing in the mineral mass. It builds channels for the movements of ions so that potassium can get out of the rock face into the soil. Ionic potassium in the soil solution is taken up by the plant and moved into the leaf. It has a similar function in both the mineral realm and the plant realm.

These qualities of potassium that can be traced scientifically are also foundational to the method of analogy. The pictorial rendering of the scientific facts is a valuable part of this work because it then allows you to say, "Well, hmm...what picture would I look for in an animal if I wanted to make a sheath to make a new preparation that is active on potassium? Suppose I wanted to make something that enhanced that action. What would I look for in an animal sheath? What herb could I look for that would actually make that dynamic stronger?" We could even ask what the action of potassium represents in the soul. You can see the value of the pictorial method of analogy.

You can do some pretty useful thinking with this because you are not locked by the fact that the mineral is separate from the plant. We want to

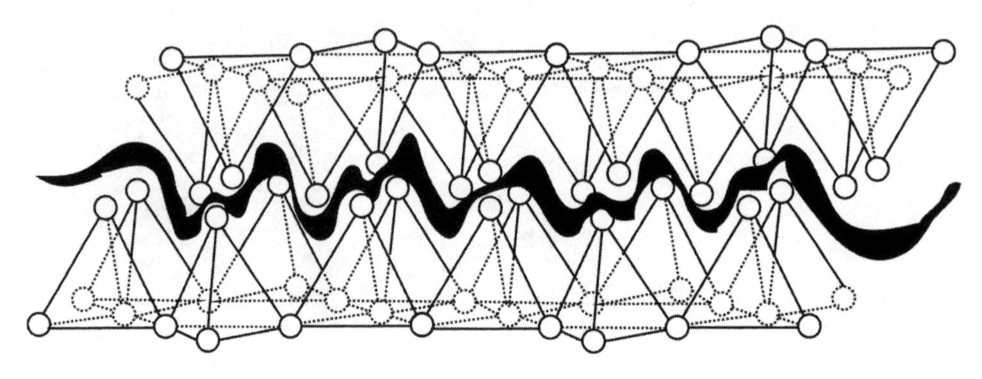

Figure 3

learn to see that the mineral is a dynamic in the plant; it is a process. It is an activity. When I look at these pictures as a scientific process, the form of the organ is a key to its activity. The method of analogy uses the language of form and gesture to look for similar motifs. In the earth realm, I have the actual mineral earth with its formal motif, but I also have plant structures similar to the formal motifs found in the earth. In the figure the forms of the grana, the leaf structure, and the stem structures of xylem and phloem all have the form of a membrane.

A membrane is the form that allows the water to get from the root of a redwood to the top. Membranes in nature are the sites of capillary action, a force associated with membranous surfaces where the water is moved by the variations in surface tension of water. The molecules in water and the surface tension of the molecules of the organism's structure interact to move fluids. Water under the influence of very fine membrane complexes can overcome gravity to an extreme height because of the fine membranes within the growing parts of trees. This allows water to rise out of the soil. When you think about it, water rising three hundred feet inside of a plant seems to go against the fundamental law of gravity on earth. The water is interacting with forces that go into another realm. Although we just take it for granted, it is really a miracle. The water used to be in the soil, and now it is growing. What is growing?

Energy is a ubiquitous term for phenomena we do not understand. We say something has energy. With life energies tracking presents problems.

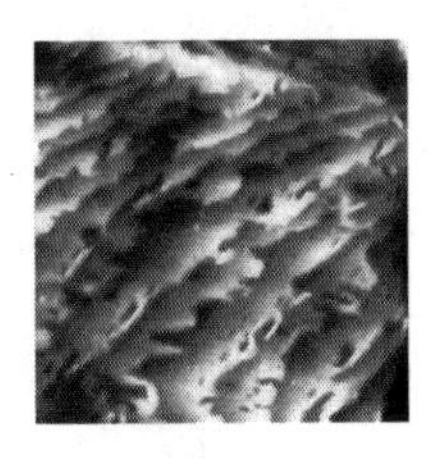

Figure 4

Movement of life energies however we can track. *Figure 3* is basically a diagram of *figure 4*, a photograph of mica, while *figure 3* is a diagram of what that looks like. There are molecular spaces between the many inner surfaces of the layers of the mica. Wherever you have surfaces very close to one another and when water comes in contact with those surfaces, there is a reaction between the forces creating the surfaces in the water and the forces of the surface of the physical object. If the physical surfaces are fine enough, water gets pulled up into the layers; it moves up into those surfaces. That is capillary action. The finer the surfaces, the more activity there is. Now, when that water goes up into those rock surfaces, its movement is very close to what happens in a plant when fluids are transferred along all of its inner surfaces. The exchanges in both the plant and the mineral are analogous actions. This is because the surfaces and the fineness of the surfaces create the potentialities for movement in both cases.

However, you could put distilled water in the mica in *figure 4*, and nothing would happen; the water would just be water. But if the water has been percolated through either the dead body of an animal or a plant or come through something in the atmosphere that made it slightly acidic, then when that acidic water goes up into that little surface, there is a pH difference between acidic water and the mineral form. This action begins to pull ions of other minerals from the rock. These other minerals from the country rock change the nature of the solution of oxygen and hydrogen in the water. Calcium and potassium get pulled out of the country rock and recombined, and then more water comes in and flushes these solutes out. When the water that has been changing interacts with a plant, it has a charge. Water that has a charge caused by the minerals in it is full of ions. Ions are interactive. When the ionic fluids in the soil meet the plant root, the plant root exchanges with the water and the inner surfaces of the root absorb what was released from the mineral because of the abundant inner surfaces of the plant.

I will present a little drama to illustrate the idea of an ionic charge. Suppose my friend David and I both get up in the morning. I am having a good morning, he is having a good morning, and we are on the subway. Neither of us have a charge. Here we are, on the subway, the door opens, and then we both go out. Now, the next morning we get up, we go to get on the same subway, but I woke up in the morning with a headache and my refrigerator motor caught on fire at six in the morning. Now we go on the subway, but today I have a charge and David does not. He steps in front of me, and I give him some lip: "Hey, that is my space man!"

What we have just acted out is ionization, the transfer of charges of attraction and repulsion at the molecular level. Alchemically, attraction is coagula, and repulsion is solve. Get thee to the periphery says solve. Get out of my space. Go over there. That is repulsion.

We are back to coagula and solve when we try to understand ionization. After my refrigerator motor caught fire, I became a little acidic, or sour, in the subway. Then another person was next to me with a neutral charge, and suddenly there is an exchange of charges. Charges and transfer of charges through imbalances create movements of "energy." These exchanges are actually a kind of a consciousness at some level of organization. An ion is active, and if an ion comes out of a rock and is suddenly in the soil, where it meets a plant root, there is an exchange. The exchange is life. That movement is life. That is what water does; it creates potentialities for life. Water carries charges and imbalances, then whatever meets that water interacts with it. Water carries certain kinds of charges and movements that result in variations in pH, temperature, and velocity.

For several years I have been researching colloid structures.[2] A colloid forms when a mineral gets to such a small size—perhaps between one and a hundred nanometers—so that when it is put into a liquid, it forms a kind of gel.[3] A particle that small, when found in a liquid, is known as a colloid. We are doing some work with minerals to try to get them to go into the colloidal

2 *Colloid* is a system of particles somewhere between 1 nanometer and 1,000 nanometers in size. A nanometer is one billionth of a meter. By contrast, a solution is composed of particles below one nanometer in size. Particles in solutions never settle out. Particles in colloids selectively settle out, and particles in suspensions always settle out.

3 Technically, the prefix *nano-* means one billionth, or a factor of 10^{-9}. In a more generic sense, *nano* may refer to extremely small particles.

state. Below the particle size for a colloid, the mineral is said to be in solution. Particles above the size for a colloid are called a suspension. In a suspension, eventually all of the particles will fall out. You can grind up a rock, and if you hit it right that rock goes into the water and just stays there. The particle floats around because it has a charge on it that repels all the other little particles. Because they all repel one another, they form an evenly dispersed net of particles that tend to lock onto each other at specific distances.

I wanted to find out what makes the particle fall out, and it took me two and a half months of bleary-eyed Googling to find someone who did a simple experiment in 1902. He worked with very subtle changes in the pH of a solution of China clay. To find an experiment done with a simple material in a simple condition, I had to go back to 1902 to find someone who did it, but it was perfect. The experiment showed that the perfect pH for a colloid is just slightly basic above the neutral seven. The experiment was done with seven precise mixtures of China clay in water. At the neutral pH of seven, the clay particles settled out. At a pH of 7.001, however, the clay formed a beautiful colloid. At 7.003, it began to settle out again and at 7.006 there was more settling. By the time the mixture got to 7.1 it had settled out like a suspension. This experiment showed the intense charge-changing potentialities for even the smallest little bit of acid or base. These minute differences create the various potentialities for the patterns of how things move through living bodies.

For a while, that experiment stalled my own experiments for producing gems in the colloidal state. My goal was to create colloids out of gems that would allow plants to take up the gem most easily. It was such a narrow range of pH for keeping a colloid up in the liquid. It turns out that a perfect colloid cannot go through a membrane, because the colloidal state is the way minerals are transferred through liquids in living bodies. However, once the mineralized colloid meets a semi-permeable membrane, the prevailing chemistry in the cell changes the pH of the colloid through the interaction with the membrane. This interaction then creates a condition whereby the colloid starts to fall out, and the ionized mineral becomes available for the dynamics of the cell.

Nature is brilliantly arranged isn't it? This kind of interaction guides the movement of minerals in your blood and cells. The minerals and metals in your blood have a relationship to the movements of those same minerals moving through the plant saps and in organs of animals. Alchemically, a colloid represents the balance of the forces of levity and gravity in your body. Thus, we are back again at levity and gravity. These qualities change the relationships of the surfaces interacting in your body. These movements are the earth, water, air, and fire elements of alchemy. Once again, it is important to understand the activity of surfaces in the natural world. This is why the pictures of minerals are included here. Surfaces are where the activity is focused in the natural world, especially inner surfaces.

The degrees of fluid viscosity go from tar to water. Tar is coagula, and water is solve. Water is the universal solvent. It is the universal solve, and it can phase shift into coagula when it becomes cold and turn into ice. However, water can also go beyond its liquid state and intensify in solve into the Sun, or today we would say *into a gas*. The major physical components of the Sun are hydrogen and helium.

Figure 6 shows a silica crystal known as a phantom. the molecular structure of this crystal is such that, when the crystal is building layers, there is a phase during which something comes with the fluid moving through the rocks. Consequently, mineral inclusion is deposited in the crystal as a "phantom." This phantom allows us to see that the geometric structure of the crystal is carried all the way through, right from the very beginning. The crystal formation process adds layers from the outside. The archetypal crystal form causes the formative process to add material from the outside. Those added outer layers are based on the initial tetrahedral arrangement. The outer layers are added as surfaces, and those additional surfaces from the outside become inner surfaces in the crystal. The inner surfaces are the pathways for energy transmission; they are templates by which nature becomes ordered. Much of what we do in biodynamics has to do with strategies for working with inner surfaces.

Silica and a cow horn share fundamental motifs for the addition and multiplication of surfaces. A cow horn is an analog of a silica crystal,

Figure 6: Phantom crystal

because the inner surfaces of both are formed in similar ways. In terms of actual substances, a cow horn and a crystal are very different, but the analog process of the formative motif is very close in both. This is because of the way the formation process in each takes place. Although they are very different substances, the molecular structure of the materials is similar because of their similar formation processes. If you look at a microphotograph of the protein structure of a cow horn and the molecular structure of silica, you will see that they are images of each other as formal, or formative, motifs.

This is a good example of what I mean when I say that science can support many of Rudolf Steiner's insights. Nevertheless, we need to understand the similarities and differences in life forms. When we use the method of analogy, it is absolutely necessary to be very discerning about formative processes. It would be an error to say that a cow horn is made of silica; it is not made of silica but a protein. Nonetheless, it's structure and formative morphology are very similar to a silica crystal. Many proteins can be crystallized when they are highly purified.

In our alchemical mandala, the earth realm is generally considered to be only representative of the mineral realm, which has gestures that can be seen as archetypal motifs we can apply to plants and animals. Our work in the mineral realm is to understand the inner structure and formative motifs of the minerals—granites, micas, and feldspars.

Some people say that in preparation 501 you can use only certain types of mountain crystal.[4] However, in his agriculture course, Rudolf Steiner says, "You can also use feldspar." He means that we can play with this idea and work with it, but we need to understand what we are doing if we use

4 Biodynamic preparation 501 is made of crushed, powdered quartz. It is prepared by stuffing it into a cow horn and burying it in the ground during spring and removing in autumn. See, for example, Klett, *Principles of Biodynamic Spray and Compost Preparations.*

something other than quartz. We need to understand the differences among plagioclase feldspar, orthoclase feldspar, and a silica formation. The structure of each is very different. What we use depends on whether we want the calcium to come on or want potassium to be present. These two feldspars are silicates, but their inner mineral structures have very different gestures when it comes to the way the charges are arranged. Thus, the ways that the structures of the minerals are arranged will give a different picture of the various plants and animal organs that serve as good sheaths. I think there is research that could be done in this area. We could try feldspar, amethyst, or tourmaline. What sheathes would they require, and what would their action be when sprayed on plants or on the ground? How do the forces in these different forms of silica move?

In the earth realm, the mineral carries the gestural possibilities for plants and animals and then the human being. There is a kind of consciousness that the mineral represents. As consciousness, the mineral represents all that is lawful and ordered. It is useful to study the mineral realm, especially the inner structures of minerals and, most especially, the inner formative processes of the silicates, because they have such profound effects on agriculture and soil production. It is useful to study the molecular formative gestures of the silicates as gestures of other things. You should find out how an amethyst crystal is different from a quartz crystal and how a tourmaline grows. Although they are all silicates, they have very different formative gestures in the life body of the Earth, because they represent different forces of attraction and repulsion. The molecules arrange themselves in different orders that reveal themselves in the form of crystals. Our experiments have shown that if you take the crystals of, say, an amethyst and put them in a cow horn to make a spray, you will get a very different gesture in the sprayed plant than if you use quartz.

The amethyst has properties that I would describe as a tendency to hold back and prevent shooting out to the periphery. If you look at an amethyst geode, you can see this holding-back gesture. The quartz crystal, by contrast, shoots into a long scepter form. When used to make a spray, this formative tendency drives plants to move toward the light. This is why Rudolf Steiner prescribed quartz as an amendment for the problem of rust in wheat caused by the cold summers in Europe. There is a climate anomaly in European summers when a jet-stream loop out of the north penetrates far to the south and keeps Europe generally cool during summer. This creates problems with crop maturation and ripening, especially in the north. Grapes from the north make a very different wine from grapes grown in Monterey because of the colder growing season in European summers. This cold creates problems with fungus and mildew.

To ameliorate the cold and damp gesture, something needs to be applied that represents the light pole. Steiner said to take the pulverized quartz crystal and put it into a cow horn, then mix it in water. When you spray that water on a plant, it will bring the light to the plant. The silica pushes the plant out toward the light at the periphery. In doing this, the quartz acts as a prophylactic against the mildew, because fungus and mildew do not relate to the light. Quartz crystal brings this gesture of pushing up and out when sprayed on plants. However, if you spray this during a California or Midwest summer in July and August, you will get a lot of burning because it enhances the light. Enhancing light is what the quartz does.

Years ago, I took some amethyst, pulverized it, put it in the horn, and then sprayed the amethyst preparation. I was looking for a moist light instead of the intense light of the silica. Among gemologists, amethyst is known as a "Brazil Twin." The tetrahedra that form the crystalline scaffold do not grow in long chains as they do in quartz. They go right and left, right and left, as the crystal grows. The molecular forms counter each other and that is the gesture you see in the amethyst that is in a geode. The quartz is held back and spreads horizontally and many little points form all around a common center instead of one in the center dominating. The crystal holds back from shooting out. So what is this mineral gesture good for when used as a spray? What plant gesture are we trying to influence? Lettuces and cabbages are crops that you want to hold back from shooting out into seed. You want those crops to form many leaves around a center rather than shooting out and flowering. Amethyst is a silicate, but its molecular pattern has a very different gestural motif than quartz does. As a result, it can be processed like Steiner's indications about quartz, but its action is very different from that of quartz.

Thus, we have a couple ideas that we have established here. We have solve and coagula as the two key alchemical ideas. Solve is the levity pole; coagula the gravity pole; out of levity and gravity, we have these combinations of various gestures of levity and gravity. Does this mineral or plant shoot way out, or does it hold back? There is a gravity gesture in the amethyst that is not present in the silica. The levity force of the quartz pushes the plant out toward the periphery. Next, we will take a look at the water nature from the perspective of the mandala of the four elements earth, water, air, and fire.

II. Water • Tension Membranes • Energetic Boundaries • Minimal Surfaces • Water Phenomena • Stirring • Imaginative Exercise

First, let us revisit the idea of the four elements before taking the work further, focusing on water as an alchemical element. The four-elements theory is very ancient and has to do with what today's science calls a phase shift. In contemporary language, a phase shift is a movement from solid to liquid or from liquid to gas or from gas to pure warmth. The concept has a rather mechanical quality or, as with atomic theory, a kind of geometrical quality, with bonding angles and that sort of phenomena seen as the operating principle of the shift.

When the four elements theory was developed, there were no bonding angles or quantum mechanics, and the four elements were the abodes of gods. They were the abodes of transcendent states of consciousness that acted on earth to make things happen. There were earth gods (*cthonic* gods and goddesses), gods of water and air, and planetary gods. We could say that the elements were ensouled. *Ensouling* means that the elements have a relationship to human beings, because human beings have souls, and souls are the abodes of spirits.

Alchemists would say something like this: Well, if a human being as a spirit has a soul, and if earth as a spirit has a soul, the spirits of the gods must form the soul of the earth. If that is the case, then it must be possible for gods and human beings to communicate through the soul of the earth. Human communication with the earth soul was the central mystery of alchemical practice. It was understood that the soul life of the earth, the soul life of the gods, and the soul life of human beings were interwoven. This interrelationship made it possible to transform things in the earth body that are considered impossible by today's rational science. The classic alchemical reference point is contained in the idea of transforming lead into gold. If you really understand that alchemical work, "lead into gold" had to happen first in the soul of the alchemist. Only then was it possible to enact that transformation in nature.

If you desperately wanted to make gold, guess what... you would not be able to make it. If you were unconcerned about whether you made gold, that attitude came a lot closer to understanding how to do it. Even then, however, you would need to wait until the spiritual world thought it would be a good idea. Of course, it was a matter of attitude and consciousness that allowed nature beings to trust you enough not to mess up if given the power to change things. When you gained enough power to change something, you were given the power because you had shown the beings who control the power in nature that you were worthy of that gift. However, people, being who we are, had a work-around strategy called black magic. I spent several years cruising on a nuclear missile submarine during the Vietnam War, and I saw enough to know that today black magic is called nuclear power. We look to it as a great source of power, but we do not understand its relationship to consciousness. An alchemist would have a different take on this subject.

Consequently, in today's world we have a split between the human consciousness that can understand the spiritual will nature of technology and the human consciousness that considers spiritual work somewhat passé.

Many believe that spiritual ideas are passé. They are considered obsolete and best left in the dust of the past, because today most people consider technical intelligence most important. This polarity is present also in biodynamic circles. There are those who are university trained and went from chemistry labs into biodynamics. They want to do formal assay processes on what comes out of the cow horn in order to understand what Rudolf Steiner said. That is the one side. The other involves those who could not care less about laboratories and understand that, if they simply ask an elemental being a question and go out in the woods, a tree will speak and give them the answer. These two camps are split even in biodynamics. There are those who say, "We do not need all that cognitive stuff, because we have powers to communicate with elementals." The others say, "Forget your powers; we have to bring this down into experimental methods that can compete with science."

My goal here is to see if we can bring the two sides together. It is my understanding that there are practices that enable us to develop an organ in our heart known as *imaginative cognition*, which is just as precise as what one would do in a laboratory. Goethe called this "higher beholding." Imaginative cognition means being able to control consciously one's inner picture-making process. That is the first step. The second is to prove to yourself that a picture you have in you is not a reality in the spiritual world. If you cannot prove that a picture you are receiving is only an image and not a reality, then you are off with the fairies and they are in the driver's seat regarding your consciousness. If struggling to determine whether an inner picture is a reality or an illusion prevents you from having an inner experience with the nature beings, then your experience of those beings was in all likelihood filled with personal fantasy. By contrast, a need for scientific rigor that prevents you from having a spiritual experience is not the goal either.

I am describing a very delicate space. That space has to do with who controls the inner picture that guides your human consciousness. In spiritual reality, you have the capacity to control the inner picture, because your consciousness has an element of earth. Earth consciousness arises when you are discursive and can say, "This is what I understand. This is what I see. Here's a piece of wood. Here is a leaf." When you look at the leaf and see the leaf, you are in a consciousness of earth. Here you are, and there is the leaf out there in space. The consciousness of that separation is called the Fall of humanity. In the Bible it says "And their eyes were opened, and they saw that they were naked." That is the Fall of the human being that occurred when Adam and Eve ate from the Tree of Knowledge. You looking at a leaf recapitulates Adam and Eve's expulsion from the Garden of Eden.

The experience of separation from the sensory world is the Fall. It created the consciousness of earth in human beings; "I am the one who is naming everything." That is Adam's version. I am the decider. Job's version of the same experience is, "This is me here, God. I know you are the great God but, excuse me, I did not do what is being said about me, and you cannot say that I did." This is the human. Carl Jung said that Job is the modern person saying to God, "I know you are the big guy on the block, and I know I am a sinning worm, but I did not do it."

Job is a great story because it puts its finger on this problem of what we can call "mystical experience" on the one side, which is the ultimate solve in the realm of consciousness. This is the influence of Lucifer and mystical wanderings. This is the belief that the trees talk to you because you are so special. The other side of this diabolical coin is Ahriman—the intellectual discursiveness that discovers the difference between calcium number two and calcium number three because of the molecular weight difference between these two forms of calcium. This influence is the opposite of mystical wanderings in one's search of being special. Ahrimanic consciousness is in search of power over nature. That influence is Ahriman and, alchemically, coagula, the conviction that I am absolutely, terribly certain this reality that I cognize is the only way that this can be viewed. In the Book of Job, the devil is Satan, a combination of the two. Modern science revels in the potential for power over nature.

Today we see water as one of the biggest of all mysteries. It is such a big mystery because it is based on two gasses—hydrogen and oxygen—that combine to form a liquid and retain the power of being a universal solvent.

Here we have our earth, water, air, and fire mandala. It is divided on one side into the gravity/coagula of earth and water, because they are attracted

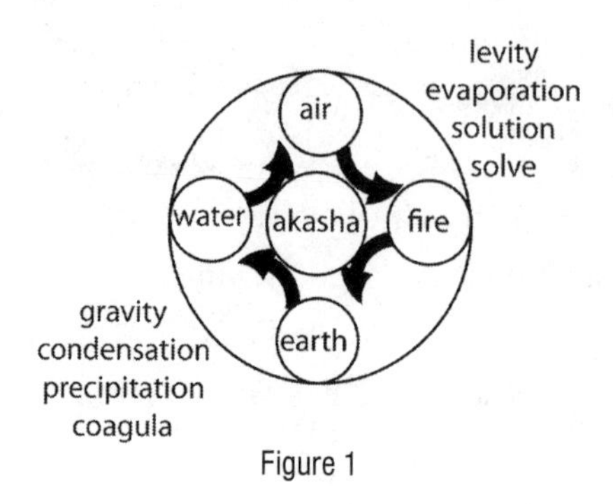

Figure 1

to gravity. On the other side, air and fire are connected to solve and levity. These two are attracted to the periphery. Gases go toward the periphery. Warmth goes toward the periphery. This is the solve side. Look at these little symbols; water has an arrow going up and down, and air has an arrow going up and down. It means that water can be in either a gravity state or a levity state. When it is in a gravity state, it is ice and considered a mineral. When it is liquid, it is no longer a mineral. It is very convenient for science to do that. However, it also can be a vapor and go into the levity side, because it has an abundance of levity in it. Why is this? It is because water includes hydrogen and, alchemically, hydrogen represents the force of fire.

Here is our earth, water, air, and fire (*figure 2*) in a slightly different configuration. We could put carbon (C) in the position of earth. We could put oxygen (O) in the position of water. We could put nitrogen (N) in the position of air. We could put hydrogen (H) in the position of fire. In this mandala we see the four universal constituents of the carbohydrate structures of life. These fundamental substances are divided by the alchemical mandala into solve and coagula. This division helps the heart eye to see the hidden relationships that form the basis for plant chemistry and soil science. This mandala of the four substances is a very good way to think of fertilizers. It is a very good way to think about cation-exchange capacity, in which ions change capacities of the clay content of your soils, allowing uptake of essential minerals for plant growth. This mandala becomes a kind of magic decoder ring when you are trying to wade into the convolutions of formulaic relationships posited by science as explanations of what could be a mystical experience on one side and then the technical jargon on the other. A mandala such as this can help us stay in the middle of the road.

Here is an example of technical jargon. Let us consider water, a substance that has a levity force balanced with a gravity force. The gravity force in water is oxygen. The formula for water is H (hydrogen) $_2$O (oxygen). In alchemy, oxygen brings a kind of gravity force into the compound because it is on the coagula side. Oxygen does this because it combines with things. Antoine Lavoisier found in the 1700s that when things combine with oxygen they become heavier. When Lavoisier proved that air has weight, that discovery killed alchemy. Oxygen oxidizes substances, such as when it combines with burning or rusting metal. Lavoisier found that the calx, or ash, of the metal left over after the burning process is heavier than the original metal. That was his great experiment, which opened the door for all analytical chemistry. His reward was to lose his head during the French revolution.

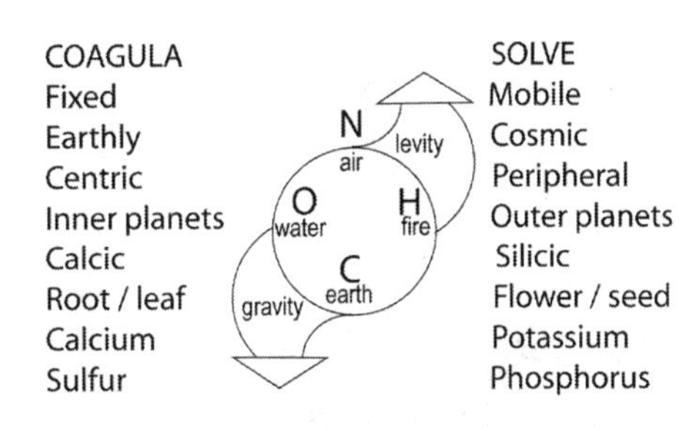

Figure 2

Alchemically, oxygen brings a gravity force, coagula. Hydrogen brings the fire. Hydrogen is the lightest of the classical elements and has a great secret. It is the basis of the whole periodic table, and all electron shell theory on the periodic table is based on hydrogen. However, hydrogen violates the law of electron shells because it has only one electron, and the law says it is supposed to have two on the inner shell. Therefore, it is kind of a dilemma for science that the very linchpin of the whole system violates the principle it establishes. Science has to assume that there is a virtual electron that shows up for the hydrogen when needed and then it goes back to wherever virtual electrons go when they are on break.

It is said that hydrogen has a weak bond because it needs something to come from somewhere, though they do not know where it comes from. The quantum people say it comes from the quantum field. In the past, they would have said "God." However, we cannot say that, so we say that it is a field of energy oscillating and jiggling into and out of existence at such an extreme rate that we cannot begin to understand it. We say it is where all the energy in the cosmos is, but we cannot see it; we cannot weigh it; and it is driving us crazy. So we just talk about it as if we know what it is. That is called "the field," where all the energy sort of leaks into this side. Hydrogen oscillates between being on that side of the field and being on this side of the field. Whenever we need an electron, one shows up.

This is water from a chemical standpoint, and it actually explains a lot about water, because water forms weak bonds but is always forming bonds with everything. It is the universal solvent because it pulls fire into combination with more manifest compounds and releases fire, all at the same time. It connects and disconnects. This is water; things that come into the vicinity of water get pulled into the dance to connect or not connect. Everything is oscillating according to quantum theory. Quantum sees everything oscillating or interacting with a huge field of energy that permeates most of the universe. However, we cannot measure the field or see it.

According to current science, the oxygen atoms and hydrogen atoms are fixed in minerals, and even in organic compounds. In most organic compounds, all of those H's and O's are locked into other things. In liquids, however, we get to the great mystery of the cosmos; here is a force called interfacial tension. This force in water is the source of the interfacial system oscillation. I will quote how science describes this so readers can get a jargon headache. It has to do with what happens when you put manure into water and stir it. We are doing something that is changing the whole quantum field in the whole universe according to current theory. You are not just out in your garden getting poop on your boots, but doing things in the whole cosmos. In water, this ability to unite, connect, and disconnect is paramount. Water is a "crystal in the liquid form." It is constantly forming interfacial systems that oscillate into and out of the quantum field. I am bringing this to show the types of science that come from technical imaginations that cannot quite keep up with the life sciences and the majestic ambiguity of life systems.

> Differences in the extent of protonation of functional groups lying on either side of water–hydrophobe interfaces are deemed essential to enzymatic catalysis, molecular recognition, bioenergetic transduction, and atmospheric aerosol–gas exchanges. The sign and range of such differences, however, remain conjectural. Herein we report experiments

showing that gaseous carboxylic acids RCOOH(g) begin to deprotonate on the surface of water significantly more acidic than that supporting the dissociation of dissolved acids RCOOH(aq). Thermodynamic analysis indicates that >6 H_2O molecules must participate in the deprotonation of RCOOH(g) on water, but quantum mechanical calculations on a model air–water interface predict that such event is hindered by a significant kinetic barrier unless OH^- ions are present therein.[5]

Does that shed a bright light on the problem of stirring manure in water? What they are saying is that within water is this huge ability to relate to other things and to form structures, inner surfaces, interfacial tension structures within the molecular structure of the mass of water, to unite with substances, and to model them to form membranes (we could say), tension structures that oscillate with other tension structures, creating in a drop of water, coherence, out of pure water. In fact, there is no coherence, because the hydrogen atoms and the oxygen atoms are just slipping by each other, oscillating in the quantum field, in and out, in and out. Until that water comes into contact with something (such as calcium) that changes the pH to 7.0014. Suddenly, the inner structure of water is reacting to that calcium, creating inner structures that are modeling what the calcium is doing, especially if the calcium is at the nano, or solution, level. The mineral is starting to restructure the water into the form of the mineral. It is something like infinite Play-Doh. It just becomes whatever it touches or whatever touches it. Water models what it touches.

Differences in temperature, velocity, viscosity, pH, and movement all create differential membranes within the structure of the water. These are called interfacial systems. If there are enough of them and if the gradients are strong enough, you get at a macro level what is called laminar flows or turbular flows.[6] At a micro level, the water is responding to whatever it comes in contact with in order to model it, to model the formative motif of whatever is in it. Since most of the things it contacts are some form of

5 From http://www.pnas.org/content/early/2012/10/24/1209307109.abstract (accessed March 2013).

6 Laminar flow is like the smooth flow of air around an airplane wing; turbular flow is more like the turbulence of air hitting the face of a flat surface.

mineral nature, and the mineral has these very clear structures, the water is continually interacting and changing with the pH of the air, acid rain, or organic compounds. Whatever it comes in contact with creates inner boundaries within the water that the water is modeling. So you can have the slightest difference in temperature or velocity or water flows moving at two different rates, and between the differentials some form of energetic pattern arises that causes interaction between polarities.

When I was in the Navy, I was on a submarine in the Mediterranean doing Spanish war games with the Spanish Navy. We would pretend to be a baddie, and the Spanish destroyers would try to find us under the water. On the sub, we had sensors to tell us when the water temperature changed. The captain would look for a temperature curtain, which is a current in which there is a big difference in temperature. When we found a temperature curtain, we would shut off the engines. The destroyers would be pinging with sonar from above, but suddenly we would disappear from their sonar scopes. Their pinging would go down along the temperature gradient and go off somewhere else, because we were hiding in a thermal curtain. They couldn't find us because the differential between the two temperatures of the current had created an energetic membrane. The sonar signals were unable to penetrate that energetic membrane, and the pings would curve off somewhere else. We would suddenly disappear.

That curtain formed because the difference in the temperature and the difference of the movements in the current had created a boundary, an energetic boundary. This happens at a macro scale in the ocean, but it also happens at the micro scale in a drop of water that comes into contact with a rock face or a particle of sulfur from a chimney suspended in the air. The drop of water falling through the sulfur picks up the particles and the pH of the sulfur changes the structure of how the water reacts. The form of the water is different after that contact from what it is when it comes out of a cloud with no contact. Whenever water is disturbed by something, it reacts. If the disturbance is strong enough, the reaction is sustained. The water has a kind of memory. If you have a fifty-gallon drum and a stick and push the stick to stir the water, it takes a bit of effort to start stirring. Then the miracle of

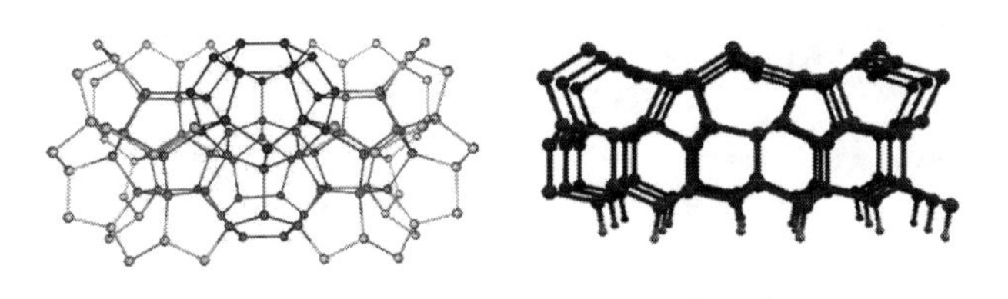

Figure 11 Figure 12

water happens, and suddenly it is much easier to stir, and it takes an effort to stop the water from going around. This is a huge mystery, but it can induce a yawn when we have to stir for an hour.

The reason stirring becomes easier is because the fifty-gallon drum has become an organ. The water has become *organ-ized* by our stirring movements. It has become organized through the discursive nature of the different velocities within the drum. The water on the outside of the drum is interacting with the tensions of the actual physical surface of the drum and slowing down. As you stir, however, the water becomes organized in layers within itself until the inner layers are rotating in a frictionless environment. Therefore, they have a much higher velocity with no drag. This happens in every streambed. The water in the center of the stream moves much more rapidly than the water hitting the banks of the stream. In fact, you can notice that the water hitting the banks of the stream (if you really pay attention) is actually going backward.

We used to go down to the river with the class and look out onto the river. I would ask my students to observe the direction in which the river is moving, and everyone would wonder what I had for breakfast. Of course, it is moving upstream toward downstream. However, as we observed the river, I would take a stick and throw it into the water along the bank, and it would move upstream from downstream. The resistance of the bank causes a backflow that moves slowly upstream along the banks.

Figure 12 shows a theoretical construct of the way the molecules in silica look. You see a kind of very beautiful honeycomb structure. *Figure 11* looks very similar to it, but it shows a theoretical structure of water molecules. There is a similarity in the structure. However, in water those linkages are always oscillating, and in silica, they have been fixed because they are earth. The open structure of water, however, has all those molecular forces adjusting to each other.

If we were all in the room and suddenly six tigers entered from all sides, the people on the outside would press against the people in the middle because of this alien situation. Those in the middle would be bonding to each other, but the people on the outside would not respond as much to the others next to them as they would to the tigers. The water on the outside of a drop form is less able to link to the air and more able to link to the water in the center of the drop. These linkages form a surface on the drop, resulting in surface tension. That surface tension runs from the surface, where the water is in danger of going out into the air, toward the center of the drop, where it is part of a community. The gravity nature of water rules at the center of the drop, and the levity force of the air on the outside is weaker than the gravity of the center. Water is at home on the gravity side of the mandala. So the drop of water has much more integrity to itself than it does to what it is outside in a gaseous state. However, if it is forced by heat into a stronger relationship to levity, it forms many smaller drops until the surface of each drop is larger than the mass and it joins the air in the form of vapor. Vapor is infinitely small drops of water with enormous surface areas that relate to the levity forces of the periphery.

In nature there is a great law called the law of minimal surface. This law states that the most efficient relationship between the surface and mass of any form is a sphere. The most efficient form to hold energy has to have a surface whereby it does not discharge energy into its environment. In nature, the form that discharges energy is tapered sides that come to a point, like the tip of the blade of grass or a point on the margin on the leaf. Every point in nature is discharging forces from the earth into the atmosphere. These ideas are regular old standard science. Points on earth are the sites of discharge into the atmosphere, except when a cloud passes over the earth; then the points under the cloud induce a charge from the atmosphere. When a cloud goes over, the point is pulling energy out of the atmosphere into the earth. This is a constant balancing act between the atmosphere and the earth. (Air will be discussed in more depth later.) The sphere, on which there are no points available for discharge, is the opposite of the point, which is a site of discharge. As a result, the ideal form in nature to hold a charge is a sphere.

The law of minimal surface says that the ideal sphere has the most efficient relationship between the surface and the mass for holding on to energy. Water will always seek the spherical form because it is the great purveyor of energy in the natural world. The smallest droplets of water carry the greatest charge. When minerals are pulverized and become so small that they form colloids, the water becomes highly charged. Water at this scale contains the ability to interact with these minimal surfaces of minerals in colloidal forms, because its interfacial tension systems are all minute, molecular-level oscillations of surfaces—molecular surfaces. Molecular surfaces arise from whatever the water is reacting with.

Think of oxygen as a gas merging with hydrogen–the lightest gas we know. Somehow they come into contact with each other and suddenly we can wash our hands in the new compound. It is kind of counterintuitive that water falls out of the air, whereas water is seen, alchemically, as the corpse of the air; in other words, water is the salt of air. If I have a solution of salt in water, and I look at it, I do not see any crystals floating around in there. It is a solution. But if I evaporate the water and send the water into a higher levity state, which it can do because it can go up into air, the salt that was in the solution suddenly appears again and falls out as a precipitate. In alchemy, that action is called salt, or *sal*. As a vapor, when water gets cold enough, it precipitates. People on The Weather Channel always talk about precipitation. Rain is the precipitate, or salt, of air. This is the way alchemists would put it, because it shows that the levity forces of the gases are coming into the gravity condition of *coagula*. They know that rain is not table salt. Rather, their idea of "salt" is a process, an activity of manifesting out of a rarified condition. When Rudolf Steiner is speaking in his agriculture course, he is using this kind of meta-language and hopes that people understand him.

It is difficult to comprehend some of his concepts because he speaks in the language of alchemy, a language of metaphor and analogy. I have been studying these concepts for forty years, and I have made it a hobby to look for the places in scientific literature where scientific findings can substantiate what seem to be strange statements by Steiner. If we can understand the method of analogy, there is indeed good science behind the agriculture course. However, it is not science as you would get from Chemistry 101. Nevertheless, those who are on the cutting edge of science are running into a dilemma; the objects of their studies are getting smaller and smaller. The substances are in such small quantities that the scientists are no longer discovering "stuff" but only forces. The next frontier, after they go more deeply into forces, is the frontier of consciousness. However, physical scientists cannot go there, so when they find forces beyond contemporary understanding they give them labels such as "dark energy" or even anti-something—antigravity is a good one. Scientists have determined that seventy percent of the known cosmos is made of antigravity. You cannot see it; you cannot measure it; we have no idea what it looks like; but it has to be there according to calculations. In this way, calculations lead scientists to the spot where the honesty in their calculations creates a crisis of belief. The tendency then is to call what we do not understand the anti-something of what we think we understand. They call this dark force "antigravity" because they cannot call it levity. Levity is a funky old term from alchemy, but that is how this strange force behaves. It is antigravity.

I bring these things because such concepts represent a dilemma in science, which is coming closer to discovering the spiritual in nature that forms the backbone of Steiner's view on agriculture. Nevertheless, the great thing about today's science is the rigor of its methods for exploring the unknown. In biodynamics, I think it is a mistake simply to dispense with the need for scientific rigor. Without rigor, biodynamics will come under attack from modern science, which is based on weight, number, and measure. We cannot afford to come under attack with contemporary science, and the laboratory work needs to be done. However, if we have the right imagination, laboratory work can form the basis for a spiritual view. Steiner asked that the laboratory bench be transformed into an altar.

I must have the ability within myself to form and dissolve an inner picture at will. That practice builds in the heart, an organ of perception that can make a bridge between my personal mystical experience and a generally accepted scientific concept. However, this bridge is not built automatically. I need to make the bridge through imaginative meditative work. I need to bring the mystical and analytical sides of work in nature together into what Rudolf Steiner calls the realm of the Christ being.

When I first go into the space between analysis and mystical experience, no answer is available. This silence creates anxiety, and I have to train myself to form pictures and give them away and then listen into the silence. This practice constitutes a question for the spiritual beings that animate nature. If I do a practice such as this before I go to sleep, images begin to approach me from outside, from other people during the day. When my heart organ is functioning, I can recognize the unconscious contributions from others in my daily life as partial answers to my questions. In this practice, the other people with whom I am in the world become the source of the true imaginations I need to corroborate in myself. In spiritual reality, these kinds of exchanges are happening all of the time, but if my heart organ is not capable of seeing, then the answers go by me without recognition. On the other side of the coin, I can do an exercise like this and believe that the images arising in me after I do the exercise are the answer to the question. This, however, is an error of inflation. If I corroborate imaginations only within myself, I run the risk of the great error of inflation.

Inflation runs rampant in the political and economic systems today. Inflation is like bodily inflammation, but it is an inflammation of the soul. I have to guard against inflammation of the soul when training my heart to see spiritual events. In my inner life, consciousness has the quality of the fluid element. Consciousness starts to model anything it contacts. I just explained this about the element of water, and human bodies are water—bags and bags of water.

When Rudolf Steiner was young, he made money as a mentor to wealthy families in the city where he lived. He was a private tutor to the sons and daughters of wealthy families. One of the families had a young son who had hydrocephalus, what used to be called water on the brain. Rudolf Steiner worked with him as a young man, and the things that Steiner learned by working with the young man set the stage later for Waldorf education. Steiner gave the young man exercises, practices that allowed him eventually to become a doctor, even though he was hydrocephalic. An autopsy was performed after the young man died, and it was found that all he had in his brain were membranes that resembled the structures of the brain, but the membranes were filled with water. There was no myelin; there were no neurons, dendrites, axons, or glial cells; only the membranous structure was there. There was just water, and he became a doctor.

His brain had the electrolytes needed to sustain life, but there were no actual tissues. Recall that earlier I gave a picture of the action of hydrogen in H_2O. What is H_2O, really? The chemical formula gives me only a small slice for understanding what water really is. Water as an alchemical idea; it has the ability to model, to make interstitial flow systems, which is just what our neurons do, but if you study neurology, the size of neurons is remarkably small. There are hundreds of thousands of neurons in a piece of brain tissue half the size of a pinhead. However, those neurons are embedded in glial material. *Glia* comes from the Greek word for glue—*glial* means glue, as "brain glue." Glial cells actually support much the plasticity and remodeling capacities of the brain as the nerves are used. It is the gluey, liquid parts of the brain that are very capable of remodeling. Remodeling, or neurogenesis, is the basis for learning and creativity.

The famous physiologist D'Arcy Thompson formulated a rule stating that germ plasma can never act as matter alone, but only as a seat of energies. This means that tissue receives energies but is not the source of energies. Esoterically, we might add to this the assertion that germ plasma receives consciousness as well as energies. Rudolf Steiner calls the energies and consciousness that germ plasma receives *imaginations*. Imaginations come from the spiritual beings that animate nature. Steiner even goes to the extreme of saying that the hollow spaces filled with fluid in the brain are the areas where we form inner pictures. The membranes that line the third ventricle in the embryo are the sites where brain cells that migrate to other parts of the developing brain are formed. This process of brain-cell migration is called *neurogenesis*. In the adult organism, neurogenesis occurs when there is a shift of consciousness. This shift often happens when we have emotional events. New neurons are created so that we can learn from our experiences. In our development, our potential to grow and learn comes from the membranes that line the hollow chambers. This process of brain-cell regeneration in sub-ventricular areas of the brain is neurogenesis. Learning involves a transformation of consciousness.

I can characterize consciousness by recognizing the different states of consciousness. We can use the elemental mandala as a key to doing this. The earth state of consciousness we could characterize with the phrase "I am terribly certain this is true." That is earth. A typical statement of the consciousness of water is "I will get back to you on that." In the mandala of the four elements, the earth element automatically begins a process of moving toward the water element. The same is true for the elemental mandala of consciousness. Once I make the earth statement that I believe something is absolutely true, my consciousness—automatically through the mandala—moves it to water and says, "Well, what about this?" Anything I can say is exactly dead wrong from the opposite side. There will always be someone who has an opinion exactly opposite to mine and who is qualified enough to prove to me that I do not completely understand what I think I do. That is just the way the world is. However, if I train myself to move between earth and water consciousness, then I move into a central place where my heart opens up and receives imaginations from the people around me. I can take pictures of the phenomena I am studying into sleep; then I listen to my fellow human beings until they provide for me the question that allows me to know that my task in the research is a good one. You have to find someone who can be a foil for your unfounded beliefs.

I mention this because, otherwise, I am just in my own little space making things up. Alchemically, I need to form a vortical pattern in my water consciousness. I need to take my water consciousness, which is modeling everything, pulling it in, becoming that, dissolving, and becoming something else. I need to organize it but not fix it into a belief until that belief is tested in the world. I need to stir the content of the water consciousness to focus it, but not fix it.

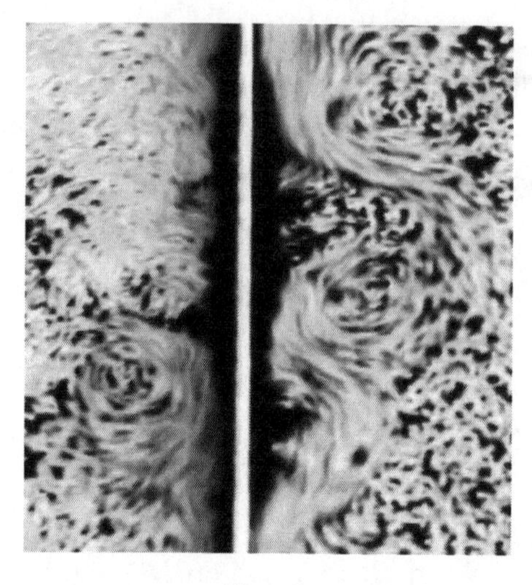

Figure 7

The vortex image for consciousness is the perfect model. *Figure 7* shows the action of the vortex. I will explain what you are looking at in the picture. You are looking down from above at a long, narrow trough of water on a table. The trough has a string stretched down the middle of the trough and parallel to the long axis. The vertical white line on the center of *figure 7* is the string. The string is suspended in a way that allows the surface of the water to touch it but not submerge it. The string is just in contact with the very surface of the water in the trough where the meniscus of the water is just attached to it. The string is being bowed by a violin bow. When you look at the string, you cannot see the vibrations. However, when lycopodium powder is sprinkled on the surface of the water, the vibrations of the string come into view. As the string vibrates, you can see little vortical swirls. The whole length of the string is organized into vortical centers by the vibration of the string. This is caused by the fact that the water can pick up the vibration and that it becomes organized through flow and resistance patterns into a vortex. As soon as I create a flow, one part is moving faster and another part is moving slower due to resistance.

Imagine that you have your bucket and you start stirring. You are creating energetic interchange places in the water where the different velocities of the water are creating interstitial tension systems. We could call them energetic membranes, although there is nothing there but the gradient between the velocities. However, if you know your physics, a gradient and velocity creates temperature changes. Your bucket from the outside to the inside is filled with innumerable inner surfaces. It has become like an organ.

Look at *figure 8*. It shows a jar of water that has been stirred, and then dye was dropped into the water. Look at the structures. Are there cloudy and diffuse patterns there? No, there are very discreet energetic membranes, thousands of them, from the outside to the inside of the vortex. That is what is happening in your bucket. We could call the water in the vortex an organ, because the water becomes organized by the stirring motions. We are causing a water organ to assimilate whatever you are putting into the water. We call it stirring, but the water is modeling along those inner surfaces whatever is present in the water you are trying to get into contact with the water. You are stimulating the water to model that substance by creating a huge amount of inner surfaces of interaction. The stirring is creating many, many more coherent inner surfaces that are maximizing the interstitial flow. Instead of allowing the water merely to continue going in and out of the quantum field and oscillating on its own, all those surfaces sliding over each other allow the form to become coherent. That is what we call stirring 500.[7] In reality, however, the water is becoming an organ modeling whatever is put into it—either manure or silica.

Figure 8

Chaos is a simple word for describing an extremely complex situation. What do we mean when we say "chaos"? When we reverse the stirring direction, the greatest impact is the creation of micro-bubbles. A bubble is a hollow drop form that has multiple surfaces. Each bubble has an inner surface and an outer surface. The law of minimal surface dictates that the smaller the bubble, the greater the action of interacting with the fluid that the bubble is in. This realm of huge bubble surfaces is called micro-bubbles. Micro-bubbles are of huge interest in science today because their surfaces are so interactive. The more micro-bubbles you have in a fluid, the more surface area of interaction. When we create a vortex by stirring a fluid in one direction and then breaking the direction of rotation, we create tremendous

7 Biodynamic preparation 500 is made by filling a cow horn with cow dung and burying it in the soil during the cooler months, generally from November through February.

numbers of micro-bubbles. Then by stirring in the opposite direction we organize the fluid again. Now all the little micro-bubbles break, and the local temperature in the micro-bubble goes up to the thousand-degree range for an brief instant, and the sound of a micro-bubble bursting at that scale has the decibel equivalent of a 747 parked in your backyard. A little vortex shoots out of each bubble, creating a sea of micro-vortices in the water at that scale. You have sound, temperature, and a new vortical motion to reintegrate the molecules of the water in this form of a micro-bubble.

The micro-bubbles in the water are moving toward the air state. This is called "cavitation." We will look at cavitation later, because it is a great mystery in embryology. The formation of an inner surface from what was a mass is the basis of what Rudolf Steiner calls the *Astral Principle.* When we get to the Astral Principle, we begin to approach the relationship between plant and animal. Alchemically, that level of interaction represents the element of air. The formation of a micro-bubble in a fluid is known as cavitation. This happens when a propeller goes around underwater; you can see little bubbles coming off the propeller. With the rotation of the propeller under water, the local temperature becomes so great that it separates the hydrogen and oxygen back into gases within a very, very small space, and then you see bubbles trailing out of a propeller underwater. The propeller is not sucking air from the surface. It is actually the huge temperature changes of the micro-bubbles that are separating the gasses and creating a condition similar to boiling water.

The consciousness you use while stirring makes a difference in the stirring, even if you are using a machine. Rudolf Steiner suggests that if you relate to your tractor in gratitude, it will work better for you. Any mechanic will tell you that. Curse your power take-off, your PTO curses you.

Figure 8 is a picture of a consciousness that has been integrated into a vortical configuration. That vortex breathes in a rhythm by alternating phases of being longer and slower and then shorter and faster. The alternations become perceptible after you stop stirring. That vortical breathing is an image of the interaction of flow and resistance, of solve and coagula in the forces at work in the vortex. After you stop forcing the little vortex, it

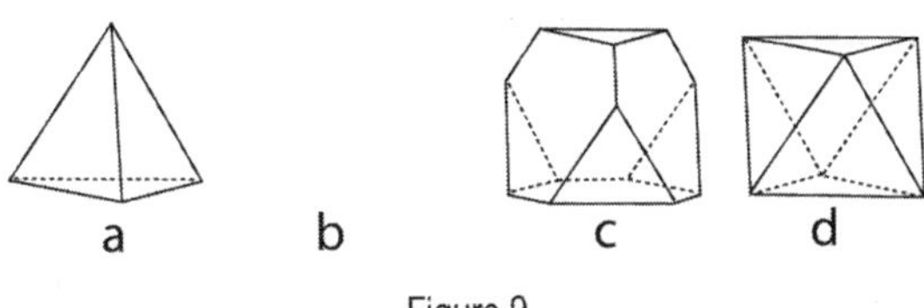

Figure 9

does not just sit there, it gets deeper and then shallower, and then it expands and contracts, oscillating in the quantum with all the little molecules being organized by the motions.

Vortical breathing in the realm of consciousness means I form coherence in my consciousness by forming a picture, and then I let the coherence go by dissolving the picture. I form it, and then I let it go. That rhythm is the model for how to work with my consciousness when I want to go from water to air in the realm of the soul. I need to be able to the form a picture, hold it, and then let go of it and listen.

Look at *figure 9*. In sleep or in meditation, you actually can pulse the picture into and out of focus. For instance you form an inner picture of the leaf of a plant about which you have questions. Then in your mind's eye you dissolve the picture of the leaf. Then you repeat the forming and dissolving process several times. Now in forming the picture, it is very useful to find the spot in your organism where you can easily form inner images. Everyone has a little different place to form inner pictures. The location of that place has a lot to do with your feelings about whether you can allow pictures to be retrieved or must keep them sealed in a concrete bunker. Some people tell me that they never have inner pictures. If I keep working with them for a while, a day comes when they do the exercise and suddenly realize that they have been forming inner pictures their whole life. Everyone has inner pictures. Not everyone has access. Some people have a dial-up connection when it comes to inner pictures. Other people have fiber-optic inner picture access, and others are Wi-Fi.

Here is an exercise to find the space where you can grasp the inner picture. Look at *figure 9* on the left. The image to the left is a tetrahedron. Use your water consciousness to become the tetrahedron. Earth consciousness is "snapshot consciousness." Look at the tetrahedron on the left, close your eyes, and try to see inwardly—see the tetrahedron. What happens when you try to do that?

In your inner eye you had to move around the form. When you try to recall an inner image of something you saw, your life (etheric) body tries to keep the inner picture from fixing. What if everything you ever saw were present all the time. That would be pretty weird. Therefore, we have a failsafe in us so that when we see something and start to lock on it, our ether body, or life body, acts to dissolve the snapshot kind of memory picture so that they do not get stuck in our memories. The method of successful inner picturing centers on experiencing how my snapshot inner pictures are always fading from memory. To deal with this so that I can hold and retrieve inner pictures, I need to have a way of tricking myself into realizing where the hidden memory force is in the sensory impression I wish to remember. You have heard of "air guitar." We want to do "air art," air drawing. Look at the tetrahedron in *figure 9* and imagine that your finger is moving along the edge of the form of the tetrahedron. Move your finger along the form as you observe it. Instead of looking *at* the tetrahedron, we call this process looking *with* the tetrahedron. Look with the tetrahedron and draw it in the air while you are looking at it—whatever you see there, as if you were tracing it in the air. Repeat the air drawing of the tetrahedron a few times by tracing the form again in the air with your finger. We want to get your movement organism—your water organism, or life body—to pick up the form and model it. Now, close your eyes and try to picture the tetrahedron. Is there any difference between this inner picture and the snapshot memory?

It was probably more stable, because you put it in your life body as a movement. When you try to remember it, your life body provides a memory form that is recorded in those movements. Esoterically, we could say that after the life body picks up the movements, your soul is moving with the formal motif of the sensory object. Instead of merely looking at the sense object, you are now *looking with it.* It is much easier to find the inner picture space when you have an inner movement you have structured. Your memory functions in a different way from the way it does when you try to remember

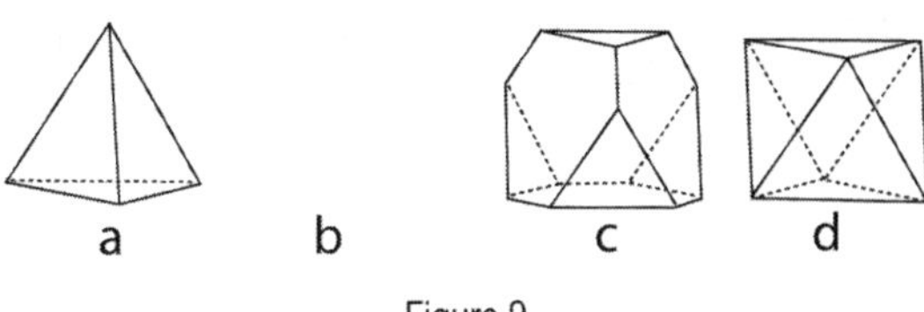

Figure 9

a grocery list. In the water of your body, there is a kind of built-in ability to flow with the thing you are looking at. You can train yourself to access it by air drawing.

However, it is a deeper secret that you do not even have to move your hand. Look at the tetrahedron again and just follow it with your eyes, as if you had a ray or a pencil coming out of your eyes. Just track the edges of the form with your eyes. Now it may not have moved at all, because you have just engraved the fluids of your body with your ocular motor muscles. This method takes you close to an experience of the essential nature of the tetrahedron. If you get really good at this, you can then take that inner image of the tetrahedron and rotate it.

When you do these kinds of inner exercises, you are actually building an organ in your organization that can look with the essential formal motifs of trees and cows and landscapes. With this organ, you can see the music in them, the harmonic structure of the life in them. This kind of looking allows you to see the ley of the land (the ley lines). If you develop this as a capacity, imaginations come from the beings who live in the forms. These imaginations provide insights, which is great, but if you do not check those insights, the human organism is such that our souls want to believe that everything coming to us as an imagination is valid. We have to have a check on that tendency.

Now go to the right-hand side of *figure 9*. You see a form labeled *d*. Count the number of sides of the form *d*. There are eight. It is an octahedron. The figure just to the left of that is labeled *c*. Form *c* is a stage in development leading to an octahedron. From one point of view, it is a separate form. From another, it is part of the sequence—a morphological, phenomenological sequence. The question is how to get from form *d* to form *c*. You also notice there is a blank spot labeled *b*. Now imagine what form *b* would look like, and do a little air sketch of that. You are following the evolution of forms. The struggle you feel is the pull to move out of earth consciousness into water.

If the figure you draw for *b* is correct, you should be able to make a little movie going sequentially from *a* to *d* and back. Form *b* is one stage going from *a* to *d*. The forms represent a coherent morphological sequence. If you feel the tension in trying to make the correct form for *b*, that tension is just the thing you need to feel to correct any imaginations that come to you based on merely allowing things to flow in your imagination.

What you are feeling when you do this is the tension between the feeling that you know and the feeling that you do not know. This tension is exactly the right place to do the kind of research we need to do in biodynamics. It is one thing to feel that the beings of the cow horns are speaking to me from the heaven that they live in. Great. Now what? Now I have to take that experience and submit it to an inner questioning process. When I slow the imaginative process, there is a space where the analytical side of my mind and the fantasy side of my mind come together into a controllable imagination. In that space it is possible to find that the molecular structure of keratin is very similar to the molecular structure of quartz, even though they are two completely different substances in completely different realms. When I am able to balance analysis with fantasy, I can find resonances through the formation of functional analogies. When I perceive resonances through functional analogies, something in me becomes much more settled and I start to find a way through the maze of my own belief structures into a space of objectivity. If I do not do this, I usually go off with the fairies. Carl Jung worked with sensitives—people who "know" things. At a certain point in life, Jung said sensitives get very nervous because they are one hundred percent correct fifty percent of the time. The fifty percent of the time they are correct is what forms their reputations, but the fifty percent of the time that they are not correct drives them into therapy and, as a corrective for the inflation of the feeling that they have the hotline to God, they have to encounter this death force of really not "knowing."

When analysis and fantasy come into balance in the center of the soul, we have what Rudolf Steiner calls imaginative cognition. When personal fantasy gets corrected by conscious acts of cognition, the result is imaginative cognition. Conversely, when analytical cognition becomes enlivened by fantasy, the result is creativity. One can feel the tension here. We want to make a form out of our imagination and somehow it has to fit in to a cognitive scheme. This is why teachers in the ancient mystery schools always used geometry as a tool for training the kind of thinking you need to enter the spirit. We cannot BS geometry... well, you can and some people do, but there is a certain space in the mathematical realm where certitude can be achieved. A balance between terrible certainty and creative fantasy is the goal. Goethe called it "higher beholding."

III. Air Consciousness • Fields and Vectors • Corpse and Resurrection • Crystal Heavens • Antennas • Cavitation • Astralization

Now we will cover air, my favorite subject. The difficulty with understanding air is that you never really understand it, because when you think you have finally zeroed in on it, air always ends up being something else. In thirty-five years of weather research, I have come to value this, because the way you make mistakes is by always thinking that you understand something. Working with weather is a very good way to learn that what you think you know is always only part of the true picture. The consciousness of not fully knowing is *air.* When you have that moment—when you realize that you may not have penetrated what you are saying—that is the air element in alchemy.

Alchemical air always brings something of the mood that what counts most is out there where the air is rare. It is out there where you cannot quite grasp it. When we feel as if we are trying to work with something but cannot quite understand it, this is air. When you think you understand something, but somebody asks you a question and you realize you do not understand it, this is alchemical air. Alchemists would call it the *feeling of reversal.* Reversal will put you up against the wall of your own misunderstanding; it is a primary way that the hierarchies help us learn. Air is learning by being able to tolerate reversals of beliefs.

Those who think they do not make mistakes are on the road to self-delusion. The invisible wall that you run into is your own belief system; this is alchemical air, which arises where there is a gap. There is a separation out there that I am not quite understanding because my belief has just been reversed. Such a reversal is actually a gift because, if I can experience reversal as a gift rather than a threat, then the door between the worlds will open to me for learning. There is a door; it swings open and closed in the great cosmic wind, and my task is to learn the rhythm of that opening and closing. This is my task as an esotericist. It includes entering the consciousness of the elemental beings in the water collected in my backyard rain barrel and the beings in charge of planetary orbits. The consciousness I need to understand them as an alchemist is one tuned into rhythmic reversals. I have to change my consciousness from thinking "I know" to one in which my focus is on the gaps between the things I know. This requires training.

When we work phenomenologically, we believe we are seeing something. I believe that this thing is separate from me. A contemporary physicist would have issues with such a belief, because they would say that part of the forces that allow the thing to appear as present are connected to the belief that it is out there. An esotericist would say that the thing appears out there so that one can say, "I think it is there."

The belief that what I call "there" is separate from what I call "here" is a vectoring quality of human consciousness. The perception of "here" and "there" as separate places is a force of consciousness that makes the world fall into seemingly separate entities. This fall into separate entities is the consciousness of the earth element. Those separate entities go through time and change; that is the water element. Then when they change, they eventually change so much that they appear to be the exact opposite of how they appeared at the beginning. This is the force of reversal in the alchemical consciousness of air.

All forms come from somewhere, and they go somewhere. They come from somewhere as potentialities for being, and then they *earth.* After they earth, they *water.* After they water, they begin to reverse and go somewhere into the realm of potentiality again. The realm of potential, or what the Greeks called the *pleroma,* is where entities come from and where they go. The door to that potential place is through the alchemical element of air. The rhythm of coming from the pleroma and returning again into the pleroma is fixed in the formal language of their body. It is fixed in how they appear and in how their body functions. It is revealed in the rhythms of digestion and circulation. The journey into and out of incarnation is reveled in the pattern rhythms of the leaves on a plant stem. Spiritually, the plant you see is not the real plant but a corpse that illustrates the rhythmic incarnation and excarnation rhythms of the spiritual entity that we recognize as a plant. The rhythms of the forms are actually interacting with your "*sal*" consciousness in a vectoring way.

A *vector* is a force that influences another force indirectly. If I am flying an airplane and I want to go from Sacramento to San Jose and there is a fifty-mile-an-hour wind moving north to south down the valley, I have to plot a course to the north of San Jose to factor in the wind force that will drive me to the south. If I plot a course directly to San Jose the vectoring wind will push me south and I will end up down in Monterey. The wind that influences my original course is a vector. It is a force acting on another force that is not the original force; but it must be factored in, meaning that the organism has to adjust to something that is not its own thing. That is a vector. Vectors illustrate air consciousness very well. Everything that causes an organism to modify into something else we can call air. Air is reversal.

In biology, however, a vector is not necessarily directional. A deer tick is a "disease vector" that can carry Lyme disease organisms from one host to another. In bodies, the idea of vector becomes more subtle. Picture a leaf growing. It has an overall form that is not directional, I think. Actually, that form represents forces streaming through the leaf to make it grow. The forces flow from the mid-rib out through the veins toward the periphery, or so I think. Do I know that for sure? I can do an experiment. Take a tiny leaf and watch it grow into a big leaf. What is difficult to see is the time signature of that growth. Therefore, when I look at the leaf as a finished form, the forces determining the growth of the leaf do not appear to my senses. However, when I use my heart to look at it and can get into the rhythm of the air consciousness, I can see that the leaf has a becoming, or a growth, gesture. The form of the leaf becomes directional when I grow it in my mind.

Nevertheless, the question is still this: What are the vectoring forces in the form of the leaf? Do you want to do a great exercise? Take a tiny little leaf and a magnifying lens; then, using an X-Acto knife, cut a tiny square in the new leaf and leave it on the plant. Then every couple of days, observe it and record how it has grown. You are going to see something very interesting about how the leaf grows. You think it is growing one way, but it has a whole other trick up its sleeve. In fact, it grows holographically. The leaf becomes larger, but the square hole remains square. It does not grow from the midrib and push out from the veins. Every part that grows exists in the leaf from the beginning and expands in context with every other part. Here we see the remarkable forces of the air element in the alchemical mandala. Growth in which every part grows in context with every other part is an example of *field properties,* which are examples of alchemical air.

A field is composed of potential vectors. Say I have Jell-O with a marshmallow in one side of the Jell-O and some raisins in the other side, and let's say I am partial to marshmallows over raisins. When I dig into that Jell-O with my spoon to pull out the marshmallow, do the raisins move? Of course the raisins move, but I am not touching them, I want the marshmallow, but the Jell-O and the marshmallow, and the raisins are all part of the field. Every time you move something in a field, something else is moved by the vectoring of the forces present in the potentialities of the field. The vectors are not all manifest; rather, they represent potential pathways of force within the field. They are subject to many forces that are not all manifest in the field, but they are present as potentialities that act on the field from the outside. The potential influences on the field of those forces from the outside are very difficult to compute. This is vectoring. I begin with a specific intent and suddenly I end up somewhere else, because everything is connected to everything else through invisible webs of relationships. Those invisible webs are field properties, and the field is what alchemists would call air. It just means the stuff you see is only the smallest part of what something really is. Air is a difficult alchemical concept to understand, but it is very important.

I will give you a picture of a physical reaction in water that an alchemist would move into the air position in the mandala. There is an experiment with *surfactants.* A surfactant is any substance like a detergent, which alters the surface tension or interfacial tension in a liquid. The experiment involves two big flat trays that hold about five gallons of water each. They are both fixed side by side to a common rod. The arrangement makes it possible to move the trays in sync with each other. There is a trigger on one end of each tray, and when you press the trigger, the ends of both trays drop about a quarter of an inch simultaneously. When the trays drop, the water in them pulses to the opposite end of the trays. The camera is set so that you can see the wave propagated when the trays drop. The waves go through the trays and hit the opposite ends of the two trays and come back. Now there are two waves moving through the trays in sync. Now, the experimenter adds a vector into the experiment. In one of the trays with five gallons of water, one drop of detergent is added. In the beginning of the experiment, the waves moving through the two trays were in sync. However, after the drop of detergent is added, the first return wave in that tray is seen to be slowing down, and by the third wave in the detergent tray, the whole five gallons has stopped moving, while the other one is still oscillating. This is the effect of one drop of detergent in five gallons of water. The detergent altered the field properties of the water. It provided a vector to the forces of the water that interacted with the whole field and changed its configurations.

The water is a field with specific qualities of forces. One of the forces in the field is the capacity to flow. Alchemically, is the flow of the water air or water? An alchemist would say that the flowing property of the field is linked to the element of air, but that the water itself as substance is water. Water as

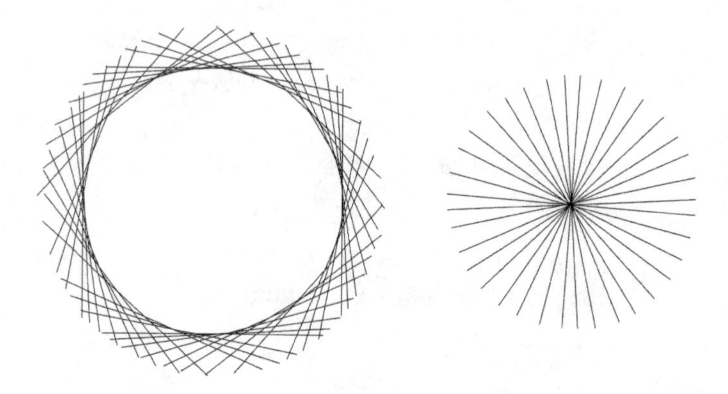

archetype seeks to become a sphere because of its inner gravity structures. As a result, still water is water, alchemically. It is active in solution and chemistry. Flowing water is becoming more like air. Water has an inner kind of surface tension that holds it together. Air seeks the periphery; it flows to the periphery. Its boundaries and surfaces are more energetic than actual. Thus, the flowing quality that creates the abundance of inner surfaces is the action of water moving toward the element of air. Water that is moving in the other direction, toward its destiny with gravity, is losing the air quality and forms drops. Air dies in water. Water is the corpse of air.

Air flows, and all parts of the air are connected in the field of the whole atmosphere. This leads to compression and rarefaction. Water can flow, but essentially the wholeness of the field of the water element is connected to the wholeness of the earth through the force of gravity. Even molten rock is water. Flowing molten rock is moving toward air.

Above are two diagrams. They are fundamental, esoteric diagrams for alchemists. Both are circles. The one on the right has a point in the center with rays going out from the center. The one on the left has planes on the periphery moving toward the center. I can make a circle out of the one on the right with points moving out from the center, and I can make a circle out of the one on the left with planes moving in from the periphery. Which is the true circle? Neither of them, because they represent two polarities of the breathing process linked to the formation of a circle. The breathing process from the center out is more in the gravity realm, and the breathing process from the periphery in is more in the levity realm. Water can visit the levity realm but has a deep affinity for the gravity realm. Air can visit the gravity realm and turn into water, but then water turns back into air because it has a deep affinity to fire. Air can visit gravity but is not gravity. Water can visit levity, but it is not levity.

They both share the tendency to flow, but water prefers to move toward the gravity pole, or coagula, but it contains great properties of levity. This is why water is the universal solvent. However, it is the planes within the water that allow it to be the solvent and to model everything. Those planes are really the action on the air, because water is two gases that have come together and fallen into the liquid state. While they are gases, you cannot wash your hands in them, but when the air element dies in sacrifice to us, it can take on everything. Then it becomes water.

Air then lives in the outer realm and only visits here. Water lives here but visits there. Rudolf Steiner calls the peripheral forces "planar forces." They represent planes of light coming from infinitely distant space, from all the stars and all the suns in all directions, our limitless surroundings—gazillions of stars in all directions radiating light to our dear wonderful little planet.

Science tells us that light travels in what are known as wave planes of energy. A plane that is perpendicular to the movement of the wave carries the wave front, where the energy of the sunlight has impact. They are infinite planes, invisible, that weave through us continually from all directions. This is getting close to what science calls quantum. The living state of air is light. Air itself is the corpse of light. Thus, we have *form* as a noun and *form* as a verb. I can form something, and when I form something, I have a form. The noun is a corpse and the force of becoming is the verb. The spiritual entity behind the form is a verb. The true human being is a pillar of *doing* made of light that speaks and understands. The form here on the Earth is the corpse of that light and warmth. We make the corpse live while our light and warmth occupy it. We resurrect the corpse composed of the physical earth every time we wake up and face the day. That is levity, and we do it with light, or consciousness.

When I try to understand air, I have to understand that the forms of things here are pictures of how they came out of the light of the cosmos to become a corpse. Moreover, the rhythms that are perceptible in the forms I see in the sensory world can be used by me as a human being to resurrect the corpses and make them live again. I resurrect them in the form of inner pictures I have that are formed in my imagination. Those inner pictures are made of *light.* My imagination is inner light. That is where my pictures are. My imagination is the tool, *sin qua non,* of working with the natural world, but it has to be tempered. My imagination has to be made coherent as a laser, so that the light that I give back to nature through imaginative cognition is not fouled by my beliefs. Remember, air is the corpse of light.

When I do that act of giving light back to the natural world, I can harmonize with the beings who create the forces and laws of nature. The beings who made nature cannot resurrect the corpse as aspects of nature. The only beings who can change nature are us, and we are the densest ones on the block. Nevertheless, we have something that the spirits who maintain nature do not have, and that is freedom. There is the rub, but also the gift. When I move into air consciousness, I connect with "me not knowing." As a human being, I cannot lift my waking state into angelic consciousness and say, "Hey, how are you doing?" My consciousness is numbed when I address an angel, but I have to structure my consciousness here through cognition and understanding so that when I do go into that place, I have my begging bowl well-polished. If I have tempered my imagination, then when I come back from an encounter with a spirit who stands behind a natural form, I am able to look in the begging bowl and see something that has come back with me. To do this, I must have developed a protocol in me for sustaining an aspect of my earth consciousness, even when I enter a higher consciousness. I learn to form that protocol in air consciousness. I have to have a protocol that allows me to check the validity of a spiritual experience that I believe occurred. If I do not, then I run the risk of inflating because I am always in air.

We need to know what our beliefs are. When I do, I can recognize that my beliefs are corpses of my experiences. To alchemists, a corpse is what it is all about because, for Rosicrucians, the whole purpose of life is to resurrect the corpse. Rosicrucians work on resurrecting the corpse of nature by transforming the senses from memory-producing capacities into creative-imaginative capacities. Once I do that, I can understand that the Earth is becoming a corpse since it is mature enough to start thinking about retirement. It will never be able to make 40,000 feet of plant material as it once did to make the coal reserves. It takes a couple of hundred yards of plant material to be compressed into a quarter inch of coal. You can go into the coal regions and find a seam of coal that is thirty meters thick. How many plants do you need to pile up to make a coal seam that thick? That was the young Earth. This is Gaea saying, "Hey, you want plants? Here are some plants." "You want a gazillion little animals in the ocean? Here you go, have a ball." We do not have that degree of life today. We have corpses of that today, and it is getting more corpse-like even as we speak, because of people doing things with technology.

Nonetheless, we take poop and dead things, pile them up, and then stand back and watch the miracle of the resurrection of corpses. Rudolf Steiner asks us to make a personal relationship to the manure. We need to understand the profound attraction that corpses have for life. This is the great alchemy of polarity and reversal found in the alchemical mandala as the element of air. Therefore, when I want to go into the next level of energetic work in nature, I have to enter what is corpse-like and lift it, with my consciousness, into new life. Lift it from earth to water, get it to move in me, and, from water to air, make it so I can form a picture of it and dissolve it. I move it back into light and then let it go into the spiritual world as a seed for new life. Christ returned to teach the mystery of the resurrection of the corpse. To do that, I have to go to the next step of being humbled by the fire—that is, issues such as climate change. We are going to be humbled by the fire.

The planar forces coming from infinitely distant space represent the pole of levity. Light and fire are the representative forces of the pole of levity. The whole key to transforming substances is to understand the stages through which things move. That is the whole premise of the alchemical mandala of earth, water, air, and fire. This mandala gives us a way of understanding how things move from being spiritual beings with infinite power but no freedom to being manifested as spiritual beings packed in little corpses and destined to die. However, spiritual beings have total freedom.

In other words, how do I get from being a spirit to being a New Jersey punk? How do I do that? Better yet, *why* do I need to do that? Why do I come down stage by stage from being a spirit to facing the death process? I have to understand the death process to understand the life process. I have to understand how pig manure is different from cow manure, because both are manure but the life of the organism that creates the manure makes a soul impact on the corpse of the manure. The impact comes through the desire of the pig and the desire of the cow. These soul forces cause things to happen to the calcium and potassium in the earth body by concentrating certain minerals in the manure. The cow does not get down on its four legs and stick its nose in the dirt and plow around. It just does not do that. That is part of the desire of what can be called astral body, or air body of the animal.

Where Rudolf Steiner uses the word *astral,* we can read *air.* Astrality is air consciousness. It is awareness and sensitivity. Sensitivity is the characteristic of air. The cow has one whole set of desires, and its digestive apparatus accommodates this. The pig has another whole set of desires; it is a rooting animal. The desire to root brings a whole different quality to the manure. What I am describing is called the soul body, or air body, of the animal. It is where they have their awareness of the field, the tribe, the group, and mom and dad. The field, or field properties, are air qualities. The water properties are life properties. Water relates this to that. However, when I understand how this changes into that, I move from water to air, because I am now in my consciousness, or astral, body rather than in my life body, where I just float along. In water, I do not have to worry about cognition, because everything is taken care of by the spirits who govern life. In water consciousness, I am suckling in a world where everything makes sense.

Some traditional indigenous cultures have that. The people live in a world of nature. If nature starts to make less sense to them through such things as death or suffering, they invent something called religion in an attempt to relink to a world in which life makes sense. They wish to relink to the beings that made nature to find out why nature no longer makes sense. Nature ceases to make sense as soon as we separate from it as autonomous human beings. When we separate from nature, we separate from a whole lot;

we separate from our community, from our family. Then the world seems to be filled with the corpses of what was once living nature. Since that has happened on a grand scale, we yearn for the past when we all felt interconnected in community. Now, however, if we make rules for everyone in community, everyone complains about the restrictions, since everyone expects to be treated as an individual.

Air is the realm in which I become aware of how the distant forces of the field are acting. In water, the fields are all connected up very clearly. If I put a stick in water, it goes over here. The vectors are easy to see in water. The water takes things on and models them. However, air is a little bit more difficult to see because it is so diffuse. Nevertheless, the great value of diffusion is that everything needs to diffuse; otherwise we would very quickly solidify in our own toxic waste such as carbon dioxide and such. If we could not get rid of the carbon in our bodies, Rudolf Steiner said we would become beautiful stalactites.

Therefore, we need diffusion as part of our work. Diffusion depends on the amount of surface available. Thus, we are back to surfaces again. Surfaces determine the ability of an organism to interact with the environment. The rule—the great law—is minimal surfaces. The most excellent surface to mass ratio is found in the sphere. It is the most economical form. That economy is the reason most eggs are spherical—except when they come through the vent of a hen, and then they get a little aerodynamic. Basically, they start out as a sphere because it has the most energy economy; the relationship between the mass and the surface is balanced. However, as soon as that sphere grows, it starts threatening that economy. As the volume increases, it grows in three dimensions: height, width, and depth, whereas surface grows in only two: height and width. The surface cannot keep up with the mass. Therefore, either it has to rupture or, if it is living, the cell has to divide. The cell divides and makes a clone, and the two sister cells share a common membrane. That is cell division.

The common membrane between the two allows the two daughter cells to reduce greatly the overblown relationship between too much mass and not enough surface. This condition makes it difficult for the contents in the cell to communicate with the environment. The ideal is to have enough surface area to allow the contents of the cell to communicate easily with the environment, but not so much that the contents of the cell are drawn out and dispersed into the environment. That happens when cells have more surface than mass. Getting large is not the ideal, nor is getting small the ideal. This law, then, governs the optimal size of cells in living organisms. However, there is a strange element in the law of minimal surface. This can be illustrated with a reference to biodynamic preparation 501, the quartz preparation.

I may take a crystal, and pound it up, I run it through a mortar and pestle, and grind it to a fine powder on a glass slab. Originally if I dropped the crystal it would fall to the floor and no one would think it was strange. Nor does anyone think it strange if, after the grinding process, I blow on the powdered crystal and it floats around the room defying the law of gravity. Why does the crystal in this form float around the room? Because its particles have been rendered below a certain threshold in the relationship between mass and surface. Now they have more surface than mass. How did that happen?

Imagine inflating a balloon. It has a limited surface but, as I blow it up, the mass of the balloon grows more rapidly than the surface. The mass grows by three dimensions and the surface grows by two dimensions. As the sphere of the balloon becomes larger, it threatens the integrity of the surface until, eventually, the surface ruptures. Now imagine a sphere similar to a balloon, but now imagine making it smaller. Again, the mass will become smaller more rapidly than the surface. There will be a dimension whereby the mass will become smaller than the surface, just as the mass became larger more rapidly than the surface when the balloon was getting bigger. When the sphere becomes so small that the mass becomes smaller than the surface the surface interacts with the environment in very powerful ways.

Surfaces interacting with the environment on the periphery of the organism are the signature of the forces of levity. Levity acts on surfaces that are on the periphery of living things. Levity is a peripheral force. The facets of a crystal are the signature of planes in the gravity pole. If you look at a gem, you see a facet, or plane. In a gem, the facet of the face is the corpse of the gem and its lawful crystallographic geometries. It came from somewhere "out there" as a ray of light. It comes into existence in a field of invisible forces that guide the molecular structure to that specific angle.

The signature of levity is the planes of a crystal, whereas the signature of gravity is the points. Gravity works by having a point that goes out to gather and attach stuff to itself. This is how rain happens; you have to have a microscopic bit of salt at 30,000 feet or perhaps super-cooled soot from a jet that attracts super-cooled water to form the nexus of a crystal. You need earth way up there if water is to condense to the point where it forms a crystal. Then the crystal comes under the influence of gravity, starts to fall through a warmer layer, melts, and goes back into water. Then we say it is raining.

That earthy bit of dust high up allows the field to make a corpse and form a seed for a crystal. The seed crystals for rain had been salt or dust, and now they are pretty much unburned hydrocarbons from jet fuel. You do not need conspiracy stuff about "What in the world are they spraying?" Just consider your most recent airplane ride; you put tons of particles in the air at a microscopic level. Those unburned hydrocarbons from the jet fuel now serve as ice nuclei for precipitation. The more ice nuclei from jet fuel, the more water vapor will condense around each seed nuclei. Thus, you get large amounts of clouds that trail ice down. Essentially it is raining up at 30,000 feet, but it never reaches the ground. That is actually drought, because there are too many seed nuclei for the available amount of water in the upper atmosphere.

Water enters liquid form as it first rises from the water here on Earth, but as the vapor rises it cools radically. By the time that vapor reaches the top of a cloud, it will have cooled into ice. Those super-fine ice crystals then fall, melt, and turn into little, spherical beads of water. Then the surface of one bead touches the surface of another, and boom, boom, boom, they start merging. Suddenly the little droplets of vapor are actual droplets of rain; finally it starts raining. However, all rain starts out as ice crystals that have precipitated or condensed around some form of particulate nuclei. The most abundant nuclei used to be salt from the ocean. Today there are many synthetic particles that serve as ice nuclei. Today's science calls these earth particles "aerosols." The whole issue around the use of fossil fuels is that they put too many earth particles in places that create challenges for the air.

Now we turn to the air analog in the mineral realm. Where is the element of air among the minerals? In the mineral realm, it is evidenced by the fact that minerals form with planes. Do those planes originate in the mineral or somewhere else? That is an alchemical question. If I have a salt and put it in water to dissolve it and then look for evidence of its physicality, I cannot see any salt in the water. However, when I evaporate the water, suddenly the salt crystals come into view. If I look at the salt crystal before I put it in the water, it is a perfect cube. If I look at it in the water, I do not see any cubes. I do not see little lines where there would be cubes. When that water goes away, the salt comes back in the form of a perfect cube. Where is the template for the perfect cube held? The cube we see is the corpse of the archetypal cube that dwells in the land of "cubeness," but activities of the land of cubeness are not the corpse that appears when I evaporate the solution. Archetypal cubeness is an ordering principle in the spiritual world. It is the geometrical rigor of planes of light that have a particular relationship of 90° to each other. That angle is a huge vector force that we use all the time. It is the key archetype in an activity called making a building.

The minerals that form into crystals are found at the top of the mass of rock known as the *pluton*. A pluton is a huge lens of rock buried in the earth. It is divided into coarser rocks at the bottom, granites in the middle, and gems near the top. The finest parts of the fluid rock migrate to the top of the pluton to form intrusions where the liquid gem materials crystallize. The gem structure in the pluton at the top is under the influence of what used to be called the "crystal heavens," the starlight coming toward Earth from the periphery of space. The crystal heavens were considered to be planes of wisdom coming from the periphery that created the ordering principles of matter. The representative of cosmic order was seen to manifest in the ordered structure of the crystal realm. Today, we call this sublime realm of cosmic wisdom "crystallography," in which we analyze the axes, angles, and facet relationships of crystals. These geometrical laws, according to science, are evidence of the molecular relationships in the bonding angles that form the shape of the crystal. Are the bonding angles in the crystal a result of the molecules, or are the molecules getting a cue from somewhere else? Years ago I went to a book from Harvard on mineralogy to look for the root of a crystal form. According to Harvard, the source of the crystal form is something known as a "space lattice." The text explained that the space lattice gives rise to the form of the crystal, and that the form of the crystal is a manifestation of the space lattice. Okay, so what space—outer space or what?

"Space lattice" is the concept contemporary science offers for the remarkable ordering seen in the realm of the crystals. That is as good as it gets. The space lattice is made of particular theoretical bonding angles of the molecules. Okay, where did the bonding angle originate? If it were like the rest of the natural world, there would probably be more variations on the theme, but it always comes back into form with geometrical rigor. Yet within that rigor are infinite variations of form caused by vectoring forces in the formation of the crystal in the pluton. A crystal is a corpse of the highest form of consciousness, because, in the crystal heavens, what Rudolf Steiner calls higher Devachan—the consciousness of the beings who live in there and create the actual fabric of our whole earth, the whole body of our earth, coming out of silicates—their consciousness is perfectly ordered to hold the mineral world in harmonic order. Moreover, they hold that so the earth can be consistent when it falls into a corpse. The corpse retains the potential to mirror that high level of ordered consciousness. Thus, I can take a quartz crystal, put a screw on it, and tighten down on it and an electrical charge will come from the crystal because I have altered the space lattice. In addition, when I alter the space lattice, the energy used to form the crystal moves out of it. Where does it go? Does it go to the local space lattice depository? It goes to the periphery unless I capture it in a device and use it to call home.

Alternatively, I could use the rigor of the crystal heavens when I crush it up and then stir it into water. When we make a silica prep, we are playing in the realm of the hierarchies, interacting with beings at a high level of consciousness when working with that crystal. That is the realm of the future moral capacities of human beings. It is known as higher Devachan.

Higher Devachan, the crystal heavens, is a realm in which nature is going perfectly according to plan. It is the guiding form principle of the mineral realm and the reason mountains endure. The minerals are not messing around with the Divine plan. However, according to that plan the life in that mineral is in the farthest periphery. The forces of consciousness in that mineral are at the farthest periphery. As human beings, if we could actually go to that farthest periphery, we would probably come back very ill if at all.

The physical crystal is a corpse. The life principles that guide the formative processes of the crystal reside in Higher Devachan. A life principle is the consciousness that a being uses to say, "I exist." The crystals on Earth are corpses of the beings who say "I exist" at the periphery of the cosmos. The physical crystal is a transceiver of the sublimely ordered forces streaming from the highest realms. The form of the crystal is an antenna for those highly ordered forces. Every corpse is an antenna that is a picture of the signal. Now, let us look at antenna theory.

Figure 1 is an image of what is known in the trade as a "log-period array," or Christmas tree antenna. On the left is what's known as log-period array in electronics. It is exactly what you have on your roof when you want to pick up a signal. Looking at it, you can see why they call it a Christmas tree. The purpose of those horizontal bars is to make an image of a signal wave. The vertical bar simply holds the horizontal bars in place. Each of the horizontal bars has a screw on it that makes it possible to slide each separate horizontal bar up and down the vertical bar. The horizontal bars are meant to pick up a compound wave coming from a transmitter. Those compound waves are composed of large waves with smaller waves embedded in the larger waves, so that you get a full signal rather than just little pieces of things. All of your signals are organized this way—smaller waves embedded in larger waves. The ideal antenna can pick up the peaks of the waves, so that you get the maximum from the signal.

Engineers have designed the type of antenna depicted in *figure 1* so that, when the wave comes in from the signal, it hits the shorter bars first. The wave then goes through the middle ones and then to the big one in the back. At the back, the signal is bounced back again to the front. The signal then bounces back and forth in the antenna creating a standing wave, which allows you to get good reception. The peaks and troughs of each wave need to be harmonized to be the same. If they are not, one of the little crossbars is not in the right position. To fix this, you go up on the roof, unscrew it, move it along the middle bar, and shout down to the house, "How's that now?" Alternatively, you use an oscilloscope, and move the crossbar until the signal becomes coherent. This is called trimming the antenna for gain. It is just basic antenna theory. Trimming for gain allows the wave that comes in to be rectified with the antenna form. You want to have an antenna whose form is a picture of the signal. If the antenna is not a picture of the signal, your reception will be full of static.

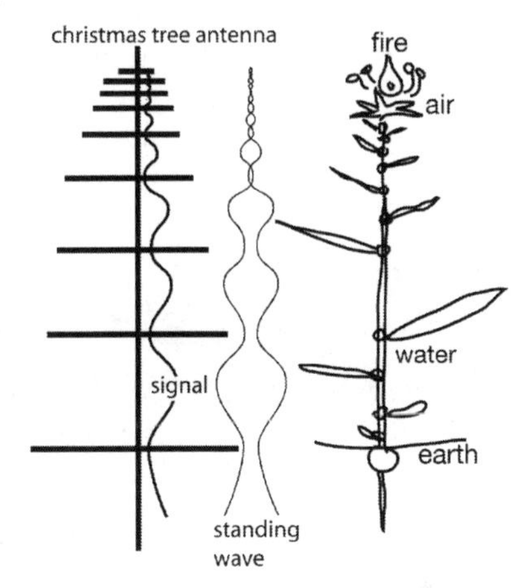

Figure 1

Looking at *figure 1,* you can see that there are certain waveforms in which the periods between the waves become rhythmically smaller. This rhythmic form makes the antenna look like a Christmas tree. Christmas trees are antennas for receiving signals from the Sun in the form of solar radiation waves. They have that form because a waveform is a logarithmic period wave. Mathematically, the waves have an inner logarithmic harmonic that keeps all of the different amplitudes in sync with one another. Each part of the wave is listening to the whole, just like a leaf, a tree, or your arm. Your arm is an antenna for your spirit. That is what temple dance was about in the past—even the temple itself—and the songs sung in those temples. These different waves of energies were intended to harmonize the temple, song, and dance with the energy waves coming from the deity to whom the temple was dedicated.

Silica in the earth is the great transceiver or antenna for cosmic forces. It pulls in and reflects light from the stars. An analogy can be made for the form of silica crystals in the earth and the eyes of organisms. Silica receives light from the cosmos and reflects it into the forms of plants and animals. Esoterically, that is the function of silica in Earth's body. The silica crystal is seen as an image of the signals of starlight that bathe the Earth as they come from the whole crystal heavens. The silica crystal is made up of highly organized molecular planes. The planes in the crystal rotate light and direct it within the molecular structure of the crystal. This affinity for light is seen as the link between the mineral and the plant worlds. Plants take up the light and in their formative motifs they make little models of the wave periods of solar energy. Look at *figure 1*. You see that kind of vase-shaped thing; it is a diagram of the standing wave in the antenna. Once the wave goes forward and backward in that log-period array, it becomes coherent and makes a form of energy in that antenna that allows you to get a good signal. Look at the plant. What can you see? A plant going through its vegetative expansion and reproductive contraction is a log-period array. Intervals of nodes and internodes go through a logarithmic sequence in growth that allows the plant to receive all the variations of wave properties in the sunlight in order to go from making leaves to making seeds. It has to have a whole field of wavelengths rather than just one. In its growth pattern, the plant is building an antenna for sunlight. The smallest unit, the leaf, is built to receive light. It is a dish antenna. Then its inner structures are used to receive that light. In fact, magnesium in the plant sap lies in the bottom of what is known as the light antenna. It is a series of pigments ranging from green to yellow to orange to red. As the light comes in, the wavelengths go through a transformative process until the wave of sun energy reaches the magnesium at the bottom of the light wave antenna. The magnesium in the plant sap allows for photosynthesis. Science calls that process the "light antenna." There are three or four different plant pigments, and those plant pigments are what you find as fall arrives and you also see the different levels of the antenna. We see this in a ripening peach. The array of colors and pigments allow energy to be received and transformed within the plant organs. Human beings use the color sequences of ripening as a signal that it is now time to harvest the fruit. Search online for "photosynthesis light antenna of plants" and you find this amazing relationship between plants and sunlight. This also has to do with the way metals are utilized and so on. Basically, it is to receive a signal in the form of an energy wave.

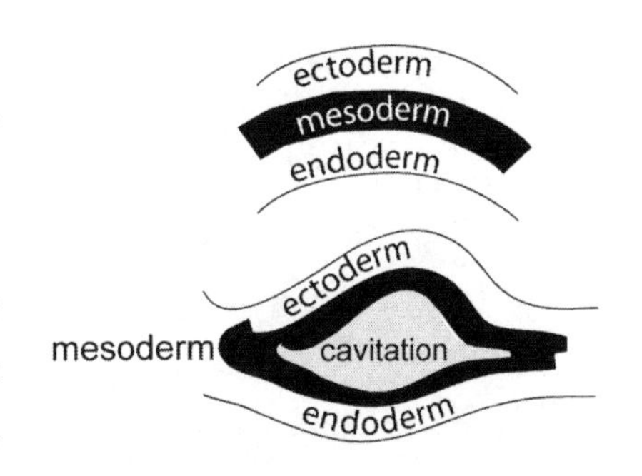

Figure 2: Cavitation

Light is the signal. Air is the corpse of light. The plant is a model of all the wave properties of sunlight. When we get to the animal, the animal takes light in, while the plant has its organs spread out in the air and filled with air, constantly taking it in and out because of breezes going across the plant—it is not like breathing in and out with a lung. The atmosphere and atmospheric pressures are breathing for the plant. With an animal, the air in the animal creates a system in which membrane after member creates an organ adapted especially to deal with the air.

There is a special process in animal embryology called cavitation. In *figure* 2, we see at the top the ectoderm, mesoderm, and endoderm, the three layers of an embryo. In an embryo, the ectoderm is the outer skin, the endoderm the inner skin, and the mesoderm the middle area where the outer skin and the inner skin "talk" to each other. Mesoderm arises in the interactive secretions between the outer skin of an organism and its inner skin or an organ. In higher animals there is always a place between the ectoderm and endoderm where the two are speaking across a space by means of secretions. When that in-between space becomes strong enough, the inner space takes its own life and becomes something higher. The way it does this is by creating a cavitation—that is, a hollow or bubble. The bubble inside the two layers is sealed off from everything outside and inside; the layer between proliferates into other spaces are used by the organism for specific functions. This is the evolutionary plan of your liver, lungs, kidneys, heart, and all of one's life organs. The life organs are separated out from the big flow and form an inner space based on what is called cavitation.

If your organs were organized just to arise from masses of cells, they would be in danger of becoming too dense. The spaces in the organs allow them to breathe with the environment, even if they are hermetically separated from it. If the organs were to become denser with no cavitation, then every organ development would result in the formation of a tumor. Therefore, in the evolution of organs, there is a failsafe. This means that, once the organ has become dense enough, a hollow forms on the inside. This hollowing out of an animal organ is what Rudolf Steiner calls "astralization." The air in the animal—this hollowing process of the air in an animal—creates cavities that allow the animal to interact with its environment while maintaining the integrity of its own organism separate from its environment. It has to be able to do both. This separation from its environment and interaction with its environment needs a mediator. The great archetype of mediation in the organism is the creation of a surface. In physiology, the surface is known as a semipermeable membrane. Membranes form along the field lines of the organs present when the juices are flowing in the embryo. Membranes are deposited like sandbars in the flow of a river. An organ is a picture of the flow that goes through it. Does it pass right through? Does it get curled around? Does it get stopped? Depending upon the desire of the animal, the air becomes very articulate, making the inner structures of the various organs, depending on the astral body of desires of the organism. The astral body is the air or soul body of the organism.

Air as an alchemical element is the formative process where hollowing processes permeate the matter of the organ with functional spaces. Cavitation is an elemental air process when spontaneous cavity forms are created in a developing life organ. The inner space is alchemically an elemental air structure. Basically, air means the gaseous state of existence. The word *gas* comes from the Greek word *khaos* (meaning "great void" or "abyss"). When Jean-Baptiste van Helmont (1579–1644) first found *khaos,* or gas, he said, "It is like chaos. Nothing is there, but then stuff falls out of it. I think I will call it 'chaos.'" We say "gas." Gas is chaos, but chaos means potential.

Animals take air from the outside and move it inside. This process is continued in each of the life organs as they create particular "atmospheres" in which they use inner spaces to unfold their functionality within the organism as a whole. Each organ creates its own kind of functional atmosphere. My soul makes an inner cavity in my body, which gives me an inner surface where I can interact with the world. However, that surface also creates a space in me whereby I do not have to interact directly with the world. The little part of me that determines the degree of interaction with the outside is called a soul. Alchemically, your soul body is your air body, because you take a part of the great field of the shared atmosphere and you make it your own field within your organism. You take the great field of potential from outside and form it as an inner image.

We call this a lung or kidney. We could also call it air, but we could also call it light, because we know that air is the corpse of light, and light—or "air," in alchemical quotes—has the connotation of a field of activity rather than a point. This gesture of the inner air is my soul, not my spirit. My spirit is fire, and my soul is a place where I "breathe" between my spirit and my body. Alchemically, that breathing process is a function of the air element. This is an astral principle. To alchemists, the astral principle is based on desire. My desire body, or air body, creates organs in me that allow me to interact with outer phenomena and take them inside, simultaneously establishing my own inner soul atmosphere. I go out and come in, out and in, as a process of soul breathing that forms the basis of life.

In ancient times, a yogi would breathe air through the nose in special ways in an attempt to hold onto the soul element. Rudolf Steiner gives a picture of how this practice of dealing with the air, or soul body, has changed. There is a new yoga; Steiner called it breathing light through the eyes. This is accomplished by paying attention to what happens when you observe a thing phenomenologically. Breathing light has to do with training one's soul to tolerate its fire. When your spirit comes in contact with your soul, the usual response is "Ow!" Of course, your spirit is on fire by having to understand why you decided to incarnate. If you train yourself correctly, you can answer the spirit and admit that you do not know why you chose to incarnate. This is the beginning of understanding your reason for incarnating and discovering your mission in life.

One's soul breathing must be made rhythmic. When soul breathing becomes rhythmic through meditative practice, it becomes a way of releasing ourselves from the tyranny of our desires. From a psychological point of view, releasing the self from the tyranny of desires means integrating the dysfunctions in my soul into my whole being instead of keeping them bottled up in one part of my being. My desires create a shadow-like element in the light of my soul. Initially, I just want to get rid of my shadows. The question is this: Why can I not just decide to do that? The answer is that you are not supposed to get rid of the shadows; rather, you are supposed to understand them. You have to shine the light of spirit on them. Then you integrate the shadows and get the spirit fire to unfold your mission in life. Fire simply means this: I have the will to do what I thought I would need to do before I came down here and deal with the corpse. If I, as a spirit, have to go and live in a corpse; suddenly I have to get up in the morning and have a cup-o'-joe just to get the engine started. Then, suppose there is no cup-o'-joe in the morning. I am not a happy camper. What is this reaction? It is desire. Desire is the key to understanding my spirit. I do not want to get rid of it, but put it into the crucible and set it on fire so that the shadow burns into a pillar of light. I can then take the ash of the shadows as a remedy.

Thus, I have my physical body, which contains the shadow of matter. I have my life, my ether body, which is given to me by gracious hierarchies so that I can maintain my life. I have my astral body, or consciousness, which I have separated from the rest of the creation. I have my spirit, which is the repository of my deeds and my mission. My spirit is in contact with the consciousness of everything. It lives in the realm of akasha, ether, space, consciousness. When I am in my spirit, I am one with the Lord of the Elements, who is the Christ. When I go from earth to water, I go through akasha. When I go from water to air, I go through akasha. The key to the Rosicrucian element in the agriculture course is the need to train one's consciousness to participate in the world in a way that the world unfolds. This is the work in phenomenology. I train myself to check the pictures that arise in me when I am out surfing among the elemental beings serving life in the world. Then I need a way to access the fire so that it does not have to burn me all the time to get me going. To do this, I need to understand how to take shadow pictures into sleep.

IV. Fire Tradition • Cooking and Destroying Fires • Alchemical Mandala • the Senses • Phenomenology and Art • Ethers and Elements • Mystery Wisdom

I should say here that *earth, water, air, fire,* and *akasha* are all alchemical code words. They do not mean what people ordinarily think of when you say "fire," "air," or "water." They are code words that include all forms of fire, all forms of air, and all forms of water. We could say that these words include the elements as universal archetypal processes. Earth as a universal archetypal process is a process of manifestation. Things that manifest are Earth. Earth as a process of manifestation has the quality of a verb. In the element of earth, things appear in space and time. That is what earth does to spirit; it makes it appear in space and time. When you come to Earth, you are a spirit clothed in the earth so it can appear; otherwise it has no appearance but only being. However, the being is not revealed. Therefore, to an alchemist the earth element represented anything that was coming into manifestation, whether a thought, an embryo, a rock, or ice crystals.

Therefore, alchemically, water freezing begins to reveal its earth nature, but water unfreezing is participating in the water process. It is moving away from strict manifestation; instead of manifestation, it represents transformation and change. The signature of that transformation is the fact that water models everything. It takes on whatever is placed in it and transforms into that. This is because water as an element has the internalizing property of taking a gestural motif from anything that has manifested and transforming it. Everything unmanifested yearns to be manifested. Everything manifested is yearning to be unmanifested. This is called being a human being. Leonardo daVinci said, "Oh, human beings. They spend their whole life hurrying toward their death."

What is manifest has a yearning to become unmanifest, and the first step, in esoteric language, is "loosening." Whether it is a chemical procedure in a lab or our meditative practice, the first stage is loosening from the status quo. In our inner life, *loosening* means to free oneself from our common, everyday consciousness through acts of meditation. We are loosening and separating our sheathes through meditation to become more absorptive of our true Self. An alchemist would call the process of attempting and establishing such a practice *water,* or *loosening.*

The whole field of existence is patterned, but some of the field is unmanifest and some is manifest; it is all patterned. It is "what it is." We have a feeling in chaos that it is unpatterned and kind of crazy. However, chaos is not without pattern; it is a totally ordered potential for anything to happen. All possibilities are contained in chaos, but they are all in potential. Here, on the manifest side of the cosmos where I like to feel my wallet in my back pocket, chaos seems to be a little iffy. However, to the cosmos it is the only way that everything can get done. In alchemical thinking, everything exists as potential in the spiritual world. The realm where the archetypes of all things exist in potential has been called the *pleroma.* The word means "fullness," or the *entelechy.* Pleroma, or the entelechy, is the realm in which the archetypes live unmanifested but patterned.

This goes back to Plato and to the seven Holy Rishis. It goes back to a worldview held when people lived here on the Earth less than they lived in the spirit, even though they were actually on the Earth. The pleroma is the chaos of the total-all that is tremendously ripe for generating manifestations. Therefore, when I go from earth to water, the water loosens the manifest earth, and the earth once again opens to pleroma. It becomes open to receiving patterns, but the water element does not do the patterning,

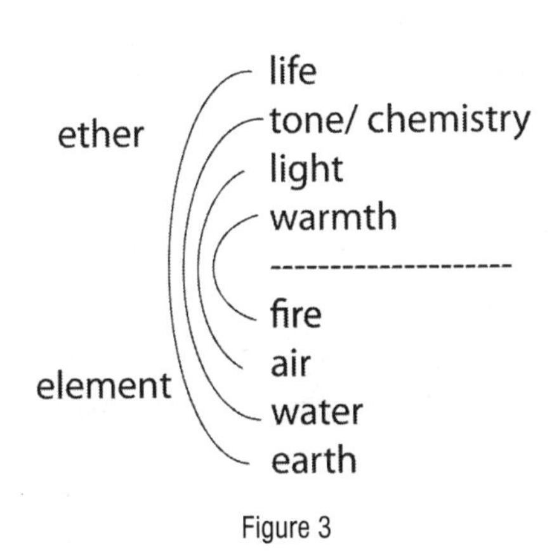

Figure 3

since the water element is part of the chaos structure of the archetypes. Everything is completely patterned; the rocks are patterned; the water is patterned; the air is patterned; the fire is patterned—even though they are all unmanifest as archetypes. These kinds of distinctions are hard for us to understand, but this is the source of our idea of molecular attraction.

I will give you ten minutes to go out and come back with a species. It will not happen, because a species is a spiritual entity, completely organized, but unmanifest. Therefore, when I move from water to air, it is an even deeper loosening, because now I am in the periphery and starting to move way beyond manifestation. Lavoisier proved long ago that air has weight. Oxidation changes the weight of a metal; when it burns it becomes heavier. This idea turned hundreds of years of alchemy on its ear.

Alchemically, there is another side to the elemental world; it is the ether world, and fire is the flaming door between the unmanifest and the manifest world. Looking at *figure 3,* you see fire on the elemental side and warmth on the ether side. The important point is that *unmanifest* does not mean unordered. I have to keep saying this, because we need to understand that, although I cannot see the intelligence behind the phenomenon, it comes into manifestation from a being that has intelligence. It is structured and ordered, and there are levels of organization already in action as it emerges from the unmanifest realm. As part of the unmanifest organization behind world phenomena, the elemental beings are a lower order of spiritual consciousness. This order rises from elementals who live in ponds to exalted beings who make worlds, but they are all connected. The elementals are the children of the higher hierarchies. This is the way Rudolf Steiner describes it.

Elemental beings are the children of the hierarchies, because human beings have a profound habit of believing that the manifest is separate from them. You perceive that your pencil is not you, and this is a perceptual habit. Science would say there is more space between the particles in your pencil than there is stuff, but tell that to your pencil and your fingers. So, science has reached the point where it recognizes that fields of force have action and are ordered and that the amount of actual stuff in the cosmos is much less than we thought. Therefore, your pencil appears to be manifest because it is embedded in very strong fields of force that are in fact invisible.

I am just talking regular science. This is Einstein. We live in an age when field consciousness is available to us. Field properties are important to how your cellphone works and what cell towers are about—fields. Our field consciousness extends from the manifest out to the unmanifest. In the properties of fields, the manifest and the unmanifest are connected by the line of emergence of the manifest from the unmanifest. The manifest and unmanifest are connected, and our job is to find out how they are connected by building analogs in the day's conscious state, which I can manipulate with my inner picturing capacities. Once I can form some pictures of what I believe is a line of emergence, I can develop the practice of dissolving the pictures into sleep. I do this as an experiment to see if my inner pictures are moving in a way that the archetype wishes to manifest along a line of emergence.

What I just said in two sentences is the heart of the Rosicrucian work in Anthroposophy. I have to find inner picture analogs in the natural world for what I think the line of emergence may be for a given phenomenon. I manipulate the inner pictures to make the phenomenon seem to develop in my mind's eye. I give those inner pictures to the spiritual by dissolving the images before I go to sleep. I listen into the silence as the picture dissolves, and then I wait for morning to see if somebody in the visible world comes to speak with me about the picture with which I have been working. The ideal is that they have no idea that they are speaking about those pictures. However, if I establish this as a practice, I will know that they are speaking about my pictures, because I am working so hard to get rid of my beliefs around them. This is our next topic—how to cross the bridge between the manifest and the unmanifest, safely without inflating into self-delusion.

Whether we go from the manifest to the unmanifest or the opposite way, it is our birthright to be able to do that. However, it is like clarinet practice; we have to practice a lot so the outcome is not ugly. It presents a problem if we think with our Google consciousness—that in a hundredth of a second we will get five thousand hits on what it means. In our manifest belief that we are the center of the universe, we assume that this is what it is, and this is the doorway through which the adversaries try to lead us into perdition. It is a battle for human attention. Moreover, it is right in that spot where, if I wait until it comes from somebody else, I know Christ had a hand in it. If I am content to arrive at an answer myself, I am never really sure that I am not gaming the system.

From whom do I usually learn about my mistakes—myself? No; usually other people point out these things to us. Other people are our greatest check against self-delusion; this is why they are provided to us. Other people provide the fire of the spirit to help us to steer toward our mission in life. Today we live in a fire time. Many traditions from the past are being resurrected as people discover that the answers they thought they had were in fact only temporary, because things are changing so rapidly. The fire of spiritual change is cooking all the old beliefs and systems of thought, and only the vapors will be carried forward; the rest will be rendered out for compost. Not every spiritual idea or practice brought from the past needs to be carried forward into the future. However, everything that was a practice in the past needs to be brought forward for review, because all of humanity is in a fire time, and it all has to be put into the pot. The vapors of past belief structures that have carried over from the past bear the essential truth of what was. There is always an essence of the truth from the past, but it has to be brought forward, purified, and harmonized with future demands. The purification of these past practices is alchemical fire.

The alchemical fire principle has two fundamental and connected gestures. One is *cooking* and the other is *destroying.* Cooking occurs when separate things become more intimate. When I make soup, I put together a carrot, potato, and water; then I cook it over fire, and the carrot becomes more intimate with the potato and both become more intimate with the water. That is cooking, and it is one side of fire; it is benign. However, if I keep cooking the soup long enough and do not use my consciousness, Mother Kali shows up and says, "Ah, you know what guy; I think you are going to drive the levity off. The water's going to go somewhere, and you will be left with a very primal state of carbon that you will no longer be able to call a potato or carrot." That is the destruction side of fire as an element. Fire has this double side—the cooking side leads beings to the threshold and the destruction side pushes them across. When we are led to the threshold by the cooking fire, everything becomes more intimate. We are all being led through a threshold here in the fire age; we are beginning to realize that we live on a very small planet. Take out your cell phone and call Budapest. No problem. That is a consciousness shift. Something has changed, and we have all become much more intimate in the cooking fire of technology.

Everything can become more intimate, but it carries with it the destruction side. The cooking soon goes through to the other side, where things are combusted. At present, dictators all over the world are finding the destructive side of the cell phone. A cell phone is cool if you want to call your girlfriend, but if you are a dictator you will tell the army to shut down all the cell towers, because the people are talking to one another and putting the regime in the hot seat. This is the nature of fire; it is nice until it is not.

Fire is a threshold between the worlds. Where you see this in the mineral realm is in the formation of gems, in which the metallic nature and the mineral nature become so intimate that there is a great transformative force. In the mineral, the water and air support the structure of the cleavage planes of a gem or a mineral. However, the metals in the solution are in constant flux, creating all kinds of variations within the silicates. The different metals in solution give rise to all the various gemstones that come from the magma. Alchemically, a metal is a kind of mineral that has its own fire and flow. Metals are ductile, malleable, electrical conductors, and more "living" than rocks. In soil, solution metals such as potassium and magnesium give a force of fire in various degrees when active in plant saps. Depending on the relationship between how much coagula and how much solve are in the metal, the fire in metals bring about various degrees of transformation. The transformation of lead into gold occurred through the action of fire. Each metal has a different solve and coagula relationship. For example, lead is very susceptible to solve to fire.

Alchemists therefore say, "It is easier to make gold than to purify gold." This means, if you start out with the gunk in the bottom of the pond first and apply the fire, the fire will move the gunk to water, then to air, and then to the fire itself. That is the key to the art of transformation, and this is what happens in your compost keep. You take the gunk and fire it into gold. What is the first thing that happens when you put all the gunk together in a heap? It gets warm. What is the source of the fire? We might say that it is the metabolic rate of the microbes assimilating the nitrogen in the urea compound. Sure, okay, but fundamentally it is *fire*. We see that, just by piling things up, the mandala of the four elements becomes active and starts to move earth to water and then to air and then the fire comes. Every time you set up something, you think it is fixed, but it suddenly becomes something else. This is because, as soon as you say something is fixed, you are saying to the cosmos, "I do not think you can do anything about this." And the cosmos replies, "Hey, little guy, check this... Boom!"

The mandala of the four elements is a universal process that moves from a fixed state to a flowing state to a dissolving state to combustion. That is the mandala, whether we are making compost or pancakes or trying to solve a problem. It is the same thing. You take the problem that you experience as fixed, then you put the problem in the retort of your consciousness. You loosen it by trying to form a picture of it and moving that picture rhythmically. The fixed nature of the problem starts to loosen into the possibility of a reversal; this is the air step of taking the loosened image of the problem into sleep. The chaos of sleep and dream is the fire evolution of your inner work.

Now, do you know what happens when it goes into the fire? The cosmos says, "Okay, I think it needs to go back again." Then it comes back to you as air; this is called a dream. Then, suddenly, the dream goes into your life body in the form of a response from your glands and you wake up. When you awake, this is earth; the evening and morning is one breathing process of the consciousness mandala.

Elemental fire is represented by the gem-forming process in the mineral realm. You can take a gem such as a quartz crystal, pulverize it, and then grind it up to make it extremely small. Basically, what you are doing here is moving it into fire. Then you put it into water, essentially cooking it.

Wherever you are, the key to working with the mandala of the four elements is to know where you are. I need to ask, "Where have I picked up this process in the becoming of the archetype?" Once I know where I picked it up, it is easier to move it along in its process. That is the whole beautiful benefit of the mandala after you learn how to use it as a tool for imagination. The mandala is a map that tells us where we are in the transformation process and the next step. Rudolf Steiner uses this kind of thinking in the agriculture course and in the chapter on ancient Saturn in his book *An Outline of Esoteric Science*—that is, the origin of earth, water, air, and fire.[8]

Figure 4 shows a cross section of the layers that form an eyeball in an embryo. In a), we see the brain, the brain tissue of an embryo. In an embryo, the brain sends out a little pseudopod of tissue toward the surface of the skin at a certain point, and the brain tissue touches the surface. The surface responds by touching back. In a), the ectoderm is the surface. The brain continues to push on the surface and the surface pushes back. In this interaction, a kind of folding of that ectoderm layer takes place. This creates the lenses of the eye. The lens of your eye is just very compacted epidermal layers, one on top of the other, until it becomes b), a kind of crystal. This is how the lens of your eye is formed; it is a crystalline morphology.

However, the poking out of the brain causes a back flow from the surface that pushes the lens back toward the brain. This is what we see in b), the second part. The ectoderm folds, and the lens is pushed back toward the brain, and then the brain reaches out, surrounds the lens, and makes a hollow

8 Steiner explains the planetary "incarnations" in chapter 4 of *An Outline of Esoteric Science,* "Cosmic Evolution and the Human Being." Explaining the possibility of describing events in such far-past eons, Steiner has this to say: "Talking about evolution in the sense intended here is meaningless only for those who do not acknowledge this hidden spiritual element in the present. For those who do, previous stages of evolution are present in their perception of the present one, just as the one-year-old child is still present in their perception of a fifty-year-old person" (pp. 125–126).

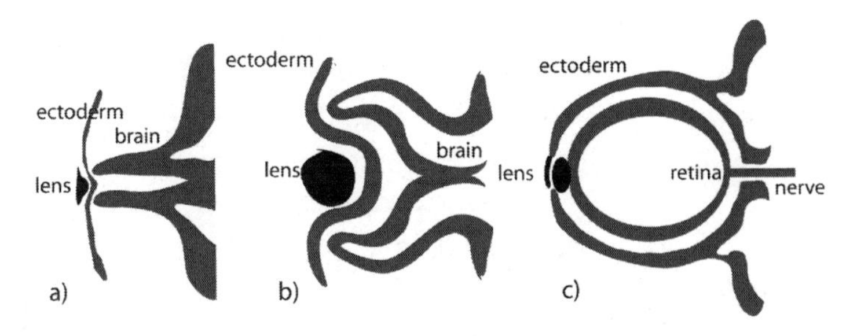

Figure 4: Layers of the eye

space; that is your eye. In c), the lens has embedded. The retina becomes the enclosed space; the back of your retina is the front of your brain. I put this in here because it shows the action of how Earth comes from a centric spot, moves out, receives a response from the periphery, pushes back, responds, and then pushes. It is this push and pull right at the threshold that is elemental fire. Fire is pushing right up at the edge, and then there is blowback that comes from it.

Fire is the farthest periphery, where this whole process is held as an archetype. The other three elements are the processes that, together, constitute the fire principle. To go from earth to water, I must use fire. To go from fire to air I must use fire. To go from fire to warmth ether, I use fire to enter the spirit. Fire is pure periphery and potential. Air is almost pure periphery with a bit of manifestation. Water is not quite such pure periphery, but it brings the periphery in so it is more manifest. Earth is, I think, where we have fallen out of periphery. The elements are always a dynamic, always a process, always a becoming. Where one element reaches a conclusion, the next in line brings about a transformation through a fire principle. The fire element is the essence of complete change. It governs each change but is not limited to each change. Fire is intimate in all things until it leaves and flees to the periphery. This is a great mystery. When the center pushes out to the periphery, the periphery meets it and pushes back inward, and they reciprocate. That reciprocation to achieve balance of opposites is the element of fire.

Recall that I said fire is the essence of a field. Fields in the spiritual realm are ordered but unmanifest. This is hard to understand. In the unmanifest realm, coagula is the potential for manifestation without a manifestation to accompany the potential. In the hierarchies of spiritual creative beings, there are spirits with only the idea of manifestation (kyriotetes). Then there are the spirits that sort of think about the action of manifestation (dynamis). Then there are other spirits actively working on the forms that will be manifest (exusiai). There are also spirits who like that manifestation and think they can make a world from it (archai). In addition, there are spirits who listen to the one saying, "I think I can make a world out of it," and they put all their intentions into little boxes so that their attention goes to the spirit wishing to make a separate world that has little to do with the others, and this spirit is called Ahriman and his workers, the Asuras, who are fallen archai.

Thus, we go from spirits who represent the idea of something that could be manifest to spirits who have made a career of making things manifest; this process is called the Fall of humanity. Just because a being is a spirit does not mean it is always nice. Just because it is a spirit does not mean it has nothing to do with matter. If this were the case, Ahriman would have no interest in the Earth. However, Ahriman is deeply interested in the Earth as a manifestation separate from spiritual input, because this is his sandbox. When we say "spirit," we need to understand that coagula is an important part of the unmanifest spiritual world.

This drama of the Fall of humanity is woven intimately with the formation of sense organs in the body. In our body's every input from our senses we would be fried by sensory activity if we did not have scar tissue to modify those sense impressions. The creative light of the cosmos would fry our souls. We have to take in sense impressions through filters and kind of condense and control them. Our eardrum is a kind of scar tissue that filters every sensory impulse that is coming in. We have an eye to receive light, but the lens of our eyes forms from the ectoderm in the same way that scar tissue forms. We have our skin, which is basically dead tissue in the outer layers. It is a scar that prevents us from being flayed alive by every little breeze.

We surround ourselves with scars in the places where we get the maximum input from sensory experience. Basically, the formation of that lens in the embryonic eye is like forming a scar. The scar is separated out from our consciousness. Thankfully we are blocked from direct experience because we have fallen and have scars over the places where the cosmos comes into us. Those scars are corpses of the creative activity of the hierarchies that we call our sense organs.

The fact that a great spiritual being has decided to come from a condition of total omnipotence to being blasted away by ignorant human beings is love. This is what I am talking about in the Fall of humankind from paradise. This archetype is the process of coming from a spiritual condition to being here and getting sick. That is great sacrifice. It is love. Love is allowing ourselves to become so vulnerable that we allow the world to impact us. I allow you to see me. That is Steiner's definition of love. He does not often make a definition, but that is what he says. "Love is allowing ourselves to become so vulnerable that we do not have to protect ourselves from others seeing me." Everything else is not that. Everything else is having my little thing that I expect from love. Expecting reciprocation for my output of attention is not love but just being human. Love is the hierarchies' participation in the endowment of human beings with sensory organs.

My body is a crystal of my spirit. My physical body is a crystalline form of my spirit. Your bones, your cells, are all liquid-crystal displays. What you think is this kind of fluid vehicle you run around in is fluid only because there is a spirit within it and keeping the fire going in the furnace. When the spirit leaves, it turns into a crystal— a corpse, dirt. The spirit fire is the only thing keeping us animated. Spirit fire is the fact that we are attached to being—a spirit packed in crystal. When I finally understand that I have to become vulnerable to the structure of the world—this is what Alchemists spoke of in the alchemical wedding. They said that you have to take the world as it is manifest, take it in through your senses, and form it inwardly. You need to marry the physical to the spirit by changing the way your senses work into moving pictures, because your senses have death in them. Your senses give you the impression that you live in a world in which most things you encounter are corpses and whose real life is active in some other dimension. Because your senses were put in place to protect you from becoming too vulnerable. If you did not have them, you would be so vulnerable that you could not complete your mission of transforming a dying Earth back into spirit existence.

Think about it. You have to do strange things to accomplish your mission of transforming the dying Earth. You have to turn the world upside down through that little scar tissue in your eyeball. You take it and give thanks to the world for being out there, but to deal with that power you have to turn it upside down. If we saw the world straight up as a spiritual entity, we would just evaporate. Therefore, we have to focus our sensory experience of the Earth and tweak it so we can take it in without it dissolving with its creative power. We take the world in and turn it upside down a couple more times, and then we call that process seeing. However, if we were actually seeing, we would not need our physical eyes because the eye of the soul is in the heart, not in our head.

When we work with inner pictures and transform them, we are developing an organ that Rudolf Steiner calls a "heart eye." This is the purpose of phenomenology. When I learn how to control the way I see something, I transform the world. I do this by first looking at the thing without adding any content. Once I can do that, I need to take the next step and look at my looking. I need to realize that every time I look at something I have a bias about what I think it is that I am seeing. Owen Barfield[9] said that percepts are concept-saturated and that fact prevents us from seeing the true nature of what we are sensing. To make progress, we have to check the difference between the thing out there that we are sensing and compare that experience to what we think we are seeing. This checking of ourselves in acts of perception is phenomenology. A higher form of phenomenology is watching our soul reacting to being checked by ourselves. This can begin with the experience of looking at a leaf. It transforms into the experience of watching ourselves looking at the leaf. Between the two, there is a breathing process.

9 Owen Barfield (1898–1997) was a British philosopher, author, poet, and critic. He was born in London, received his graduate education at Oxford and in 1920 received a first-class degree in English language and literature. Barfield was a longtime student of Anthroposophy and, as a member of the Oxford Inklings, had a strong influence on C. S. Lewis and, through his book *Poetic Diction,* an appreciable effect on J. R. R. Tolkien.

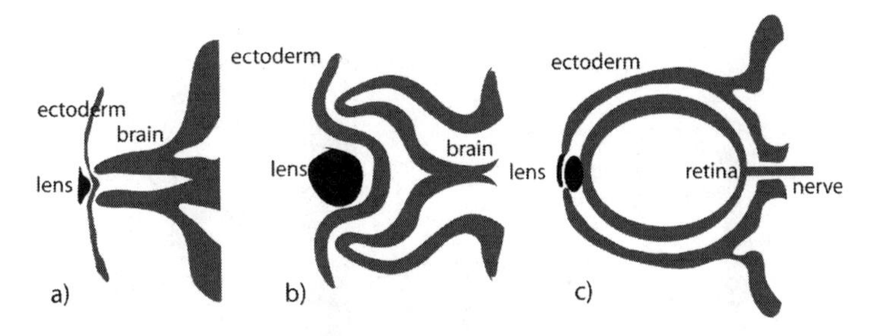

Figure 4: Layers of the eye

As we work with these breathing polarities, we start to use the sensory world as a barometer for what is happening within ourselves. As I understand it, this is the true purpose of phenomenology.

Once phenomenology of the soul becomes possible, the soul world opens up to reveal the situations in which I have a belief, which may not reflect what is actually occurring in the world. This is the great training of all esotericists—to eat my own belief and find a way to balance myself in a world that suddenly has no room for blame. This is where we get back to having another person to represent fire when I am sending images into sleep. Having the patience to wait until another person comes to me with an image or comment allows me to know that I have not manipulated that insight. Then I take that insight and Google it to take my understanding further. The other person has shown me that my imagination is guided by an inspiration linked to a spiritual being.

In *figure 4,* we see that there is a pulse to the periphery in the fire element that guides the incarnation of physical bodies, and then the periphery pushes back. Over time, the pushing and pulling of the polarities in the fire mode creates what we could call "form." However, it is a form with quotes around it. This "form" is the potential for a particular form (without quotes) to manifest.

That is how organs are created. Look at *figures 5* and *6* (next page). The first is the cross section of a fish heart. Notice the little arrows and such in the upper part, which would be the atrium in a human heart. It is called the *bulbus arteriosus.* There are vertical currents flowing in that fish heart. Then

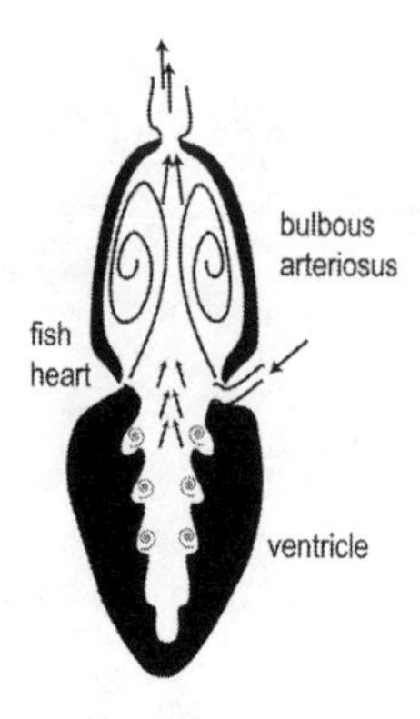

Figure 5: Fish heart

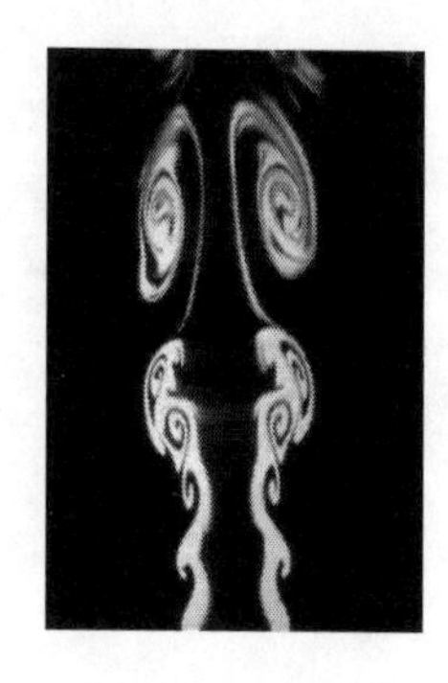
Figure 6: Vortical flow

there is a kind of intake vent where the arrow goes into the chamber, and in the bottom the ventricle is the part that contracts as fluid goes up into the *bulbus arteriosus,* from where the blood goes out into the body. You also see tiny spiral currents in the loops of the ventricle. The gray part is the tissue of the heart that has "little fingers" in it. When the heart contracts, little vortices are created that go up through that little orifice, or venture-like form, into the *bulbus arteriosus.* This heart action happens just as the fish moves.

There is a hierarchy of inflow of the forms of little vortices that evolve into the big vortex, then the blood begins to move out. It is kind of like a venturi, or supercharger. That is the fish heart. Hearts in general take form as a result of fluids falling out of activity into manifestation. These tissues are like sandbars in the river of your blood; that is how they form. They go from a solve to a coagula.

Figure 6 is a photograph of water coming out of a tube into a tray of still water. Dye has been put in the water. As it comes out of the tube it forms the same type of little vortex trains and eddies that are in the fish heart diagram. The small vortices are followed by a venturi-like form that generates an upward surge into a larger form. Even a superficial look at the form of water and the flow in a fish heart shows that both have the same pattern.

Although one is part of an organism and the other is part of the elemental world, they share the same formative pattern. How and why does that happen? We could say, according to Rudolf Steiner's picture, that the elemental world—the elements of earth, water, air, and fire—are receiving imaginations from the unmanifest world. In organisms, the elemental beings at work in the bodies of living things take up those imaginations, which create the life bodies of the organisms. A stag bladder, cow intestines, and an animal skull are each imaginations of a particular relationship of sets of forces that create an organ through a specific formation process that results in the form of the organ. Once the organ is formed, it maintains a connection to the forces that formed it. The patterns of forces that formed the organ become the organ's function once it has formed and takes its place in the body. The form of the organ stays in contact with the forces that formed it. This means that we can take that organ from the organism and put other things into it to take advantage of the formative forces still contained in that organ. This is possible because the forces that guided its formation are still connected to it. Those forces of the organ's formation are the unmanifest, or etheric, side of the manifest mandala of the elements. Rudolf Steiner calls unmanifest imaginations activities of the etheric world, the creative imaginations of the ether world and its spiritual beings.

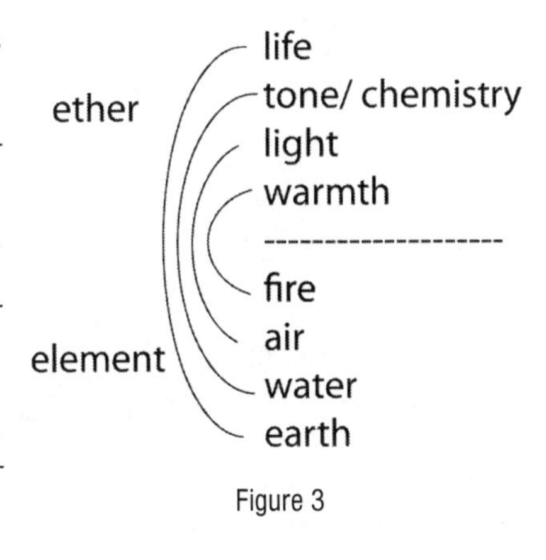

Figure 3

Figure 3 shows a list: earth, water, air, and fire. It starts with the most manifest, the earth element; then the less-manifest water; then even less manifest, air; and finally the least manifest element, fire. Then a little crossing point goes up into what Steiner calls the etheric sphere. The little membrane between the elements and the ethers divides the elements on the manifest side from those on the unmanifest side.

We see that fire, the least-manifest element, is a coagula of warmth, the most manifest ether. Warmth is the potential and fire is the manifestation, even though fire is the least manifest of the elements. I could draw a little connection between fire and warmth to illustrate that the element of fire contains the more subtle state of warmth. Esoterically, warmth represents enthusiasm and will. When enthusiasm becomes manifest, we see it as a fire. In ancient cultures, fire was a god; it still is for some people. There are articles on the web about *Agnihotra,*[10] an ancient Indian and Central American ritual in which people burn cow manure when the sun is on the horizon as they sing their ritual Agni hymns. They take the ash and spray it on plants, which grow like crazy. In an experiment, they put the ash on one plot and not on another. The plants in the plot that received showed remarkable growth. In fire and warmth, the ether and the element are almost the same thing, but not quite.

In the realm of the soul, enthusiasm and will are related to the ether of warmth. Therefore, with my consciousness I can enter into that warmth and the warmth of the fire is the agent of transformation of the Earth into the water, and the water to the air, and the air to the fire. In this work my akasha, my consciousness, carries an enthusiasm for being even when I am sick or else I would not be here.

When I lose my enthusiasm for being, I am on the way out. In fire and warmth, the element and the ether are very close to each other. Warmth is the "enthusiasm for being" of the hierarchy of the Thrones. The "will to be" is the warmth of ancient Saturn. The "willingness to be," or enthusiasm to be, is the warmth of ancient Saturn. This then goes through a process of wisdom, from the Thrones to the Kyriotetes, the beings of consciousness of light. When ancient Saturn goes to ancient Sun, the will warmth of Saturn becomes the light of wisdom of ancient Sun. Light is the wisdom of the enthusiasm to be. It is the consciousness of the wisdom of what it means to be rather than just the will to be. Light is the awareness of what it is to be. Therefore, in the cosmic periphery, the light ether represents the ability of the tone ether to manifest. The tone ether is even higher than the light ether. The light ether is a corpse of the tone ether.

It is difficult to understand how Steiner organized the ethers without knowing what happens in the transformation of one into the other. Each ether, as it rises, includes the one below it. Each element, as it goes down,

10 See agnihotra.org. *Agnihotra* is a healing fire from the ancient science of Ayurveda.

excludes the one above it. When I go up to light ether, I include warmth ether, but it is warmth that ignites and becomes light—not just visible light, but also the light of the wisdom of understanding, the gift of the Kyriotetes to humankind.

Light is the force of transmission that reveals how the cosmos creates the great field of energy. When seen with the background of projective geometry, the field of energy of the universe can be seen as composed of interweaving planes of light. This is an image of what Steiner calls the weaving of the light. According to science, light streams from the stars as planes of energy. The innumerable sources of light from the stars continually weave planes into energetic points of great potential. Geometrically, three stars produce planes of light and create potentialities for the planes to meet in a point. However, there is then another star right next to that one, creating another point, and another point, another point, an infinite sea of potential points of light. Points of light are everywhere, fluctuating continually from one position to another, appearing in space, and then wiggling out into another dimension. In quantum mechanics, physicists call such motions the jiggling of the quantum field—that is, a manifestation of photons always moving in and out of existence everywhere. It is the quantum field—the source of the energy from which we get light and warmth. Rudolf Steiner called this process the self-procreation of the ether world.

Every point of light and warmth is burgeoning and growing. These movements all create the great sea of weaving light. The ethers fashion the potential for this *pleroma* of forces, and the elements make the potentialities manifest. The states of earth, water, air, fire, warmth, light, tone, and life each represent the state of consciousness of a spiritual being—either that or they are just dead forces we cannot seem to pin down enough to measure with instruments.

Continuing with the ethers, when we move even further out from light, we see the light and interweaving of the planes, and it has a kind of geometry. In the past, that geometry was called the "music of the spheres." The music of these spheres is the infinitely distant weaving of all the light into ordered patterns. Rudolf Steiner calls that realm the tone ether, number ether, or chemical ether. The number ether includes the warmth and light ethers, but takes the potentialities a step further toward the infinite. With the tone ether, we approach the crystal heavens, where all is ordered and the number ether represents potential for all phenomena to be ordered, even in the state of ultimate chaos. Remember, chaos means potential. Tone ether could be called the Logos Word. *The Word,* with the capital "W," is the logos nature, the numbering of things, ordering, the great cosmic chemistry, the great cosmic music of potential, and the word-like nature of everything.

We covered the fire element as the corpse of warmth, and air as the corpse of light. We will do the same with tone, number, and chemistry. We end up with our friend water. Water is the corpse of number. That seems strange until we recall the potential of water to work actively in all the chemistry—in all combinations and transformations of one thing into another. Water is the corpse of tone. Air is the corpse of light. Fire is the corpse of warmth.

I am speaking alchemically about the "corpse" of the unmanifest as something that has formed. Therefore, the number, tone, chemistry of this ether becomes as its corpse what we call water, but it is a very high ether because it is a universal solvent. Alchemists used the word *corpse* to denote this falling from spirit into manifestation because their central task was to reanimate, or resurrect, the corpse of substances back to their potential states. They wanted to resurrect the corpse of substances so that the original creative force in the substance could be used for healing. To alchemists, a corpse is not something to be disdained; it represents the potential for eternal life. Hidden in the corpse is a great secret of the salt, or phantom, of the potential for human sensory life as the key to transforming Earth into a star. Do not be put off by the term *corpse.* I use it consciously in keeping with alchemical reasoning.

Nevertheless, we are not done yet, because we just have water as an element and not the earth. We have to bring the ethers down to earth, and we have just chemistry and tone. What guides the chemistry? Life—universal life ether—sacrifices to manifest as earth. The lowest is the highest. This is the "as above, so below" of Hermes's emerald tablet. It is the second day of Christian Rosencreutz's alchemical wedding. It is the parable in which you are invited to the wedding feast; you go and sit in the front, and the host comes and says, "Excuse me, I think you need to go back there because there are some other people here who are a little closer to me as friends. If you do not mind, please move down the table a little." This makes you feel blue, but if you sit in the back, the host will come and say, "Why not sit a little closer to the head table?" This is Christ—the highest shall become the lowest, and the lowest shall become the highest. Blessed are the poor in spirit.

Therefore, universal life is the highest ether, containing all the other ethers and all their potentialities, and this creates the willingness to sacrifice by those beings beyond the Thrones, Cherubim, and Seraphim to bring their gifts to Earth and to create a place where the great drama will happen. This is the great Rosicrucian drama of resurrecting into spirit what has fallen into earthly manifestation and then giving it back to God.

Now we can discuss a new agriculture through which we need to be on Earth as priests, enacting rituals of resurrecting what has fallen into corpses through the sacrifice of the Earth spirit. This is the deed of Christ to move out of cosmic life and to become fixed in a body. This is the destiny of Earth, which gives life for us all to live. Agriculture can be a way of taking the corpses of past life, resurrecting them through alchemical methods, and giving them back to the Earth as a healing medicine. How do we give nature back to God in a better form than when we received it? The Earth falls into a corpse through my deep human belief that my sensory experiences separate me from everything else. *Love* is a code word for a willingness to be vulnerable to the realization that—if I am using the Earth with impunity and not connecting to the spiritual destiny of the Earth and my role as a human on the Earth—I am causing the Earth to suffer without being recognized. It is one thing to suffer, and a whole other thing to suffer unrecognized. Therefore, when I come to realize that I am treating the Earth as a resource instead of a being, it crushes me like a bug. This is the guilt that many people experience when they study ecology in school. They feel hopeless and ashamed.

What can I do? Do I just get crushed like a bug? No, I have to make a vow to myself that I am going to bring pictures to the cosmos that allow me to feel that I am co-creating with the Earth and that I recognize my small, though important, task of recognizing the Earth's spiritual destiny. This allows me to intuit ways through which I can give back what has fallen to God. The only way I can turn my soul from existential guilt into the willingness to imagine my role in the Earth's destiny is through *active imagination.* Through imagination, I can move past existential guilt over my habit of just using the Earth's resources for my own ends. Such guilt is useless beyond the first awaking step; it can be paralyzing and a great cause of despair. It is absolutely true that the one being paying for all of this freedom is the Earth. Moreover, whether we are aware of it or not, that is the great force used by the adversaries against the whole global community. Existential guilt will not solve the pressing problems; only enlightened and imaginative human cognition can free the Earth from the shame of not being recognized by humankind as a spiritual being of a high order.

We are all guilty of using the Earth as a resource, but this does not make us bad. It means simply that there is a lot of work to do. We have to transform our consciousness so that we engage the Earth and the beings of Earth, as well as the elemental beings that are part of the Earth. Then we will have prepared our consciousness to recognize their place in the great spiritual hierarchies whose forces are interwoven with the destiny of the spiritual Earth. We can work with that perception meditatively by taking images of earth phenomena into our sleep. This is the initiation of a dialogue process; it allows the spirits linked to Earth's destiny to give us pictures of the correct ways to move into the future. We need their help to imagine how we can take materials and make them more effective in dealing with the health of the Earth. How can I take natural things and make them more effective? I have to understand how they suffered to become what they are to begin with, and then I have to identify with that suffering, that death becoming a corpse. I have to become vulnerable to that drama of eternal life dying into temporal life as a being with an earthly body. This means realizing that my car key is also my key to the polluter's club. Once I can realize this, I have to transform the existential force of "Oh my god, this is too much" into imaginations of the Earth's spiritual destiny. That is a death-force kind of depression. The way I change it is through working meditatively with pictures from the Earth that seem to have a personal, symbolic relationship to me.

The personal link in working with a symbol is a way of showing gratitude to the spiritual world. When you form an inner picture of a phenomenon in which you are interested, the beings in the natural world behind that phenomenon can then relate to your consciousness. It becomes possible for them to inform you of the deeper significance of that phenomenon in the natural order. The formation of the inner picture should be accompanied by a feeling of gratitude and a sense of reverence. This is best accomplished by choosing an image that seems to have something to do with your life in a personal way. This process involves the formation of a symbol as a bridge to the hierarchies. Nevertheless, we have to be clear that this exchange requires more than just taking what comes from the spirit through us personally. That leads to what Rudolf Steiner calls "a war of each against all," in which everyone becomes a little magician with one's own little set of magical pictures and nature deities. This kind of situation is what led to the world cataclysms in Atlantis and Lemuria.[11] The Lemurian catastrophe was linked to plant forces, while the Atlantean catastrophe was linked to atmospheric forces. In both cases, magicians learned to manipulate the forces of nature in ways beyond their understanding morally. They did not have the failsafe of the Christ, since the Mystery of Golgotha had not yet happened.

Rudolf Steiner has described all of this. They were times when Jehovah ruled as God, the creator of the natural order. Human beings looked to the Creator for direction in what to do. Magicians, however, took the opposite route, doing whatever they wished without reference to the natural order. The result of not doing the right thing was catastrophe. We are now in a similar situation with our approach to technology, which allows little individuals to do really big things without real awareness of what they are doing. This may be called being President of the United States, Premier of China, or the Grand Imam of Iran. We are on the way to realizing again that we live on a very small Earth, which is losing forces. Now we need to come together

11 See Steiner, *Cosmic Memory: The Story of Atlantis, Lemuria, and the Division of the Sexes.*

again to do something about this. However, as an esotericist I have to recognize that there is a vector in the way I receive answers to research problems. That vector is having to wait with my own answer until you help me understand that my answer is really the answer to the question I am asking.

This is the path back to oneness, because when I train myself to do that I am always listening to what individuals say to me all day in a completely different way. I have to learn that the answers I seek are coming through other people who intuit unconsciously what the spirits are saying to me. When I realize that these kinds of communications are always coming through the people around me, I can still exercise my imagination and get answers, but at least I am aware that for truly serious questions there is a layer of failsafe whereby I am not just getting answers directly. Ancient magicians worked that way, which eventually led to the pollution of mystery wisdom with personal belief, dogma, and a perceived need to control. This led to the decadence of the mysteries. This decadence took over when whole groups of people got together and said, "This is the only way this mystery can be practiced." Go tell people you have a different way of working with preparation 501, and you will experience the old way.

In the elemental world, fire is a threshold that leads into a whole other dimension of what we could call the consciousness of warmth, enthusiasm for being, or the will to be. The consciousness of the beings of the light ether perceives the world as made of consciousness. The consciousness of the beings of the tone ether see the world as ordered. This is the prevalent awareness of the field of the Logos. Then the consciousness of beings linked to the life ether is that ordered consciousness is completely open, with no biases. Life contains all polarities without strain or contradiction, even when there are complete opposites. When we get to be the tenth hierarchy and the Earth becomes a sun, we will all have that consciousness. The collective human consciousness will act as a homeopathic drop in the evolutionary cycle of the great ocean of the Logos nature. Today, that consciousness exists as the consciousness of the Earth. According to Rudolf Steiner, Christ put a seed of the Logos nature into the Earth following the Crucifixion on Good Friday. Human beings need to germinate that seed as a collective spiritual deed for the future. The Christ being is now the spirit consciousness of the Earth. Moreover, the Christ being is open to all, but appears only when two meet. It is a very beautiful relationship here between the destiny of Earth and human destiny. When the two meet, the combination of their two separate forms of consciousness allows the Christ to come in, and we start talking about the Holy Spirit.

V. Nicholas of Cusa • Liberating Elemental Beings • Spirits of Time Rotation • Qualitative Research Methods

The theme in this section has to do with consciousness in relation to the transformation of substance as an agricultural path. My motif comes from the alchemical wisdom that the operator's consciousness needs to be in harmony with that of the level of operation. In other words, if you are working with minerals, your consciousness must harmonize with the mineral world. If you are making a solution, doing a distillation or extraction, your consciousness should harmonize with the rules of the water sphere. Alchemical sublimation occurs when the substance with which you are working transforms from a solid to a gas without going through the liquid state. If you are trying to do this or trying to make a remedy that will affect the flowering process in a plant, you have to harmonize your consciousness with the guiding principles of the air element.

What I have tried to describe are the rules of the Earth, the mineral, the water, the air, and the fire. Your consciousness affects the akasha linking these elemental states. When you work to harmonize your thinking with the guiding principles of the elemental world esoterically, this is called "writing in the astral light." When you gain an insight into the natural world, the elemental beings who animate nature have given you a gift. It is not you who has an idea. You are a satellite dish or antenna. In reality, *the idea has you.*

As an alchemist, you have to work to trim you antenna so that you create a standing wave in your consciousness. The standing wave in your consciousness has to be a picture of the signal.

I am speaking the language that stands behind Rudolf Steiner's agriculture course. It is an alchemical point of view. Your akasha tunes your antenna. Writing in the astral light means simultaneously reading in the astral light; this comes directly from Steiner in the little book, *Inner Reading and Inner Hearing.*[12] Writing in the astral light, in the akasha, is the equivalent of reading in the astral light. This means, when I download what I think is an idea, I have to guard against it becoming a belief. Therefore, the rule is this: you need to do your reading and writing simultaneously in the akasha. That is the failsafe against self-inflation. However, as human beings we tend to believe what we download is the answer. This problem of consciousness has to do with our human sensory experience and cognition of our sensory experience. Normally, we do not cognize our sensory experience, because it is just part of the given world. That is a tree. That is a leaf. We learn this as little children. Then, forever after, our senses reinforce that sensory experience by saying that the leaf over there is not me. However, every esotericist in the world will tell you it *is* you; it is just vibrating at a different rate. In addition, it has its own consciousness, which actually is linked to you. You just have to be able to harmonize your consciousness with the leaf to understand how it was made; this is the lesson of John's Gospel—everything that was made was made by the Logos. Without the Logos nothing was made that was made.

This is a very deep meditation on how my consciousness as a human interacts with the Logos, or the becoming of the world. Attaining this consciousness is the task of esoteric students. Goethe called this higher consciousness of the world becoming "higher beholding"—in contrast to the lower beholding. The difficulty is that I project beliefs and memories from my lower beholding into my higher beholding, and therefore "that is not me." My lower beholding is composed of all of my uncognized sensory

12 See *Inner Reading and Inner Hearing: And How to Achieve Existence in the World of Ideas,* lectures 1 and 2, Oct. 3–4, 1914.

experiences. My experiences of higher beholding occur when elemental beings come to me with insights as a result of my previous work to purify my lower beholding.

When I assume that my lower beholding is higher beholding, that is a fatal mistake as a spiritual researcher in the realm of natural science. My beliefs based on old memories create problems for me, because as soon as I revert to old memories as solutions to my questions, I immediately drop out of the realm in which I harmonize with phenomena in the elemental world, and all of the elements and ethers get cast into a corpse of belief. I "earth" my process of discovery; I factoid it into a corpse. As soon as I factoid something, I own it as a belief. Moreover, as soon as I own something as a belief, you cannot mess with it—actually, neither can I. My world becomes locked into a whole system of beliefs that I can structure around what I am doing to justify it. Unless I always challenge those beliefs, they just become a fixed system that dampens my enthusiasm for taking the risk to learn new things.

This leads to all kinds of experiences that are mere specters of reality. In pursuit of higher consciousness, I always have to test my beliefs by engaging in a process of observing the phenomenon and making sure that the inner picture I am looking at is linked to the phenomenon. In addition, a further task is to engage the process of taking the inner picture into sleep and waiting until something comes from someone else to reinforce the thoughts that this inner picture stimulates in my consciousness. Basically, this is a process of trying to keep my interactions with elemental beings open without falling into a belief.

Look at *figure 11* (next page). When I take images into sleep and wait for a response from others, my belief transforms into a process of discovery. I start to have a meditative process I will describe in a moment. That process goes through akasha again, and in *figure 11*, you see belief and process are in the gravity corner of the mandala. However, when I want to go to a higher place with my spiritual research, I have to shift my process into reverse. The biggest reversal is trying to tell others, "This is what I think I saw," to which they reply, "Huh." The other person's consciousness gives me a failsafe against my own belief, which I need to progress in

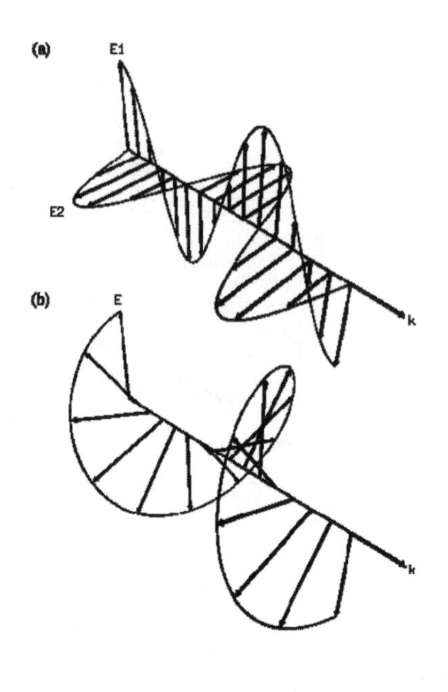

Figure 11

spiritual research. If I do not have this failsafe, I just keep playing the same tape of my beliefs again and again. In the old days, alchemists had what they called a *soror,* because they all would get stuck in their beliefs. In this instance, the alchemist was the *frater,* or brother, and the *soror* was the sister. Alchemists would know *what* to do, but they did not know *when* to do it. They needed a sister to indicate when to do something, because women tend to be much more tuned into process than men are.

Therefore, alchemists needed the *soror* to know when to do something, but she would not know what was needed to be done. She would cultivate herself to be open to *when*, but if you asked her *why,* she would not have a clue. Nevertheless, she knew *when.* Together they did the work. Today, the *soror* and the *frater* have to be the same person. That is the alchemical wedding, in which the male part of consciousness and the female part of consciousness are married and go on a honeymoon in the elemental world. Thus, I have to be able to access imaginations, but I also have to be able to check them so that my insights do not become ironclad beliefs. To prevent this, I have to take any insights or imaginations that come to me back into process. Eventually, if I do this, my consciousness moves to the exact opposite of my belief, and my heart eye begins to open. If I can work my consciousness over to the exact opposite of my belief, it is not a problem when someone shows up and tells me I am full of hot air, because I will have already seen this through my own efforts.

We will do a little quick exercise here to show you how that feels. We will use *figure 7.* The exercise comes from the work of a famous theologian of the Middle Ages, Nicholas of Cusa (1401–1464). He was a bishop

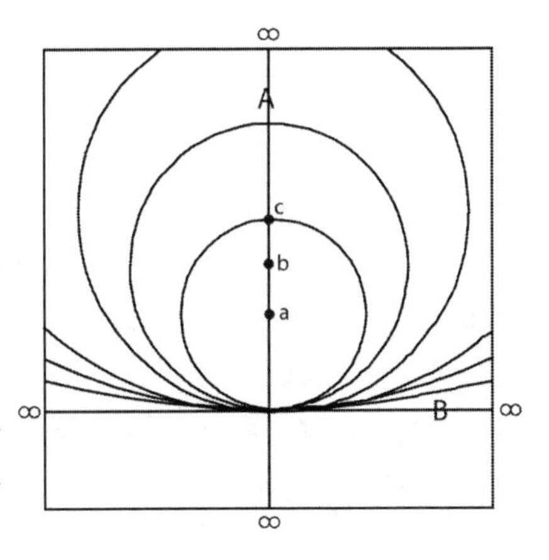

Figure 7

and a geometer who had an experience that changed him so much that Rudolf Steiner says he was the first modern person in his consciousness.[13] When Nicholas of Cusa was alive, theological arguments were rendered in the form of geometric theorems. To prove a point, theologians would make a geometric diagram. This was the way people described the relationship between God, the hierarchies, Earth, and everything; it was all geometric. If you read Kepler's books, you see that his arguments are presented in page after page of geometric theorems and drawings. It is like junior high school on steroids. The accepted language for theologians was geometry—sacred geometry. Thus, Nicholas of Cusa, being a bishop, was doing sacred geometry when he did this exercise and received a spiritual insight that caused him much doubt. In his reasoning, he came up against something that turned a thousand years of geometry from Euclid and Egypt on its head.

His discovery was a theological experience for him, but it rocked his worldview. He experienced doubt that human beings could actually know God, which is a sticky position for a bishop. It was this exercise that did it. I hope that when you do it, you will see why. I present this to illustrate the issue around one's belief that something is one way and then the discovery that the reverse is true. This kind of feeling disconnect causes a person to hurriedly shove another belief into the hole of disbelief. The problem is that, with this exercise, you cannot shove anything in there. This experience left Nicholas of Cusa in a completely different place.

In *figure 7* you see horizontal line "B." In geometry, it is a convention that lines are designated with capital letters, while points are designated with small letters. We will designate the vertical line in the middle "A." Then

13 See *Mystics after Modernism: Discovering the Seeds of a New Science in the Renaissance,* pp.71ff.

you have the small circle and "a," the point at the center of the smallest circle. The edge, or circumference, of the small circle is tangent to line "B" where "A" and "B" cross.

Looking at the diagram, the smallest circle has point "a" as the center. It is on line "A," the vertical line. The bottom edge of that circle touches the spot where horizontal lines "B" and "A" cross at right angles. Line "B" is called a tangent line to that circle, because it is just touching the small circle at one point on the circumference. Now, this is not a big deal, but what Nicholas of Cusa was playing with was the idea of infinity. He was a bishop and was trying to understand the nature of God. He thought he would make a bigger circle on line "A." The second circle has as its center "b" on line "A," but it is still tangent to horizontal line "B." In effect, he was moving the center of the circle toward the infinite on line "A." So, if you look at *figure 7* and the top of line "A," you see the little infinity symbol.

A big issue in the Middle Ages was to explore the infinite. A great question asked how many angels could dance on the head of a pin. This is a sort of "oranges and apples" question today, but back then they were exploring infinity and its relation to geometry. They asked this question: Is God still present in matter? It is a question of what is known as *fiat,* or "I create." Wars were fought over the principle of *fiat.* The theological issue in *fiat* is this: Did God make this creation and then abandon it, or is God still within it? In terms of this experiment, the question would be this: Is the infinite part of the finite, or is it separate and beyond our understanding? That was a big theological theme.

This issue had to do with the mystical body of Christ and issues in the Church relating to the problem of matter in the context of the infinite wisdom of God and the fallen nature of the flesh—we might say, the purpose of the corpse in the grand scheme of things. Thus, Nicholas of Cusa was exploring the issue of the creation with a thought experiment about God. To the Medieval mind, God is symbolized by the Circle; God is the great circle. It represents the all-inclusiveness of the divine nature. However, now Nichalas of Cusa was taking a circle and expanding it to the infinite. He expected a double dose of infinity when the circle that is God returned to the infinite. However, this is not what happens when the circle reaches the infinite.

In the experiment, we have a small circle, which expands while maintaining contact with the Earth, as represented by line "B." Lines "B and line "A" meet at right angles, the angle of the Earth. This means that the circle representing the diving remains in contact with (is tangent or touching) the Earth as it expands into an infinite dimension. This is the reasoning going on here. If the circle tangent to the Earth is moved along line "A" to the infinite, the question becomes: What happens to the circle?

Nicholas of Cusa is saying, "This is my circle; this is the divine. I am going to take the divine all the way out to the infinite, but I am going to keep it connected to the Earth, or finite; and what happens? Try to imagine what happens as we move the center of the circle along line A toward the infinite at twelve o'clock. This is what he did in his mind and what I ask you to do in your mind as a thought experiment. What happens to this circle as the center of the expanding circle reaches the infinitely distant? Just play with this in your mind and reason it out. If I look out to the horizon, do I see a curve? No, it appears flat, but we know it is a curve... or at least we believe it is a curve. Nevertheless, it appears as a flat line, and this is just the Earth's circumference, not an infinite circle.

Nicholas of Cusa found that, when the center of the circle reaches the infinite, the circle becomes a perfectly flat line. The question remains: Is it still a circle? Its center is the tangent point, but that tangent point representing the Earth is now on the periphery, while the circumference of the circle is now the Earth. When you experience this, you experience reversal. When you grab something really big like this, it becomes a great challenge to one's accepted beliefs. That is really good. When your own experiment reveals a gap between our belief and reality, that is very good because we have the experience of actually doing battle with our beliefs.

What matters here is that you can go through a reversal and still maintain integrity in your consciousness without snapping to another belief or having to hold on to your old belief. It is useful to remember that Nicholas of Cusa inherited a thousand years of belief that God is a circle with a center and a periphery. This experiment reversed that concept in his mind and created a very difficult dilemma for him. This experiment challenged a thousand years of belief.

If you do not revert to a belief when your worldview is challenged, the elemental beings kept in prison by the nature of your belief will be liberated. Christ cannot get in to your belief structure, because human belief is based on personal memory. Every doctor knows this problem. They get to a certain place in their work and have to say, "Either I am the source of the healing or something else is the source of the healing." They have to work with that conundrum as an emotional reality. For a thinker or esotericist, the conundrum is this: Am I the source of my insights, or do insights come to me through the acts of grace of spiritual beings? The whole Rosicrucian test is that you get to be a player only when the spiritual world needs you in the game; that is the rule. The failsafe in this is that I have to find some way to temper my imagination to get it to where I can imagine reversals myself. Reversing an imagination is my protection against self-inflation. If I can do that, elemental beings that are part of the phenomenon I am studying do not have to go through the extra process of correcting my inflated beliefs because, by being able to imagine the reversal of my belief, I have already made it okay for them to approach me.

There are many different levels of elemental being in nature and Rudolf Steiner has said that the greatest source of elementals is created by human beliefs. We create elemental beings with our consciousness; this is the key to the kingdom we are given as human beings. It is why nature around us appears separate. We have a particular way of filtering the light that comes in from the periphery; we cause it to be fixed into patterns. The elementals that serve the cosmos sacrifice themselves to be fixed into our human thought patterns. The elementals who have sacrificed to being imprisoned in the sense-observable manifestations of nature need to be liberated by our love and understanding. This is the great Rosicrucian root of the work in biodynamics. However, to do this, I have to work on myself so that the imaginations I put into myself include reversals. They should include the idea that the imagination I am holding may not be true. I need to entertain

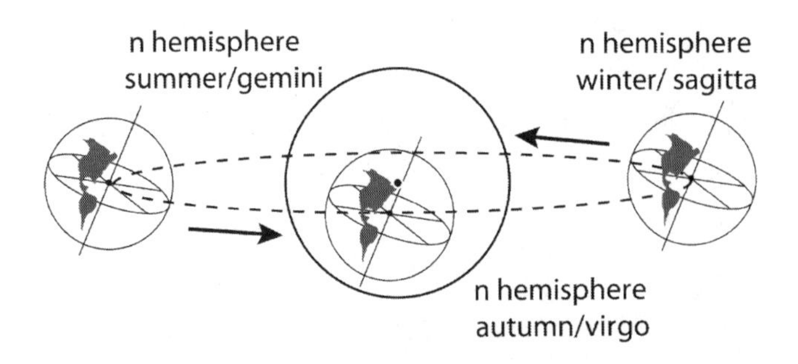

Figure 8

that thought without it causing me to lose energy. I need to train myself to do this, because the world is full of reversals. This experiment by Nicholas of Cusa is just one.

We will now look at a big reversal that is very useful as a meditative device for agriculturalists. Look at *figure 8*. This is another imagination you can use as a meditative device. I am now going into the realm of akasha. If you want to create the ability to incarnate imaginations that include reversals, I suggest studying planetary movements, because they always include reversals. This is why arithmetic, astronomy, geometry, and music were known as the *quadrivium* in the school of Chartres. Those four subjects were the four great pillars of the higher octave. The lower octave was logic, rhetoric, and grammar, which were taught first. The higher octave would expose students to the idea of the harmony of the spheres. When you study this type of reasoning, you begin to understand that it is not what you know or what you do that matters; rather, it is lining up things in time that counts and that everything reverses. With the *quadrivium*, you could experience doing something one time and getting one result, then doing it again, believing you are doing the same thing, while actually not doing the same thing because the elemental beings who guide the rotating movements of the planets are of a higher order than the elemental beings who animate earth and water.

These teachings relate to Steiner's description of three levels of elemental being. At the lowest level of elemental beings are the elements that animate the forces of nature. Steiner calls the next level of elemental beings "elementary" beings. They guide the laws of nature and have a kind of self-awareness. Steiner gave a picture of these beings as the laws of nature. Esoterically, as human beings we collectively create these elementary beings through our belief that everything is separate from us. We create the laws of nature as thought concepts in our human souls. However, every century or so since the 1400s, the beliefs of science—the accepted laws of nature—have been violated by some genius that comes along and says, "Here is a new way to look at things. Unfortunately, this new way invalidates our old way of understanding natural laws." As we rise in the spiritual hierarchy, our understanding of the laws of nature requires an approach to those elementary beings that is different from our understanding when, for instance, we manipulate forces in nature through new technologies.

According to Rudolf Steiner, an even higher level of elemental being is available to human consciousness. He calls this highest level "the spirits of the rotation of time." These guiding spirits of the natural order have an even higher consciousness. Therefore, if I wish to train myself to enter nature to understand how reversals work in order to free up my beliefs so that when I do an experiment I am not always just repeating the same stuff again, in my mind I have to practice harmonizing my consciousness with the movements already present. This gets back to where I started. The consciousness of the operator has to be harmonized to the thing being studied. This means harmonizing one's consciousness with the highest level of elemental being working above natural laws and forces. These highest elemental beings guide planetary movements. Therefore, if I want to do myself a service, I study the movements of the planets as rhythmic signatures—not just as a study, but also as a pictorial meditation. I picture those movements to myself, and this puts me in contact with a level of transformation that allows me to see substance from a whole other perspective. Any substance is a movement that has come to rest; a substance is the corpse of an activity. If I wish to make it live again, I have to understand how that substance has gone through the activity and died into that corpse. I have to understand the rhythm or the time signature of that process of becoming.

If I understand the time signature of how a substance or entity has become, I can also understand the difference between phosphorous and calcium as a set of rhythmic pulses in the growth of a plant. A study of the planets' motion creates a rhythmic signature in my heart's eye, which helps me relate the action of calcium and phosphorous to the rhythmic cadences in the music of the spheres. To form an analog of this relationship, I could say that calcium is the note, while phosphorous is the interval between the notes. It is a perfect analog for the way these two substances act in a plant's growth processes. Phosphorous is the thing that says to calcium, "Go over there." If I want to look for a phosphorous deficiency, I do not look at the bottom part of the plant in the leaf; rather, I look at the growing tip, where it becomes brown when phosphorous is unavailable. If I wish to look for calcium problems, I do not look at the growing tip; rather, I look near the leaf's midrib. These symptoms of mineral deficiencies can be seen as an image of a rhythmic gesture that I can learn to read as a script. Training in observing natural phenomena within the context of planetary rhythms allows the formation of very useful symbolic analogs for one's work on the land.

Phosphorous is the stretch between notes. Calcium is the note that sounds so that you hear it as celestial music. With this symbolic thinking, I can link calcium with the lunar node, for instance. What I am describing is "Imagination" with a capital *I*. It does not depend on me and my pendulum but on the rigor I use to control my belief structure, linking my inner picture formation process to natural phenomena by finding functional analogs. Through Imaginative cognition, we can find analogs in the natural world to guide us toward more harmonious and creative thinking about what is happening in nature. It is my understanding that it is this type of thinking behind Steiner's agriculture course.

My hypothesis to you is that you can develop this kind of creative thinking by picturing to yourself the orbital movements of the planets. You are training your pictorial eye to look for lawfulness in movement. Once you do this, questions start to come up; then you go and get a ephemeris and ground those pictures in knowledge, which opens the door to creating an activity alchemically in your imagination. In learning to look for the nodes in time, there is a huge amount of information available through electronic media today to support any insights. When imaginative insights begin to be corroborated by data from scientific sources, a feeling arises that the elemental world is being liberated by your imaginations instead of being imprisoned by your beliefs. We have to make our imagination rigorous in a mathematical sense. Rigor means "I have tested it, and here is the test." We understand the process when we make our biodynamic preparations; we pull them out of the ground and look at them, and someone asks. "Is that good?" Then someone else says, "It looks good to me." Good on a scale of one to ten is sort of good. I mean there is New Jersey good, California good, and Swiss good. Biodynamic work requires of us to build organs for qualitative research that cannot be quantified or statistically robust. However, the research can be statistically consistent and rigorous. When the old science of weight, number, and measure collapses under an avalanche of conflicting data, a new science of rhythm will emerge. In this new science, statistically consistent proof will be as important as proof that is statistically robust.

Statistically robust means that, if you are a drug manufacturer, an acceptable probability structure for an experiment has to be in the range of eighty or ninety percent to place a new drug on the market. However, a statistically significant rate for an experiment is somewhere in the range of fifty-five to sixty percent. A placebo is forty percent. This means that, if I am working with an experimental protocol that consistently repeats a fifty-percent effect, that experiment is pointing to a statistically significant effect. If an experiment fails to reach the forty-percent probability, this indicates the futility boundary in the drug industry. Prozac, or *fluoxetine,* was introduced in 1977, and its patent expired in 2001. In 2010, more than 24.4 million prescriptions for generic formulations of fluoxetine were filled in the United States alone. Now there are findings that Prozac cannot achieve the effective percentage of a placebo.

If you consistently come up with a fifty-eight-percent probability in terms of an experimental result, that is very significant if it is consistent. Unfortunately, in biodynamics we are now attempting experiments aiming at a robust eighty percent. I believe that this is not going to happen because of the way the cosmos is organized. Prozac once had the coveted eighty-percent positive result status. Now Prozac-type drugs are struggling to maintain forty percent. Success rates for research experiments will likely change when the paradigm supporting the research changes. When it finally does change, will all the data that supported your eighty-percent success rate become bogus?

The statute of limitations on the Prozac patent ran out, and then we find that it is no more effective than a placebo. Does that mean that Prozac was more effective than a placebo when it first came out? This is just the way the world operates. We live in a world in which statues have wheels. They are wheeled in and out of the hall of fame as new evidence is presented. In fact, placebos are becoming stronger and more robust. This seems to indicate a future when uncertainty will only become greater. It means that, in regard to experiments in the life realm, the burden of proof is shifting. A workable parameter for qualitative research in the life realm does not require a probability structure of eighty percent, but a consistently low level of probability greater than a placebo. I can make my experiments aim at that.

If I wish to have an imagination that can harmonize with life in all of its fluid mysteries, I have to understand the rhythm of things rather than expecting the level of experimental results that one can achieve in the realm of physics or electronics. Biodynamic work is not like testing circuit loads for transistors. Rather, it is at the level of life. Alchemically, it is thought that the consciousness of the experimenter interacts with the life in the experiment and changes the way things happen. There is a doorway there to a method by which I use my consciousness to repeat a procedure again and again to reinforce an outcome with each repetition. No matter how many times I turn on a machine, it will respond the same way until it is broken. This is in contrast to the enhancement possible by disturbing the soil around a plant every ten days to shift the growth energies from one organ to another. That is very different from having a pill that works all the time. If I understand that life requires different experimental parameters, the next step is to ask a question: "How can I train myself to be guided into rigorous methods in the realms where the experiments I do are not based on physical forces?" I need to build an inner organ that perceives and then harmonizes with the laws of that realm. In short, how can I develop an organ of perception for higher worlds?

Rudolf Steiner is very radical in this area. He says that we perceive the movement of the planets through harmony with the movements of the blood in our arms. I know that this sounds far-fetched, but the movement of blood from your heart into your arms forms an antenna for the astral realm, the planetary realm of the soul forces. Through a practice of visualizing planetary motions, we can learn to be aware of planetary orbital periods to the degree that you can take them into sleep. In the morning, if you spend a few minutes listening first to your heart beat and then to the pulse of blood in your arms and hands, during that day that listening extends to your relationships to other people, and you can learn to use them as the bellwether of where your research is going by them asking questions or by having them give off-the-wall suggestions to, for instance, see a movie. Then when you finally see the movie, you suddenly have an insight into a current research problem.

Other people begin to participate in correcting the experiment that you are doing, even if they are unaware of their role in your inner process. However, you have to be sensitive to that by creating within yourself a rhythmic practice of taking the planets into sleep, giving the planets back to God, and listening in the morning to your organ of perception of the blood moving through your arms and hands. This eventually builds sensitivity in you about how time comes into focus in particular events and, if you learn to listen to this, people will enter those nodal points in time and help you in your learning process. The whole esoteric art of this involves paying attention to the small things during the day and having the presence of mind to use them as guides to where you are going.

It may sound crazy, but this method is a necessary failsafe against personal inflation. When we start getting these downloads, it can be like Bob Dylan said: "I have got a head full of ideas that are driving me insane." Once it starts, it is very difficult to determine when you are getting the download and when you are just off with the fairies. The tendency is just to go into a belief that the fairies are your friends. They may be, but then again they may not be. Just ask Shakespeare.

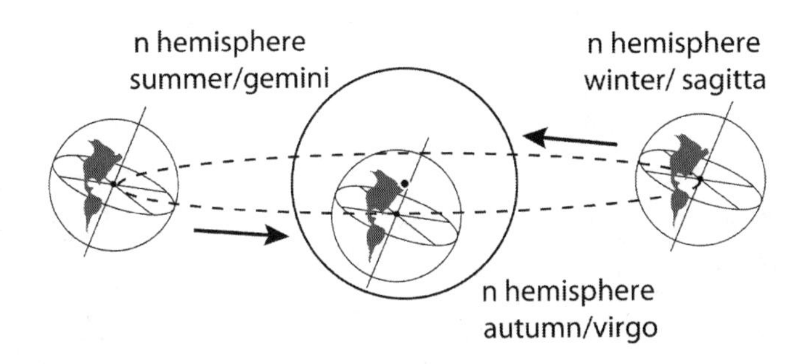

Figure 8

Here is an exercise. *Figure 8* shows the oscillation between summer, fall, and winter in the Northern Hemisphere. The big circle in the center is the Sun. The dotted oval shows the Earth's orbit around the Sun. The arrows show the direction of rotation. On the right is the Northern Hemisphere winter. It is winter because the whole Northern Hemisphere is pitched away from the Sun's direct rays. If you are on the Earth in California, that little gray thing there is North America and South America, with the Equator well within the northern part of South America.

Now find California, where I am looking at the Sun. The Sun appears low in the sky, because it is below the equator; that is winter in the Northern Hemisphere. Following the dotted line, it goes on the other side of the Sun, and I will go to the left, where I have the same position of the Earth and the heavens, but now the Northern Hemisphere is catching the direct rays of the Sun. If I am in California and look up, the Sun is now above the equator and it is summer in the Northern Hemisphere. Therefore, the relationship between the pitch of the Earth on its axis and the Sun has Earth pitched at 23° to that position of the Sun and maintains that 23° as it orbits the Sun. When it gets on one side, this 23° pitch allows the Sun to hit the Southern Hemisphere more directly. Then it is summer in the Southern Hemisphere and winter in the Northern Hemisphere.

If I look at the Sun at noon during winter, I see the star group Sagittarius behind the Sun. It says *Northern Hemisphere winter sagitta*. If I go to the other side, to the left, or counterclockwise, it is summer in the Northern

Hemisphere. When I look from California to Sun at noon, I see the Sun in front of the star group Gemini. Those two signs, Gemini and Sagittarius, are opposite each other in the zodiac. When seen from the Northern Hemisphere, Gemini is high in the sky and Sagittarius is low in the sky. If I can hold that as an imagination, run that orbit, and picture those stars, I am participating in the lawful motion of the elemental hierarchies of the spirits of time rotation. If I do that in the evening before going to sleep, those beings who are in the spirit guiding these sublime rhythms take an interest in my interest, because then I am a human being who is saying to them, "I think what you are doing is totally cool. I am studying it and participating in it with the gift I have been given to do that with my imagination. I am taking hold of my attention in my imagination and applying my imagination to study your realm in a spirit of humility and reverence." I give them the gift of my attention focusing on their work before I go to sleep. What do you think they are going to do? Will they do something constructive? Think about it. What would you do? I am harmonizing my consciousness with their playbook. Of course there is going to be something that comes back to me. When I listen in the morning to the movement of my heartbeat and the blood moving through my arms, I am picking up their response because my reading in the astral light is simultaneously writing in the astral light. They have the playbook, but it needs annotations from a being living in a body of flesh. They cannot do the annotations, because they are not separate from the playbook. All they can do is read from the playbook during the game. It is we who play in the game; they are reading the playbook. They need to know what happened when you tried to execute that play.

This is what we provide for them as archetypes. We need to share our understanding of what their imaginations look like from this side of the cosmos. Therefore, when we do this inner work, what builds in us is an imagination of points in time where I can begin to enter and work with substances in a whole new way. I start to see that, if I alter the time and put the organ into the ground earlier in the summer, I might get a better result. Such imaginations come by entering the rhythmic structure of time so that the time beings show us things about substances and the transformation of substances. Moreover, this can be taken much further so that, when you do a meditation such as this, you will be led to all kinds of insights into the Moon and Sun rhythms and the Jupiter and Saturn loops. There is a huge realm of imaginations available to us.

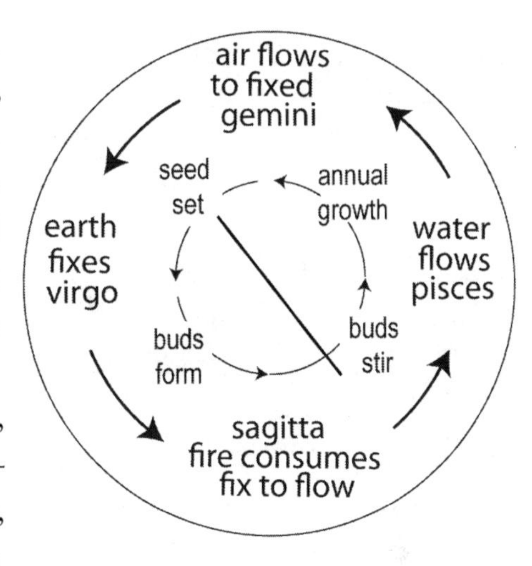

Figure 9

The imaginations come to us, but we have to wait until the doorway appears from someone else, allowing me to corroborate what has come. When it comes from someone else and the doorway appears, that is the Christ realm. You may be sitting in a lecture, tired and drifting off, then suddenly the speaker says two sentences and something in you lights up. "Alright, I get it!" That is Christ giving you a wake-up. In the space where it comes from someone else, the quickening reignites enthusiasm and pushes you to go back and study. Reading, studying, and struggling through the facts protects us from the inflation of our beliefs. When the download comes, we definitely need some protection; otherwise we just become playthings of the spirits of the true Self's shadow. The esoteric highway is full of people who have become the playthings of other spiritual beings who possess higher consciousness than human beings have.

If I do this exercise I make myself porous to the action of other kinds of laws. The formation of the exact imagination of rotations of the planets is a Rosicrucian pictorial meditation, and I am trying to rectify my consciousness by working with it. As I do it and start listening or reading, I begin to understand strange books; suddenly, I get an idea. Today I could Google that idea for corroboration. If I search outside of my own belief structures for corroboration, then I treat the inspiration in the proper way. The important point is that I do not get an idea and then confirm in myself that the idea holds water. It sounds fundamental and it holds in physical science.

However, the checks and balances of natural science that block personal fantasy are virtually missing in the esoteric work. The whole idea is that I need to develop a protocol for checking ideas so that I am sure that the inspirations and imaginations are not just coming through me. The checks and balances of natural science are not present in most of the work in the esoteric sciences. I have to make an effort to guard against inflation when it comes to inspiration and imagination, because the empirical method is often too coarse to give accurate or usable research results. Therefore, I must make an effort to establish a study protocol that provides a firm guard against inflation and fantasy when I actually begin to receive imaginations from the spiritual side of the threshold.

Imagining the motions of the planets as an inner discipline is an excellent start to establishing a study method and safeguard for spiritual research. In *figure 9*, repeating the pictures that contribute to the phenomena of the equinoxes and the solstices leads to a deeper understanding of certain patterns in a natural world. For instance, in the equinoxes on the right, when the Sun is in Pisces it is a time when water in the Earth flows. This is the gestural motif of the spring equinox. Pisces is the star group and, esoterically, it is linked traditionally to the water element, and we call it a "water sign." The nature of the solar motion during the equinoctial period bracketing the equinox has the Sun moving every day very rapidly in latitude. *Latitude* is the measurement of planetary declination, which really governs the solve and the coagula of flow structures. With the Sun at maximum north or south declination, the Earth's forces and atmosphere tend to stagnate. These are the periods of the solstices. With the Sun at the equator, the forces of the Earth and the atmosphere tend to accelerate. This is the gesture of the spring and fall equinoxes.

The Moon follows these solar rhythms, but it moves through the yearly solar cycle once a month. The Moon moves from equinox to solstice in a week. In the *Stella Natura* calendar,[14] lunar declination is described as the

14 *Stella Natura Biodynamic Planting Calendar: Planting Charts and Thought-Provoking Essays*, edited by Sherry Wildfeuer and published annually by The Biodynamic Farming & Gardening Association.

moon runs high or low; this oscillation is known as the "star moon," which moves north and south in declination following the solar cycles.

Studies have been conducted on the rhythms of the Moon moving in declination, showing strong tides in the air that oscillate with the lunar declination cycles. When the moon is at maximum north latitude, research has shown that there is a tendency for the atmosphere to bunch up under the Moon at the maximum high latitude. This effect also happens at the maximum southerly declination. This bunching of the atmosphere creates drag on the speed of rotation and changes the length of day in nanoseconds. In other words, the drag of the atmosphere bunching under the Moon actually changes the rotational period of the Earth. This has a significant effect, because the Earth spins fastest at the equator.

With the Moon at maximum high or low latitude, there is a tendency for atmospheric flowing to be reduced, because there is more drag at high latitudes and less drag on the equator. This allows the equator to accelerate, but this causes the atmosphere to move more slowly, owing to the law of the conservation of momentum. The slower action of the atmosphere occurs when the Moon is recapitulating the solstice position of the Sun.

What happens at the summer and winter solstices when you try to make compost? Nothing. However, at the equinoxes when the Sun is at the Equator, it is moving rapidly in declination each day. The natural rhythms are also in periods of great change and accelerated growth during the solar equinoxes. Research has shown that the Moon, as it crosses the equator every two weeks, also has a stimulating effect on the atmosphere. During this biweekly passage of the Moon across the Equator, research has shown that it bunches up the atmosphere at the equator. The atmospheric drag slows the Earth where it rotates fastest. To conserve momentum, the link between the rotational speed of the Earth and the movements of the atmosphere creates a condition in which the atmosphere moves more rapidly. The atmosphere on a biweekly rhythm accelerates into a pattern called "zonal flow," which accelerates the atmosphere and weather patterns. This is similar to conditions when the Sun is at the equator during the equinoxes.

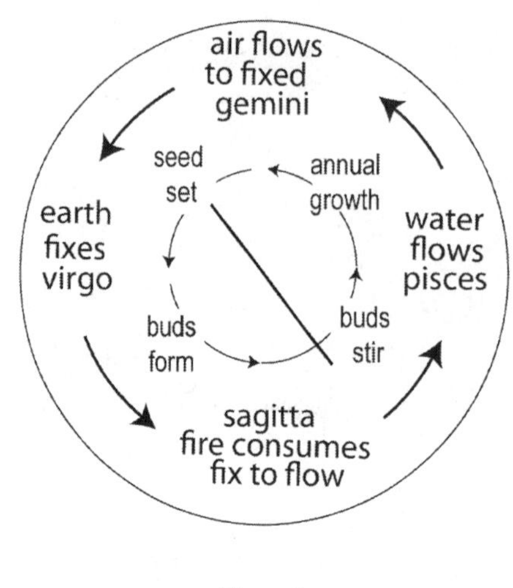

Figure 9

We know what happens when you make compost at the solstices—nothing. However, what happens when you make compost at the equinoxes? Bam, things light up and decay accelerates. Therefore, when you make compost at the equinox everything is active, with a lot of flow and change. The solar rhythm as a meditation eventually leads to insights into seasonal fluctuations. It can also lead to insights about climate shift and monthly or weekly potentialities for working with soil and plants. These are the types of imaginations that can come from a simple exercise of holding an orbital period in the mind's eye before going to sleep. The sky is the limit, because the spirits of the time of rotation are writing the playbook of how the rhythms, forces, and laws of nature operate. The rhythms of nature have their root in planetary interactions.

In terms of seasons, the real reversal is from solve to coagula as seasonal attributes change from flow to coagulation and back again. Look at *diagram 9;* you can see it is organized into polar opposites. At this level, consciousness gets very subtle. To understand the true nature of reversal, we need to understand that even reversals reverse. When I can do this kind of thinking, I get things like the dynamics of *figure 9.*

In Virgo, the Earth is fixed, but in Sagittarius the fire consumes what is fixed and begins a flow. Alchemists call what is fixed in nature "ash," the ultimate alchemical corpse. The ash process begins in Virgo and culminates in Sagittarius. All the seeds fall to Earth like ash from a forest fire. All the plants that were living in the last season go through a combustion process we call "flowering." From the flame of the flowers, the spiritual plants come down to Earth in the death process of the physical plant going to seed. The spiritual plants become part of Earth; this is "fixing." It begins at the fall equinox and it reaches a peak at mid-winter. However, as this happens, the Earth process of Virgo moves to the fire process that produces the ash, which culminates in mid-winter. Then, the ash allows what was present as a fixed form in the physical plant to be available again to cosmic forces for the creation of a new form. This chaos brings potential for a new generation, moving the fixed nature of ash into a flowing quality of potential.

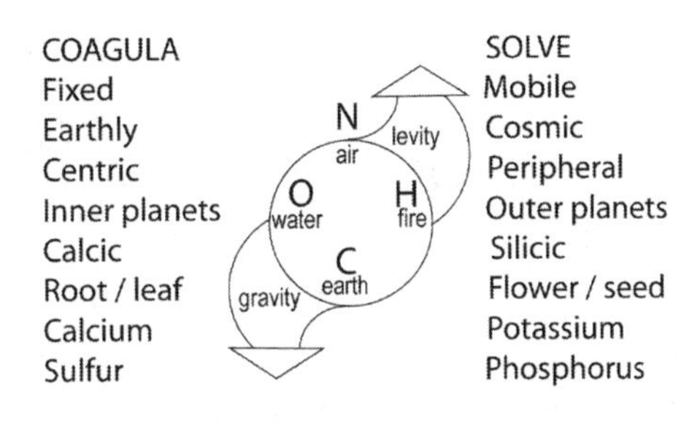

Figure 2

The potential is then taken up in spring when water and things begin to flow; the ash is taken up into the water, then the salt comes out of the ash and begins again to attract the water that carries the life, and a new plant comes into being. This imagination is just crop rotation or agronomy. However, the mandala gives it another dimension related to the Sun. Now, what follows this?

Let us review the alchemical mandala of the elements, keeping in mind that this is very subtle. In *figure 2,* we can place Virgo at the bottom, in the position of Earth. Then we move clockwise from Virgo/earth to Libra/air, then to Scorpio/water and to Sagittarius/fire. In *figure 2,* we would be in the fire pole at three o'clock. In *figure 9* the fire pole is at six o'clock. We are following the Sun as it goes through—earth, air, water, fire, earth, air, water, fire. This is how the Sun moves through the zodiac and one form of the mandala. *Figure 2* shows the rotation of the elements as they occur in nature, with earth at the bottom, then water, air above, then fire. This is the order of the elements in a puddle or in a burning candle. They are ordered as earth, water, air, fire. This is another way to organize a mandala. Suppose I take these two mandalas, put a pin in the center, and place one on top of the other. I can rotate each wheel separately to do a process the alchemists called *permutation.* In reality I need to permute these two different versions because each describes a reality that is true itself but needs to be permuted with other values to duplicate the way nature actually works. Sometimes in rotating one to the other, both places in Earth are going to match up or the airs will match up or not. The diagrams are not fixed alchemically; they made these kinds of wheels for permutations. Basically, they are the first computers. The way nature works is that sometimes things line up, and when it does not, drink lemonade and go to the mall. However, when it does line up, make sure the tractor is gassed, get up at 5:00 a.m., and go. This is the key to what I am presenting as a new agricultural consciousness. If you start to understand rhythms, you will be led to understand the principles of polyrhythm. These are two different schemas, but this does not mean they neutralize or contradict each other or that they are incorrect. It just means that, in the flow of time, sometimes they come in to sync and sometimes they do not. This is why we need a practice to build an organ that consciously perceives rhythm. You cannot just think it; you actually have to immerse yourself in it. You have to be in conscious contact with your blood, and that rhythm helps you understand the larger rhythms. Once you get it in your blood, then you have to get it in your head by studying the motion of the planets. If you try to understand this only with your head, you get a huge headache. I designed the little wheels so you could get used to the breathing of time.

This kind of imagination can be useful for such things as spray applications. If you spray a substance on a plant leaf, the plant will be altered in a subtle way. If you pay attention to where the Moon is when you spray and work in that rhythm to spray, for example, a leaf crop only when the moon is in a water sign, then the rhythms of the spraying program become active in the plant as well as the substance that you are spraying. You will not see an effect with one or two sprays, but if you spray in sequences you begin to see eventually that the plant responds. Spraying plants in a rhythm can offset things that were not optimal at the beginning.

We just put in a bare-root apricot tree. However, the timing for transplanting was not good at all, so I just heeled it in for two and half weeks until a good day appeared in the calendar. Unfortunately, it was raining like crazy on the day for the transplant. Fortunately, months before the time of the purchase I had dug the hole and put in the amendments for the tree. I thoroughly worked the soil on very propitious days, so on the morning of the transplanting the hole was already dug and everything was there. I went out in the pouring rain with my fork and popped in that little bare root because I wanted that day's beneficial influence retained in the soil of the planting hole. I worked this all with the Moon in fruit signs. The goal was to have the little apricot spread out in the soil that is programmed to support fruit production. I planned for that three and a half months before the actual day I transplanted, so that I could do all the digging, gather the right kind of compost, and put it all in without any pressure. Nevertheless, the day I wanted to put it in was not a good day weather-wise, but it was a good day that reinforced the original impulse I put into the soil when I dug the big hole. I made sure that the transplant day was in harmony with my target day. I did everything in fruit, starting two and a half months before I bought the bare-root seedling.

Spraying directly on a plant can modify this or augment the gesture. Unfortunately, we can augment some gesture in the soil to the point where it needs to be "de-augmented." I can get it so far in one direction that I need to establish a new balance. Getting that rhythm is the true power behind all this; then the door opens for you to start a little agricultural meditation practice. At that point we begin to learn about the planetary rhythms, the beings behind the phenomena. They will supply you with time imaginations of when you can do something or not, and this leads to very interesting things. There is a universe of qualities we can study with time imaginations in the natural realm.

Training our consciousness is really the great key to agricultural work, which is basically the formation of a new priesthood. As a priest, you have to be trained to be in a place where the forces flowing into the ritual you are doing is correct. It is the same for the work with the beings of nature. In the past, studying for the priesthood included a study of the planets, so that the priests would know when to do a ritual. I perceive that work on the land is a sacred task, so when I go to do something I pay attention to what is happening in the cosmos and my being becomes a conduit for the lawfulness of the cosmos to enter the soil, the plants, the animals, and ultimately myself through the food. My consciousness is the thing that allows substance to be transformed. That is the gift we have been given by the hierarchies. They gave us the substance, but they allow us to transform it. Therefore, we can call biodynamic work the rudiments of a new priesthood. We need people who are conscious about the spiritual dimensions of really good food. Food allows us to develop consciousness. Dead food feeds the corpse. Living food allows the soul to see the spirit.

Rudolf Steiner said it is food that allows people to have the consciousness needed to take esoteric work to the next level. The work we do with food production is truly vital for the future. My goal here is to present the idea that the quality of consciousness we bring to our daily tasks allows us to find the right level of work for the planet and for other human beings. The key to understanding and training that consciousness is the study of rhythm.

CHAPTER 3

MACRO AND MICRO IN BIODYNAMICS

I. Agricultural Imagination • Agricultural Individuality • Training Imagination: Observe, Represent, Silence, Record • Macro and Micro Forces as Polarities • Framework Substances • Exercises for Imagination

This topic is something I have wanted to work on for many years. During the first ten or so years, I was mentally unable to do so. Then for the second ten or so years, I was emotionally unable to do it, because of what it would take to break the mold a bit. However, I am getting old enough that I no longer care. Breaking the mold simply means that there are certain understandings of Steiner's work that are very useful when we want to work with agriculture. However, they are very subtle and difficult to understand. One of the most subtle pictures he works with involves etheric forces. There are many false perceptions and strange beliefs about what we call *the etheric,* or *etheric forces,* both in conventional science and in biodynamic circles. This is why I never wanted to wade into this, because when you wade into this you are automatically wrong. This is a kind of disclaimer—automatically, I am wrong. Therefore, if you think I am wrong, you are right.

I would like to share the fruits of my thirty years of doing this kind of work in nature—trying to understand how to experience the etheric, or what Rudolf Steiner calls the elemental beings behind the etheric world, and how to experience this and have a feeling when working with natural phenomena that one is in contact with spiritual entities. It is really important to have that feeling; otherwise, the experience we have of nature becomes increasingly abstract, and we are in a position now where abstract thinking, even in biodynamics, threatens to block biodynamic work from growing in the necessary directions. This is a personal opinion and, like I said, I am wrong, but it is my personal opinion.

In talking with many people, traveling to many places, and then working with wineries and seeing how biodynamics interacts with their operation, we could ask this question: How does their operation as a moneymaking activity based on science interact with biodynamic concepts. There is a kind of sound you hear when you ask that question—it goes "crunch." The recommendations wineries get from the agriculture school at University of California at Davis encounter the things you need to do in biodynamics. They often meet head-on with a kind of crunching sound. They do not need to, but they do; I have seen it. I have talked to the growers and spent time with them, and they all say the same thing—it is very difficult to show why a bloody cow head in the back of your pickup truck means something spiritual. The purpose of the bloody cow head is a very difficult thing to explain to people who simply want recipes. Most people just want the recipe; they do the recipe and do not have to worry. If there is a problem, they call the county agent. It is a guarantee, like an insurance policy.

However, in biodynamics Rudolf Steiner is not selling insurance. He does not give recipes; it seems to me that he was presenting principles for future research. He was giving us brilliant research parameters that had a deep tradition in the alchemical work and go back even to the mystery schools. It is about how we would interact with nature with heightened consciousness; from an esoteric perspective, the key to transforming the natural world is human consciousness, which is the fifth element added to the four elements of earth, air, water, and fire. It is *akasha,* and without akasha nature is simply chaos. However, when human beings interact with nature, things change. Human interaction with nature creates potentialities that go beyond the standard parameters of the natural world. Take, for instance, the making of a medicine. If you get a medicine or tincture made from an herbal product, you will not find that tincture in nature. It is not part of nature, but rather part of an interaction between human consciousness and the natural world, and that changes what that substance is. It changes how it acts. It makes it, we could say, unnatural.

This is a difficult thing to get your head around if you are a nature person. When you go out and plow your field, that is not natural, and if you keep plowing it the wrong way, you find out why it is not natural. You will need an agent to come in and tell you, "Go this way, not that way." It is this quality of consciousness as a force of transformation for nature that is behind the agriculture course. In its most accessible form, that consciousness could be called, "agricultural imagination," which hinges on an inner experience that could be called "a new participatory consciousness." This term comes from the scholar Owen Barfield. The old participatory consciousness could be described this way: the gods, goddesses, devas, and elemental beings in the world made you do things unless you sacrificed chickens, and it was driven by a kind of fear of the unknown of the natural world. We could call that, "original participatory consciousness." Such participation often amounted to placating beings in nature who would otherwise wreak havoc on your life.

Then came what we call, "empirical consciousness," or "consciousness of the observer." This level of consciousness discovered certain scientific laws that could be used to manipulate nature and overcome the threat of crop loss, pest invasion, or plagues. We could overcome them with "Drop Dead Twice"—DDT—which is definitely unnatural and creates patterns in the natural world whereby the observer can do things and then observe the results. This eventually led to formulas and practices that led to NPK fertilizer. If we heat chicken manure in a barrel, fumes come off the manure. Eventually, from the fumes we get crystals that form in the top of the barrel. If we scrape them off, we call the crystals *urea*. In a package of NPK, that is the "N." When people learned to do this, they took just the nitrogen component from the manure to make munitions. Then it was found that applying urea to plants caused explosive growth; the plant would exhibit rapid water uptake and growth. That is the process we go through when we make NPK, and it is a kind of alchemical/chemical process that has become codified into a formula, but we lose sight of the fact that there are beings in manure, soil, plants, and even nitrogen. There is a consciousness of the beings of nitrogen, and human beings have access to those beings through the nitrogen in their own bodies. To say such a thing to a soil scientist is the height of folly, but from an esoteric point of view this is an important concept.

In the ancient world, using the nitrogen in one's own body to link to nature spirits was the principle by which to learn how to placate angry deities. You had substances in you that could act as a bridge between you and the spiritual beings in nature. It was found that one could take certain substances to enhance the ability of one's nitrogen to access the nitrogen beings of plant alkaloids, because it is the nitrogen in those substances that alters your consciousness. Such principles started with magic and culminated in the empirical, critical analytical view of nature that drives modern agriculture. Rudolf Steiner brought a whole other set of rules to the table to move from observer consciousness into Owen Barfield's "second participatory consciousness." The original participatory consciousness was magic; the second participatory consciousness is the need to understand the rules of the observer consciousness. The most fundamental rule is that I have to understand what happens in me when I observe something. Consequently, I need to observe my observing. When I do so, I enter agricultural imagination. Eventually, I can change my consciousness to such a degree that, when I go into the natural world as a participatory observer rather than a spectator observer, my consciousness allows me to experience activities of forces as personalities, or qualities of being, instead of just abstract forces. The forces of nature are not simply formulaic abstractions.

With training, nature forces can speak to me through pictures that arise in me—what Rudolf Steiner calls "Imaginations." In our culture, we tend to think of *imagination* as fantasy, but his use of the word refers to a quality in people that has been trained to weed out the fantastical in imaginative practice and instead find the laws of the forces. He calls those laws "elemental beings." Steiner calls the patterns of forces in the natural world "elementals." Thus, we could say that the shift of the observer's analytical consciousness requires a new ability to "see," so that we can enter the laws of the natural world in imaginative participatory consciousness. This is where the work in nature becomes the work of priests. It becomes sacred rather than simply the application of analytical rules. Rules are okay; they provide a kind of foundation, but they do not allow movement into new participatory consciousness. They simply allow you to apply what already is known.

In agriculture, for those who do the work, new participatory consciousness makes possible perception of what Rudolf Steiner calls, "agricultural individuality." This is the combinations of forces and laws present in the mineral structure of your land—the climatological niche, the particular flora, the fauna, and what you bring to that microstructure. All of these factors constitute a being with a particular persona. It has a behavioral pattern that is present in the way forces flow through the land. You can go to someone else's land and have an inner experience that something is different. The question is this: Can you bring that difference to consciousness, or will it remain a general feeling? You can train yourself to bring it to consciousness. In this book, I would like to offer some indications about how this is possible. How do we train ourselves to see the macro in the micro? How do we train ourselves to see the cosmic pole in the forces and how those forces manifest as substances? Can we do this and still follow lines of reason that would make sense in conventional scientific terms? If your land out-produces the land farmed conventionally, people will have questions that provide opportunities to build bridges between what those growers have been told by government authorities and what you know from experience. You can place Rudolf Steiner's indications as learning tools between the county agent's advice and your daily work. However, if you begin with Steiner's indications when dealing with the data-driven abstractions, it just becomes "he said, she said." It is different, however, when you speak directly from your own experience.

This is why this book is about "macro/micro." The big picture in Steiner's work is that the "agricultural individuality" is a spiritual personality suspended between the laws of the cosmos and the laws of Earth. The two are not directly interchangeable, and this is the problem. There is a reversal that happens, and training one's imagination offers the ability to participate in this reversal without wanting to fix the immediate answer.

There are four qualities in training agricultural imagination: observe, represent, carry into silence, and then record. In addition to practical exercises, we will look at various phenomena and experiments and try to make sense of the "form motif" of the agricultural imagination. When I go into nature and want to attempt this work of the agricultural imagination, there is a law that the form of the thing I see is a corpse of the forces that created it. Moreover, hidden in that corpse is our key to consciousness participation in the activity of the becoming of the corpse. Our human consciousness can reanimate the corpse through imaginative participation. We can create inwardly an imagination of the "becoming of the form." We could call this becoming "growth" or "life gesture." The life gesture of a plant, animal, or mineral has definite laws, and we can participate in those laws imaginatively by observing the form and then representing that form inwardly as a picture. However, to participate in the transformation of nature, the picture we represent inwardly must be one that moves, since we are trying to make our consciousness participate in how the form came into being rather than seeing it as a snapshot.

Look at *figure 1*. Looking at this form is *observing*. I can observe it as a snapshot in my mind's eye. Look at the form, then close your eyes, and ask yourself if you can see the form in your mind's eye. This is our first exercise. This is the agricultural equivalent of learning to play the clarinet; if you cannot play the clarinet, practice is the answer. Similarly, if you cannot imagine the form in your mind's eye, the answer is practice. If it is difficult to form a picture of what you just looked at, there is someplace in your body where you can do this. I have found that when people claim they cannot form an inner picture, eventually they find a place where they can perceive an inner picture. Suddenly the picture springs up. If you can do this, it is difficult to imagine how you cannot do it; but if you cannot do this, it is hard to imagine how you can. It is all based on imagination and practice. *Imagination* contains the word *image*. It means I can magically form an inner picture. Forming an inner picture can be called "representation." This means that I take the picture of something outside, move it inside, and try and represent—or *re*-present—it to myself. I represent the picture of the outer thing inwardly to my mind's eye. This is the basis of imagination.

Here is another exercise. Draw a simple circle—it does not need to be exact, just a reasonable facsimile. Look at your drawing and see that it is a

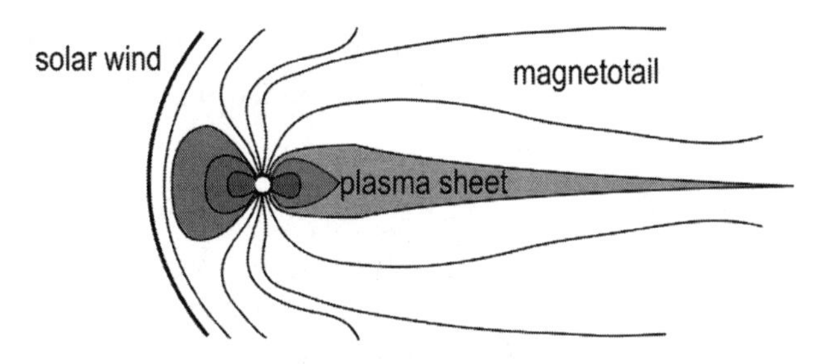

Figure 1: Magnetosphere

circle. Now close your eyes and try to hold the circle in your mind's eye for about half a minute. When you try this, as you try to hold the circle in your imagination, your life forces say, "I think you got a circle stuck in your head. Let's move that along." That is what Rudolf Steiner calls the action of the ether body. When you try to hold onto a picture, you are actually working against certain life forces that want to keep that picture moving—and thank goodness those forces keep the picture moving, because if they did not your head would retain every picture you ever looked at and everything you ever saw. You have within you a force that does not want you to allow pictures to become stuck in your mind's eye, which keeps you healthy. However, when you do wish to hold on to a picture and move into it, the ability of the life body to dissolve the picture is a key. Your life body simply wants to move it. Therefore, if you take the inner picture and represent it as a sequence of images instead of trying to hold onto a snapshot, the life body is happy and you are in control. That is a big deal. Your life body is happy because you are moving it, but you are the one who determines how it moves. This is representation.

Looking again at *figure 1*, imagine how it would feel to begin with that dark little space on the left, then expand something, and move it out to a little point. Using your fingers, squeeze on the left, and then expand your fingers as though sculpting in air. Do this a couple times. Squeeze, lift, flow, and expand. Do it until your life body starts to give you a picture.

In this way, you are training your life forces. It is basically like trying to get the attention of a cow. It is as if you have to repeat things a couple times. Your life body is like a cow. It says, "Just live." I train it by trying inwardly to duplicate the movement I would have had to make if I had made that picture. I could make the picture as a sound: "Bloop, bloo, bloooooooo" or "Xtkkk, whoo, whoooooooooo!" If you are audio rather than visual, you can do it that way, and that is also an imagination. You do not actually have to see a picture, but you do have to feel as if you are in touch with something like a picture inside of yourself. If you work in agriculture, you may be more of a tactile learner, so get your hands involved. Get two sticks and draw it in the dirt, and your life body will say, "Oh, you are making something that looks like this." You are training your imagination to participate in the form.

The important thing is that the form holds the key to the elemental and elementary beings. The form of the thing is their signature of how they come into being and how they go away. The form is a key to understanding the force, whether a plant, kidney, bladder, or mesentery. The form is a picture of the force, and if you learn to work with form imaginatively and you begin to participate imaginatively in the things you normally do; suddenly, the things you normally do start to speak to you about other kinds of potentialities. Rudolf Steiner calls that the, "elemental world." You do not see little guys with funny hats and pointy shoes eating pineapple; rather, you start to have imaginations of what needs to happen in a particular area. Gardeners call it a "green thumb." It simply means that your imagination has the ability to participate in the sets of forces in a particular area.

You all know someone who receives a plant as a gift, and a month later it has turned yellow. This person knows you are a gardener and brings it to you, and a month later it is green again. How does that happen? It is the quality of imaginations a person carries that the plant picks up because it is alive; it is made up of living imaginations. Its organs are pictures of the forces it needs, pictures of the particular areas in the cosmos where it receives the sunlight in a particular way, a particular season or time, or a certain form that allows it to create substances that make that plant useful. Forces from the cosmos, in Steiner's worldview, create the form that results in the finished form of the plant. So here we have an imagination in *figure 1*. We squeeze it, spread it out, then draw it out to make the sound or picture—or it feels a certain way to you; that is another way to do it.

Now look at *figure 2*. Squeeze it, pick it up, move it down, or out to a little point. This is a very similar form. When forms work that way, their relationship is called "correspondence." I could say there is a formal correspondence between *figures 1* and 2. *Figure 1* is an image of the plasma environment around the Earth that is generated by the solar wind and how that solar wind distorts the electromagnetic spheres into a long tail that goes out thousands of miles in space, off to the right. That little space in the middle is a plasma chamber; it is where the aurora borealis forms. That is a picture of something very big. *Figure 2* is a copepod; it is about as big as a pollen grain and lives in plankton. There are a gazillion of them on Earth. This relationship is the title of this chapter, "Macro and Micro." The question is this: Is that correspondence the basis of something we could use for a new science, or is that just a fantasy?

Figure 2: Copepod

When I begin doing the inner work of observing nature and start to see correspondences, this becomes a huge question for my own self-development. The world starts to appear different from ordinary reality. I can tell you this from personal experience. Thirty years ago I wrote the *Biodynamic Book of Moons*. I had been reading Goethe and doing inner visualization exercises with natural phenomena, and one day I walked out in my garden and all the plants were kind of saying, "Wow, wow, wow, wow." All the flowers were not just still snapshots; they were all going, "Woop! Woop!" I could not turn it off. This happened because I was very involved with exercises for perceiving the forms in nature as *becomings,* and suddenly I was seeing the flowers' forms as becomings. The plants were showing me how they grow, how they become. It was a little freaky, because I was not used to them showing me how they grow. The whole question is this: If you start to do this, can you also turn it off? It becomes a problem if you cannot turn it off. To turn it off, you need to be able to control the way the picture moves. Otherwise, it will just keep moving; then, you have difficulty going to bed, because the pictures come in whenever they decide to come in, and they do not know it is 2:00 a.m. You are just seeing this kind of twisting stuff in your head all the time, and you are saying, "Oh, Lord, get me out of this. I will never do it again."

The upside of imagination is that you have an active imagination; the downside is that you have an active imagination. In this kind of inner work, you can see how easy it is to invent things, and this is a problem when someone from a newspaper comes to interview you about the weird stuff you do with cow horns. You start to become "creative" because the juice is rolling in you, and when your quotations go public you get disturbing emails from scientific types and you regret doing that. This type of work on the imagination is what Steiner calls "exact imagination." It is not fantasizing or about making up anything, even though it does involve the imagination.

Simply because a copepod and the Earth's magnetosphere interacting with the solar wind have the same form gesture does not mean that they are directly connected. It just means they have a similar form. If I want to find how that form operates, I have to move through the *observe* and *represent* phases to the *silence* phase. Silence means I keep representing and dissolving the image I am working with as a rhythmic practice. I form the image, move it in my mind as a representation, and then I dissolve it. Then I repeat: I form it, move it, and dissolve it. I make a practice of forming and moving the sequences of pictures. I form it, move it, and dissolve it. Dissolving the picture moves it into silence. When I move the picture into silence, the beings behind the form forces of becoming approach me, because I have now shown an active interest in how they live their life. I am paying respect to the elemental beings who guide the manifestation of the form, and I do that as a human being by becoming actively, or willfully, silent. They recognize that a conscious, silent human is very rare, especially because this person just entered silence honoring their work. Elemental beings are very attracted to that. They come, enter the silent space, and say something like: You know, your imagination is cool, but man, it is a little more like this, or a little more like that. They will help you to put something else in a particular place next time—try this. The suggestion, "try this," comes when I manage to put my image making into total silence.

If you have the right heart, have formed the picture accurately, are truly interested in how the image moves in your imagination, and take the image into silence—the silence of letting go of the picture—something comes to you. The way that the macro and the micro interact with each other is simply the law of reciprocation. My every action to push something out creates an opposite and equal reaction to let something in. Every time I form and release a picture of a phenomenon in the natural world, something is released and formed in my consciousness by the spiritual beings behind the production of that natural form.

What I want to do with that is I want to *record* this dialogue. I make a sketch; I make a picture, I Google a word that has come to me, whatever. I write something out, even words that are kind of wacky. However, do not assume that what you write out or that the picture you get is the answer, because it is not; it is just the *beginning* of an answer. To get the answer, we just need to keep repeating the process, adding something on the next day or next couple hours. We need to repeat the inner imagining many times. This constitutes a practice of forming, dissolving, and recording a formal motif from nature. With time, an intuition starts to emerge in our consciousness about what we could do about that thing in nature; how we can work with a life form; or how we can use a particular substance, plant, animal, or mineral. With time, we start getting pictures of how they correspond. When this happens, we can Google what is arising as a question and discover amazing correspondences. We might call such correspondences a spiritual cosmology. This is very useful today, because science has produced a wealth of factoids that can benefit from a spiritual cosmological approach.

Spiritual cosmology places human beings within the context of the spiritual world, which is populated with beings who serve humankind by animating nature. Without spiritual cosmology, which sees us as living among the spiritual beings who animate the natural world, we have only a sea of unintegrated factoids in databases used as experimental analytical protocols. If you delve beneath the surface with spiritual cosmology, you will be led to insights that dovetail with natural scientific findings. You will find certain substances that are used for particular purposes; you have already had an imagination that this plant accomplishes that purpose, and maybe you will find a connection between a gem and one of the biodynamic preparations or a connection between a particular mineral and a plant. Then perhaps you find that an organ in the animal actually is a specialist at rending pH in the animal and that it affects a particular substance that regulates growth. There are innumerable discoveries by natural science that do not have a cosmology. However, they can form remarkable insights when contextualized as part of a spiritual cosmological approach.

Rudolf Steiner provided the fact that it is possible to have a cosmology in the new participatory consciousness that allows us once again to see the elemental world, but in a precise way—not in the old way, by which one approaches nature beings in fear. When people experience nature beings in fear, the organizing sheaths of the subtle human bodies are loosened and people start to see all kinds of fantastical things in the world. When people saw those beings, they needed others people to give them codes to explain what they were seeing. Those codes became tradition, indicating that, when people would go to a certain place, they would see a particular being and outlined what they should say.

By contrast, the new approach is to train your imagination to enter the natural world so that it speaks within you about the place where you are working; that agricultural individuality becomes a reality to you instead of a concept. It is a being that you get to know, but you do not have to put chickens out on an altar with their heads chopped off in order to placate an unknown, malevolent nature being. If you work on yourself to purify and organize your imagination, the elemental beings with which you come in contact reveal themselves in the context of a larger spiritual being—such as an archangel of all the forces where you live, including the elemental being who guides the forces of your life body. Here we enter the edge of expanded and creative human consciousness.

This is a good place to introduce a series of exercises with these pictures to get a feeling for this process of self-transformation. We can begin with macro polarities, or as Rudolf Steiner speaks of them, "peripheral forces." Macro forces are cosmic forces: micro forces are earthly forces. In Steiner's language, *cosmic* is "macro." Macro forces are the source of all potentialities. They are alive with potential. They operate at the periphery through the action of what Steiner calls, "planes of light." (We will look more closely at those planes later.) Planes of light work inward from the periphery. Cosmic forces are active in processes in which one thing is changing into another. The activity of light in the world is to create the ferment needed to move things along—this is chlorophyll. Plants are organized to utilize light through chlorophyll. It is the substance through which a plant receives light in order to become active.

The polar contrast to cosmic forces is the earthly forces, or micro forces. Whereas *macro* is potential, *micro* is manifestation. Micro forces represent "this thing here." Macro works in planes; micro with points, the center point of the center of the mass of something—what we call "stuff." Macro is on the periphery; micro is in the center. Macro is process; micro is substance. These are the great polarities. Rudolf Steiner uses these terms in his agriculture course; he sprinkles them into his descriptions of other, more familiar things such as manure, lime, and silica. Steiner is speaking a kind of meta-language that always involves the polarity of macro/micro. His imaginations of how the natural world works with things describe the interactions of these two spiritual or alchemical polarities; the system of these polarities constitutes the cosmology I mentioned earlier. Steiner's cosmology is that the macro and the micro are interacting with each other. The cause of their interaction is consciousness, whether plant consciousness, animal consciousness, stone consciousness, human consciousness, or the elementals living in your body. They are conscious of what goes on around you. You can access this, but you have to do it coherently and systematically rather than fantastically.

Here is what Steiner had to say in his agriculture course:

> There is a big difference between dead nitrogen in the air and another kind of nitrogen. This kind of nitrogen must be formed under the influence of the entire heavens—the other kind, the living kind of

> nitrogen—must be formed under the influence of the entire heavens. The living nitrogen must be alive.[1]

What does it mean to say "nitrogen must be alive"? Steiner is referring to "potential." He is saying that the nitrogen must not yet be manifest as a substance, but alive as an activity or process. It is living; it is potential and interacts with other things to stimulate growth.

> Everything that enters a body by way of digestive organs only provides materials to be deposited in what belongs to the nerve and sense system. By contrast, the substances we need for building up our bones and intestines and blood are absorbed from our whole environment by way of breathing and even via our sense organs. What comes from our surroundings insofar as it consists of "framework substance" and is pervaded by warmth and light—this is taken in by your senses.[2]

What does "framework substance" mean? It is patterns of forces that will eventually yield a particular substance. It is the unmanifest energetic template or framework of something that will later manifest as a particular thing. Steiner uses that term *framework substances* to describe what could be called "sheaths" or, colloquially, "sheaths of the preparations." A sheath is a form, or energetic form, in the natural world that attracts substances. A sheath is a framework; or in today's language we could call it an "antenna" that attracts substances to be deposited in a particular way. Steiner describes the sheaths as the actual source of a kind of consciousness. This is why we use sheaths for preparations; we want to give the substance encased in the sheath a form of consciousness. We may hear phrases such as "a particular preparation allows the plant to pick calcium out of the soil." What is this? It is a mode of consciousness.

The language Steiner uses here is alchemical, and the polarities are always macro/micro, periphery/center, process/substance, life/death, or potential/manifestation. A substance is something that has fallen out of life, but nonetheless bears the signature of life; this is the greatest key of all.

1 Steiner, *Spiritual Foundations for the Renewal of Agriculture*, p. 10.

2 Ibid., p. 11.

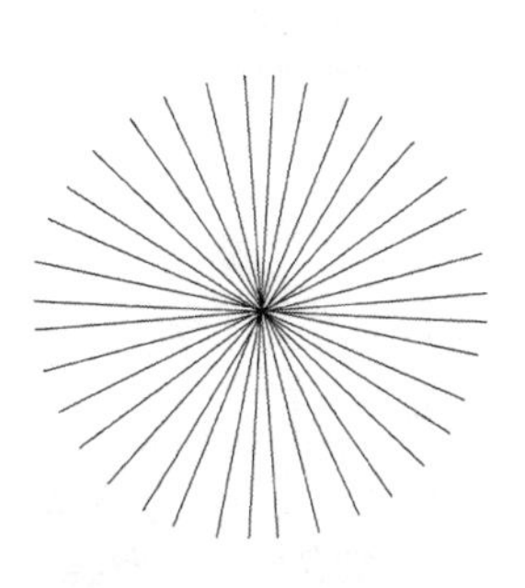

Figure 3: Pointwise circle

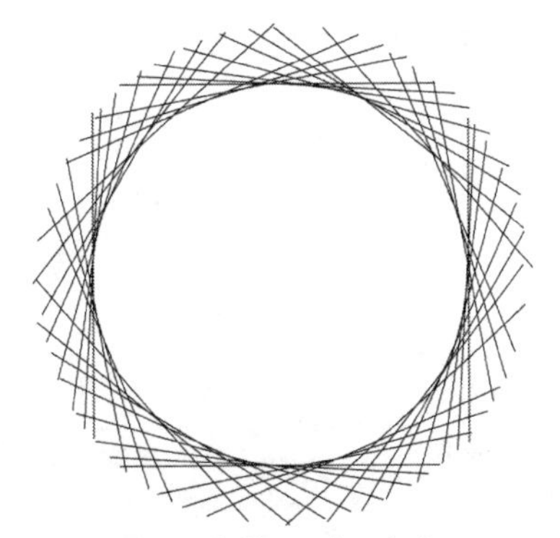

Figure 4: Planewise circle

Forms have what we call "motifs"—sets of patterns that give a form its character. The motifs of the forms are related perceptually to human feelings. Feelings circulate around how the form arises in us as a representation, and this experience is the part of consciousness we can train. It can be developed to the point that when we go out into nature and look, we become aware of how the various elements in the landscape affect our feelings. This has a great deal to do with the patterns of light coming into the senses from sensory objects.

We will now do this with a set of forms. You will observe, represent, go into silence, and then record as we go along.

Look at *figure 3* and then imagine how it would grow. If you had to make this form grow, how would it grow? Imagine that movement inwardly, and then take the image into silence—go blank, listen into the silence. Next, just write your experience of how it would grow. This process constitutes a "thought experiment" for training your imagination. Observe, create an inner image as a representation, and try to imagine how it would grow.

If you wish to go more deeply into this form, just keep repeating the exercise and modifying the inner picture with what you learn with each repetition. Eventually, the form will speak to you, and then you can go out into nature and see the forces of generation in a particular life form. This "seeing" will simultaneously help you understand the form. This is what I mean by "the form will speak to you." This inner activity in your consciousness creates a bridge between you and that energetic form in the natural world.

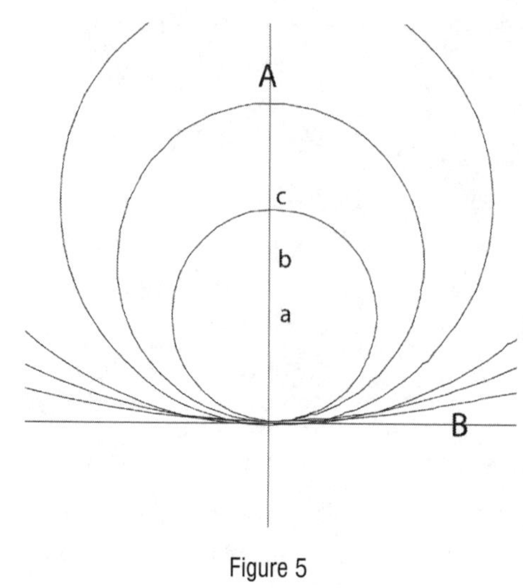

Figure 5

Now look at *figure 4*, and do the same thing. Observe it, represent it, grow it in some way (how that form might grow), and then take it into silence and record your experience.

Now look at *figure 5*. This is the diagram we looked at earlier from Nicholas of Cusa, who used it when trying to understand more deeply the nature of the Divine. In *figure 5*, you can see a circle with a dot in its center and on a vertical line, "A." The horizontal line where the circle meets the horizontal is labeled "B." Lines "A" and "B" cross at 90°. The circle, whose center is on line "A," touches at the point where lines "A" and "B" cross. That is a tangent. The little circle crosses line "A" where "A" and "B" meet. Now look at the second largest circle and try to estimate where the center of that would be and put a point there on line "A." Then do the same with the third largest circle, even though it extends beyond the diagram; see if you can put a point on line "A" that would be the center of that circle.

Now the first circle, with its point and circumference, is tangent to the crossing points; the second one is also tangent to line "B"; and the third circle is also tangent to the crossing points, even though the circles are getting bigger. This is our protocol for the becoming of the thought experiment. The thought experiment is to continue making the circles larger by moving the center of the circles vertically along line "A." When Nicholas of Cusa did this, it blew his mind, as we might say, because he experienced something for the first time that allowed him to understand a cosmic law. Expand the circles by moving the center of those circles continually up line "A" until you move the center of the circle to the infinite. This is what Nicholas of Cusa did. When the center of the circle reaches the infinite, what happens to the circle? Remember to observe, represent, go into silence, and record. All of the forms in *figures 3, 4* and *5* are related to the idea of macro/micro.

If you do this exercise with others, discuss what you discovered about these things. Listen to what the others say about what they found, and then see if something in you shifts. If it does, the quality of that shift is the key to working with the elemental world. In this work it is very useful to be aware of the times when your understanding of the idea shifts, because elementals want us to become aware of those shifts. This is because we do not understand what they are trying to tell us, because what they are trying to tell us often goes against how we think things should be—our beliefs.

Figure 3 is called a "pointwise circle," because it grows outward from a point and expands in a linear way; each point grows at the same rate out from a center. It is an "explosion." That is the picture. It is explosive. Centric growth comes from a center and moves out. However, you would be surprised how little natural growth in the world works that way. Even crystals do not grow that way, as we will find as we go along. They grow from the periphery.

By contrast, the forces diagrammed in *figure 4* are shown as a "planewise circle," which has planes coming in from outside. In fact, there is an infinite number of planes around the circle's circumference. Where those planes cross one another, we would have an infinite number of points. The strange thing about this is that the infinite number of points could also come from an infinite number of lines radiating, or exploding, from the center. Every center in nature has an infinite number of lines issuing from its center and an infinite number of planes streaming toward the center from outside. Radiation from a center represents explosive motion. Embryology shows that most things grow from the periphery.

Nicholas of Cusa found that if we expand a circle infinitely (*figure 5*)—he actually did it with a sphere—it becomes a plane at the infinite. If I expand a circle into the infinite, it becomes a line as the center of the circle reaches the infinite. We know that the earth is round, but when you go to the ocean and look at the horizon, you see a flat line. If we went to Jupiter and looked at the horizon, would it be flatter than the one on earth? At a certain range, when the human consciousness encounters the sensory world at about 100 yards, science says that it experiences all light as parallel. At the horizon every circle becomes a flat line and every sphere becomes a plane. This is called, "collimated light." At 100 yards, all light rays appear parallel.

We will finish this section with one more look at the agriculture course:

> The extent to which the soil itself becomes alive and develops its own chemistry depends primarily on the sandy component of the soil. The conditions encountered by the plant roots in the soil are greatly influenced by the extent to which the cosmic life and cosmic chemistry are collected by the rock and stone.[3]

Light. It is through silica that the cosmic factor is absorbed by the earth and becomes effective. We should never forget that silica is indispensable for life. Rudolf Steiner also said, "The soil is a child of the Sun."[4] The soil is a child of the sun because it is the mineral content that takes up the life, the potential, the activity, and then it becomes organic, but it needs a mediator. The light needs a mediator, which we will take a look at next.

II. Infinitely Distant Plane • Biological Surfaces • the Action of Fields • Gradients • Nodes • Intervals • Wave Patterns in Nature

Now we will consider the forces that operate in imaginations of the infinitely distant sphere, or the "infinitely distant sphere of planes." This idea is found in projective geometry, and it offers a key to understanding much of Rudolf Steiner's work. It is called the "infinitely distant." It is a sphere made of one unique plane—a completely flat plane at infinity. That completely flat, infinite plane infinitely distant appears as a sphere because

3 Steiner, *Spiritual Foundations for the Renewal of Agriculture*, p. 30.

4 "The soil is a child of the Sun. Above its surface, it is exposed to the near planets—below its surface to the distant ones" (ibid., from handwritten notes, p. 201).

human consciousness is point-centered. In our point-centered consciousness, that unique plane at infinity appears everywhere as the only infinite plane, but it appears as the same plane in every direction. Paradoxically, it is everywhere on the sphere, but it is only one infinite, unique plane. However, because human beings have point-centered consciousness, we experience that unique plane as the same plane that we experienced as the infinite in all directions from a point in the center of a sphere. In perceptual reality, if you get past the horizon there is no such thing as direction in space. Direction is only a kind of local phenomenon.

Point-centered consciousness has a name: "Me!" When I say, "This is me, here," the whole cosmos becomes constellated into a sphere. We could call it a background. The whole cosmos becomes constellated into the music of the spheres, and this is the action of human consciousness in the limitlessness of the cosmos; it constellates limitless infinity into a sphere. That action changes the quality of light coming from every direction and constellates it around points encircling points. We could call them buds, crystals, skulls, lungs, bladders, or little cosmoses of planes.

When we apply this language to the agriculture course, it is a huge aid to learning. The music of the spheres is a fundamental principle of the past. The spheres have to do with the fact that we are here in a relative center of the cosmos, constellating it with our point-centered consciousness into a spherical form of planes. The planes, in biological language, are "surfaces." Wherever you have a surface in nature, you have a boundary or gradient; things on one side of the surface and things on the other side of the surface interact with forces that follow laws. One might say that surfaces are where we meet the elemental beings or where we can begin to appreciate them. A surface is the product of an interaction between two different fields with different properties.

Therefore, in nature, wherever you have two different field activities, the interface between them, the boundary layer, becomes a kind of surface. Where a meadow meets woodland, you have a very active boundary where the predators and prey meet, where flora of the woods meet flora of the meadow, and where you get all sorts of fungal interactions. Walk out to the middle of a meadow, and you do not get the profusion of ephemerals that live at the edge of woodland shade. Nor do you experience a profusion of ephemerals deep in the woodland. They need enough light from the meadow to interface with the tall trees to live their life. In the natural world, the boundaries between different areas energetically represent the interaction between whatever is on one side and the other. Those interactive areas are called "gradients," and they represent a surface of interaction.

Boundary layers appear not only at the micro level, they also appear at the macro level. For instance, there is a surface at the edge of the Sun's activity field where the solar wind stops. According to astrophysics, we live inside the Sun; Earth exists inside the Sun. The ball of gas we think of as the Sun is just the center point, but the Sun and its activity go far beyond Pluto. Satellites have been sent out there, and when they get to the edge of the solar wind, those satellites must go through a turbulent zone to reach interstellar space. It is a kind of energetic membrane that creates turbulence as a satellite passes out of the Sun's area of influence.

In the natural world, surfaces are localities where there is great activity. Surfaces are where the action is. They are composed of nodes, or points, that make up the surface plane. The surface plane is made up of nodes of inactivity, interspersed among areas of intense activity. The intervals between the nodes are where the activity is greatest, and the nodes are the places in the surface plane where the substance shows up in the plane. In biology, nodes are the substances of a membrane, and the intervals are the areas where the membrane is permeable. Together, the nodes and intervals make up the most fundamental building block of the natural world: the membrane.

Therefore, the "not-stuff pattern" is where the activity is; the "stuff" is where activity has stopped. This is just standard physics for phenomena such as sound or music. The note you hear is not the true music; the music is how your soul stretches between one note and the following note. That is the music. Your soul uses the notes as stepping-stones to get over a stream, and you hear only the notes, or stepping-stones, not the motion from stone to stone, from note to note. The music dies in the note you hear with your ears, but it lives in what is in you and is moved by you as your soul tries to stretch and accommodate the movements contained in the intervals. These are all pictures of the same phenomenon. They are analogs of how the elemental world operates.

Because human beings are stretched between macro and micro, there is always an interface where macro and micro have to work it out. That interface could be large or small. It could be at the edge of the solar wind or between one cell and another. It could even just be the difference in pH between two fluids coursing through a cell in your liver. That differential between entities forms an energetic membrane. There is no substance in an energetic membrane, but it is nonetheless a membrane, and if that energetic membrane persists, eventually salts will precipitate in the space between the two pH flows. These relationships drive osmosis, diffusion, and semi-permeable membranes and are analogs of cosmic activity interacting with Earth. Whether geomagnetic force fields or cellular activity, it is all the same form motif. Form constitutes a diagram of forces.

How does science view what we would consider to be light? It is the agent of the way energy comes from the cosmos to Earth—not just the light of our own Sun, but also from all the other suns we call stars. They all send out light and this realm of starlight gets a bit crazy. Scientists did an experiment with the Hubble space telescope. They looked into the constellation of Pisces, in which there is a spot about the size of a pinhead where they have never seen anything whatsoever. Astronomers tend to have these little points, and they say, "You know, nothing is there." Therefore, when they put the Hubble space telescope into orbit, being scientists, they naturally said, "Let's just poke that in the eye!" Then they took some 350 consecutive photographs of just that one dark "nothing" spot over 350 days. They just kept shooting that sane spot, over and over. Once they had collated the images, they discovered that there are thousands of galaxies in that spot the size of a pinhead. They looked like critters swimming in a petri dish. We are talking galaxies and thousands of them in the tiniest smidgen of nothing. Yet light is coming from them and bathing us in it at this very moment.

If we simply extend that imagination out to a sphere around the Earth and think about it a little, we are talking about a great deal of light surrounding

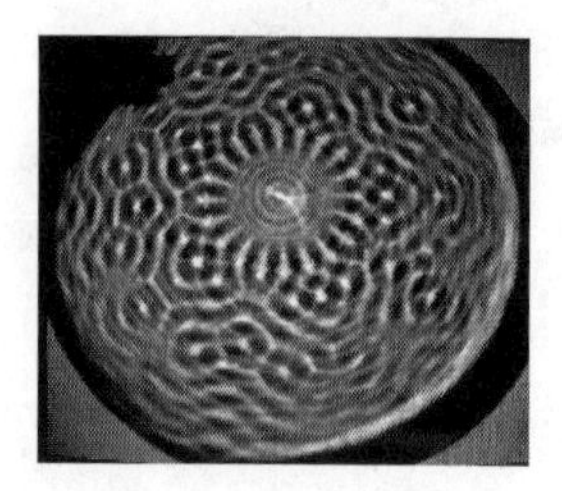

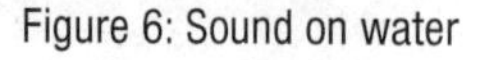

Figure 6: Sound on water

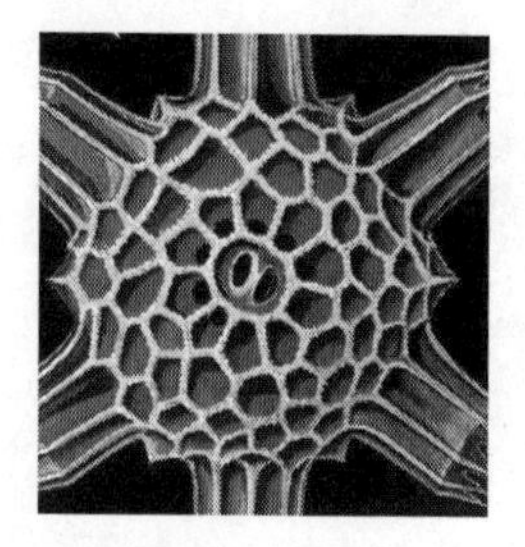

Figue 7: Diatom

us from all directions at all times—a sea of light as Rudolf Steiner would put it, a sea of ether.

Figure 6 is an image of a large, flat plate of water attached to a vibrator that can be regulated to various numbers of cycles per second. As the vibrations reach a certain frequency, they create a cellular pattern in the water. This occurs in a direction of "interference." You can see that the pattern has a center surrounded by a hexagonal or a kind of geometrical form. The form manifests as the sound waves move across and interfere with one another. Consequently, in certain places where the waves cross, a standing wave occurs. There is an increase in the wave amplitude, which creates areas where the water surface rises higher than the surrounding surface, with valleys in between where the water is not vibrating. Where the vibrations cease, you get nodal points. That image is from an experimental approach called "cymatics."[5]

Figure 7 is a drawing of a diatom. It is so small that you need a microscope to see it as it floats around in the water. It is part of the plankton; it is the shell of a tiny microorganism. Observe the correspondence between the pattern in the vibrating water and the form of the diatom. The image shows a silica structure that this animal grows around itself as a shell. The diatom has a little animal living inside, and the animal makes a shell of silica. Observe the form. Here we are researching how form operates, especially with siliceous materials interacting with light, sound, and warmth.

5 *Cymatics* is the study of visible sound and vibration. Visit http://www.cymatics.org for more images and descriptions of this interesting scientific approach to uncovering symmetry in water. See also Jenny, *Cymatics: A Study of Wave Phenomena*.

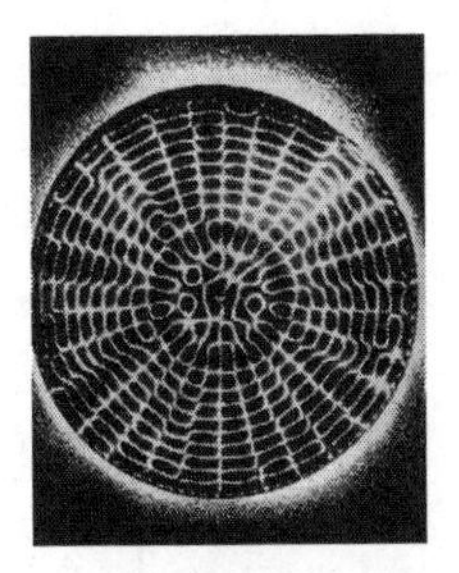

Figure 8: Chladni plate

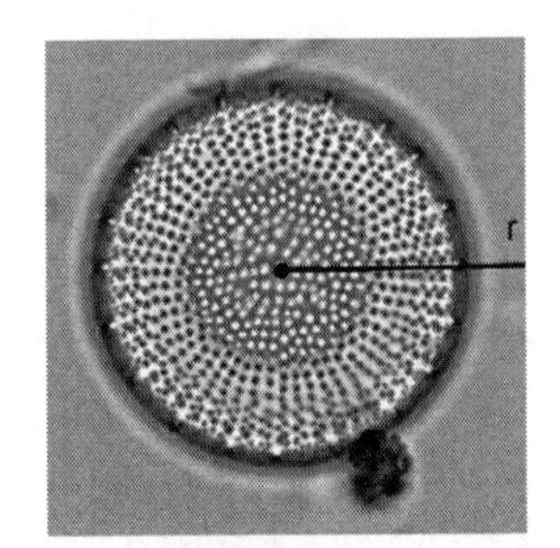

Figure 9: Diatom

Figure 8 is a circular brass disk, a Chladni plate, with boric acid crystals placed on it, after which it is vibrated with a violin bow to reach a certain pitch. When it reaches a certain pitch, all the parts of the plate vibrate to that note and the grains of the boric acid migrate to the nodal areas where there is no vibration; they move away from the areas of vibration. What you see in *figure 8* is a pattern of the pitch, or note. If you could see the sound made by the plate at that frequency of cycles per second, you would see this honeycomb pattern that is revealed by the boric acid flowing out through the air.

Figure 9 is a microphotograph of another diatom that closely resembles the Chladni plate. These pictures show the activity of sound, especially between an active principle and the nodes, where there is no action, resulting in form.

Figure 10 is a picture of two sound waves interfering with each other. One is coming from upper left to lower right, the other from lower left to upper right. Where they cross, you see that they amplify each other, producing a form principle that results from both waves, but also starts to show the kind of lozenge form we see in the Chladni plate. Where the waves of energy meet and cross each other, their interference creates cellular patterns in spaces of activity and no activity. The cellular patterns of activity and no activity result in the form created

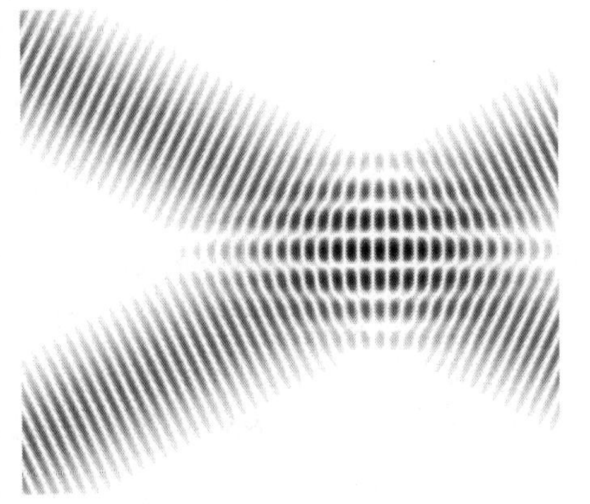

Figure 10: Sound wave interference

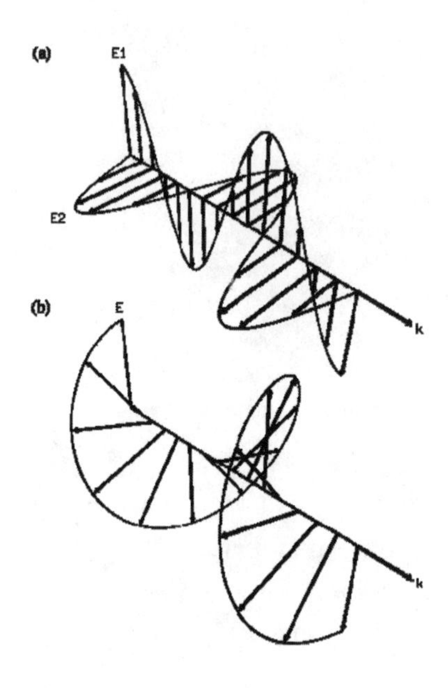

Figure 11: Light wave propagation

when substances fall into the nodal areas. In *figure 10*, we see a possibility of why the ancients said that the universe is sung by the Great Creator. Waves of sound cross each other, and where they cross the vibrations create cellular patterns revealed by the deposition of substance. There we get an energy, or activity—light, sound, or warmth. Then, where the waves of light and sound and warmth cross each other, we see the potential for a form motif to arise, which we could call "cellular"—pure forces interacting to create form. Substances fall into the areas that are void of activity, and the form of the forces is revealed.

Now look at *figure 11*. Science recognizes that light, when it comes in from distances, has properties measurable with existing instruments. One means of measuring light is called a "wave front," "plane wave," or "wave plane." It is understood that in light coming from interstellar space, all the movements of the light are, ideally, parallel to one another. This means that if we try and imagine forces of light coming in from the periphery, we see than the organization of the light is not random. Light comes to us in "energy packets," which nevertheless have, according to science, a form. That form is a plane at right angles to the movement of the energy of the wave. This is the "plane front." If I want to measure the light, I do this at that wave front, because there all the packets of energy have the same quality of measurement. This is ordinary physics that supports Steiner's idea of what we call the "planes at the infinitely distant," being the source of the energies from the periphery. He was very keen that people should study projective geometry, because he recognized that a way to understand this mystery of the etheric realm exists in the laws of projective geometry. When you work with projective geometry, the plane at the infinitely distant is the focus of very intense thought experiments that are practiced to purge the belief that the universe emanates from a point centered in one's body—that the universe, like us as beings of light, actually has its origin out there in the periphery.

According to science, therefore, light coming in arrives in the form of planes, but they are not just random planes. There is this whole issue surrounding the "propagation of light." *Figure 11* shows a diagram of the theoretical propagation of a light wave. There are two axes in the wave: vertical amplitude, Y, and horizontal amplitude, X. In the wave propagation of light, one component of the light wave is propagating vertically; the other component is propagating horizontally. Looking at the upper part of the diagram, you can see that the horizontal wave begins its cycle right at the peak of the vertical one. It is a little like the schoolyard game: "Can you rub your belly and pat your head at the same time?"

Where the X dimension reaches a node, the Y dimension reaches the amplitude of its waveform. In the next instant, the Y dimension of the wave collapses into the node as the X dimension reaches its amplitude. To measure a cross sample of the wave, the only thing we can do is cut the wave through a plane perpendicular to the line of travel for both waves. That is the wave plane. The wave plane is rotating through every point that the two waves share at a single instant of travel. The resulting motion of the wave plane is a screw-like spiraling of the plane that represents every point of the amplitudes of both waves at any given instant in time. The result is a rotating plane of light moving through space away from its source at the periphery among the stars. This is just physics and is the way science asserts that light propagates. It is planes corkscrewing through space, and the corkscrewing planes are perfectly perpendicular to the line of travel of the wave as it moves from its source.

If I have a source of light that begins to radiate in all directions like a star, planes of energy are corkscrewing out from there and continually expanding as shown in *Figure 4*. If I imagine this, then for each plane there is a star as the propagator of the light. In that diagram we can imagine the plane-wise circle as if it were a sphere. The periphery would be studded with stars, one to each plane on the edge of the circle. If *figure 4* were a sphere with planes generated by stars in the center of each plane on the circumference of the circle, the diagram could be a picture of the way physics says light travels from a source of an infinite number of stars. In *figure 4,* the sources of the planes of light are not in the center, but represent the starry realms of deep space. The planes of light are moving toward the center of the circle. The center of the circle is a point somewhere in the manifest space of the Earth. In this imagination, the light travels from the periphery toward a relative center in a spherical form, and that is difficult to understand unless you also understand Nicholas of Cusa. If I have light spreading out in a sphere from a source, like a star it will eventually form planes and then those plane waves are traveling in all directions from that star. If I have billions of starry sources and place the Earth in the center of that sea of light, I get an image of what Rudolf Steiner calls the "weaving of the light." *Figure 4,* as an image of the cosmos, is a sphere of weaving planes of light. A sphere of weaving planes of light could be considered a kind of archetypal antenna for light. An antenna for light would have the form of a sphere of planes.

If I have an infinite amount of stars around me and their light travels toward me, when it reaches my locale it arrives as rotating waves of energy moving toward every center in my locality. Consider the meristem of a plant; the void area between the leaves of that little meristem is normally spherical. It is a sphere of planes for the purpose of receiving light into that little form. The forms of trees are planes of leaves arranged into a sphere. The forms of many organs and animals are spherical derivatives or sections of a sphere. The human head is a sphere of planes of bone.

Thus, the sphere of planes is a central imagination because, as physics maintains, it is the way light propagates, and all other life forms are trying to make little antennas to receive light. Therefore, a plant working with this principle makes little planes of light to receive the starry planes of light; we call those planes of light on the plant leaves. They are antennas for intercepting and receiving energy of light, and their form reveals the way they do this, as well as the particular frequencies they want to take from the whole spectrum to do what they do. Antenna theory dictates that the antenna must be an image, or picture, of the signal the antenna is designed to receive. Moreover, those form qualities of the plant structures then result in energies that drive the chemistry, phenolics, Krebs cycle,[6] or whatever it is that you want to work with in the plant. In a plant, almost everything other than the carbohydrate is an alcohol derivative of some sort. Anything that has the suffix "ol" on the end of it is a form of alcohol. Alcohol covers almost anything from pigments to enzymes to whatever. Alcohol is a kind of solidified or condensed light. All we have to do is hold a match to it, and you can see this. Alchemists call alcohol "the essential mercury."

What *figure 11* shows is the rotating planes of light. It is a theory in physics that light is made up of immense, rotating planes intersecting and permeating every point. Light originates at the periphery of space as planes moving from a source such as a star. On Earth, these planes are focalized into a sphere, because the Earth is a point in this immense web of light. Every point receives light and is a focalizer for the extensive planes it connects. The places in nature designed to receive the most light are spherical or nearly spherical in form. That is one way to understand form. In the language of form, a spherical form in nature is saying that it wants to receive and hold onto the maximum light energy. This tendency toward spherical forms is called the "law of minimal surface." We will have more to say about minimal surfaces.

I apologize for this tough part, but it is physics and I want to build a bridge between Rudolf Steiner's idea of the etheric, in which he uses all these imaginations from projective geometry, and concepts in physics that describe exactly the same thing from a different perspective. To me, it is quite remarkable that through his clairvoyance he could see laws of physics that require a lot of hardware for other people to find. His clairvoyance was so developed and his imagination so precise that he could confirm that his perceptions were linked to physical and spiritual reality. I think it also helped that he happened to have a doctorate in physical science. As a

6 The Krebs cycle, or citric acid cycle, is a series of chemical reactions used by all aerobic organisms to generate energy through the oxidization of acetate derived from carbohydrates, fats, and proteins into carbon dioxide.

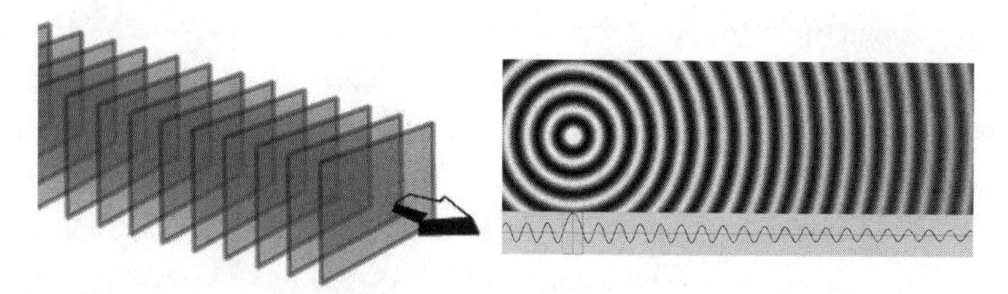

Figure 12: Planes moving through space

Figure 14: Spherical wave

contemporary initiate, he needed a foundation for his work, so he graduated from a polytechnic institute.

Figure 12 is an imagination of planes of moving through space, and the planes there are at right angles to the direction of the motion, showing what I just described. *Figure 14* shows the "propagation of a spherical wave." From any point in nature, the energies leaving that point are spherical. This is why the dewdrops on the tips of blades of grass each morning are not little cubes. From every point in nature, the Earth is actually leaking energy in all directions. From every point, huge energy, megavolts of potential, are leaking outward. Every time there is a change in cloud cover, for example, the potential changes either from going from the earth through the point on the blade of grass or from the sky down to the earth in the opposite direction. It is a constant exchange or flow of energies, and every time the energy comes up out of the earth through a blade of grass and radiates out, it does so in a spherical wave. Therefore, that spherical wave emanates from a point (in *Figure 14* we have a point on the left), but we see that spherical wave eventually reaches a place where the sphere becomes so far from the point in the center that it flattens into a plane. The great mystery, therefore, is this: How can a spherical wave become a planar wave? It just has to move down the track a little bit. Once again, this relates to the experiment by Nicholas of Cusa.

Now, because the generation from a point of a spherical wave is a portion of the cycle in the generation of light, the generation of a spherical wave from a point can happen every time a plane wave of light hits an "attractor" and is reflected or refracted from it. In *figure 13*, you have a plane wave of light coming in from the left, hitting a wall with a little hole bored into it. The light goes through the little hole, but because it is circular when light goes through the little hole, the hole becomes an attractor for the light and radiates out from the hole in another spherical wave. That is what you see in the diagram. And that spherical wave goes out and grows until it begins to become planar, and then it runs into another aperture somewhere, another hole, and each one of those holes, or attractors, generates another spherical wave that then runs out until it becomes planar, and so on.

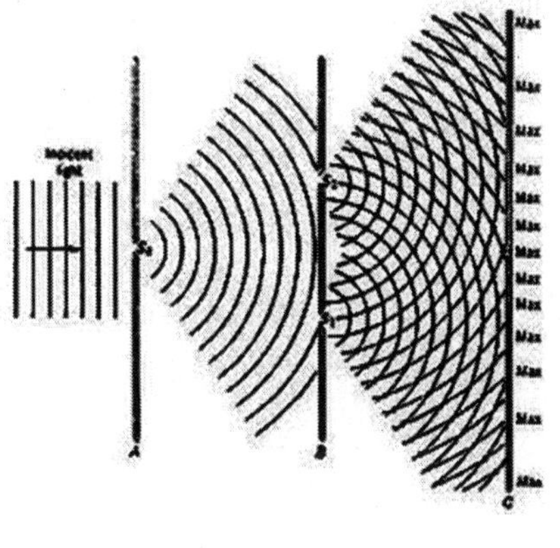

Figure 13

So you could say, "Well, okay, that is physics. Where do we see this in nature?" I will tell you a little story. One day we were at the college and it was a day of a partial solar eclipse. We were out in the garden near some tables with the sunlight shining through the trees—there is a big sycamore tree out by the table. We were sitting and talking, waiting for the eclipse to happen, and one of the children who was climbing in a tree said, "Oh, look at the wall!" The adults were talking about orbital periods and blah, blah, and one of the children says, "Oh, look at the wall!" When we looked at the wall, we saw thousands of pictures that looked like a crescent. What we saw covering the wall were thousands of pictures of the crescent shape of the moon blocking out the face of the sun during an eclipse. It took a kid to see it. Our conversation shifted from eclipses and orbital periods to "What the heck is going on here?" We realized that, after the eclipse went away, the wall was covered with sunbeams creating a dappled wall, and we realized that the dappled wall we generally took for granted actually showed projections of precise images of the Sun. Then we saw that those images of the Sun were focused by the spaces between the leaves in the tree. The leaves created tiny spaces, each a small aperture, and the waves were generated again. On the garden wall were numerous images of the Sun, each projected by a natural pinhole camera.

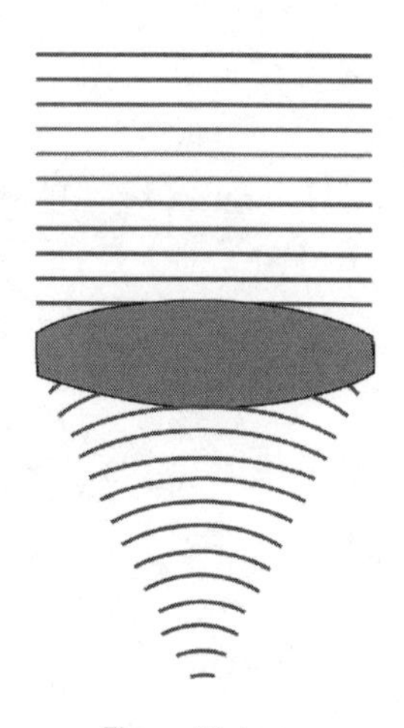

Figure 15: Lens

Figure 16: Plane wave

Therefore, the space between every leaf in the tree where the light goes through becomes a lens, a pinhole camera for the light, and each time that happens, a new propagation, or spherical wave of light, is created. Science recognizes this and calls those little spaces "attractors," because they attract the light, focus it, and then create another beginning, so to speak. In electrical language, we might call them transformers. In terms of optical systems, we would call those attractors lenses. Thus, there is a pattern in the natural world of how light moves through and is received by things. Then, through attractors and propagation, the energy of the light steps down and up within things. We call these force patterns generated by light the "life force" or, as Rudolf Steiner called them, "etheric force"—the energy of light. There transmission forces in the natural world are ways that light moves from one level to another, and they all have the same form—a sphere of planes.

Figure 15 shows a lens where spherical waves hit the lens and then become planar. This shows how these patterns can go the other way; you can also have planar waves hit a lens and become spherical. It is this plane-sphere–plane–sphere–plane–sphere oscillation in the natural world that is the basis for many transformative properties of substances. There are many variations of this fundamental pattern.

In *figure 16* we see a plane wave moving from the left toward an attractor in the center. When a flat plane wave hits a spherical attractor, a vortex arises, because the plane wave is rotating when it hits the sphere. The plane wraps around the sphere and kind of turns inside out. It becomes vortexual on the other side and starts an inner kind of reversal, creating a vortex of light. The sphere takes this rotating plane and folds it in on itself. Then it has this inner kind of quality, so in some places in the natural world where I have a sphere, the light is tweaked by it, so that as it emerges from the interaction an inner space is created rather than just another spherical wave as it would from an aperture. Therefore, rounded forms in the natural world are special kinds of attractors; holes are one kind of attractor, and a round form is another kind of attractor. We could call what I am describing "formal motifs of the light." From a certain perspective, these forms are just balls and holes, but they do things to light in the natural world. Skulls, orifices, cracks, inner qualities, and inner surfaces affect the way the light operates within organisms; the patterns they produce are predictable, and they can be used to develop an imagination to see more deeply into interrelationships in the natural world.

Now consider something difficult—the relationship between light and form. This will anchor the idea of the etheric instead of having to say, "We do not know. Rudolf Steiner never told us *what the etheric is* but only described it." However, he did leave us quite a few clues, as if saying, "Get to work. I have told you that there is something called the etheric; it operates in a particular way, but I am not going to fill in the blanks for you."

During Steiner's time, the question of the etheric was huge and not just an anthroposophic issue. It was also a very big question for scientists because of an 1887 experiment intended to find the "luminiferous aether," or aether wind."[7] (They spelled it *aether.*) It was a well-known experiment by Albert Michelson and Edward Morley.[8] During Steiner's lifetime, people said that some sort of medium was needed to carry these light waves and planes. Because of where science was at the time, they understood (or thought they understood) that light was transmitted via the luminiferous aether, a substance so fine that we could barely measure it. Nevertheless, it had to be

7 See http://en.wikipedia.org/wiki/Luminiferous_aether.

8 See http://en.wikipedia.org/wiki/Michelson%E2%80%93Morley_experiment.

a "substance" to fit the model of materialistic physics. They believed substances were necessary to transmit energy—that it could not travel through a vacuum. After all, sound does not travel in a vacuum. As a result, Michelson and Morley set up experiments based on the speed of light, which had been calculated to be a measuring device. They set up light beams to go between metering devices, from one mountaintop to another. They were testing the speed of the light in the context of the Earth's speed of rotation. They shot light beams both with and against the Earth's rotation, looking for any difference in the speed of the light. They were looking for "aether drift," or "aether drag." This was 1887. Therefore, Rudolf Steiner, who was attending a polytechnic institute, would certainly have been aware of the cutting-edge science of the time.

The experiment produced only the minutest possibility that aether drag is real. Nevertheless, though it was small; it was consistent. At that time, Einstein was developing his theory of relativity, and he said that the amount of drag could be accounted for by experimental error and was not robust enough for us to prove the existence of the "aether."

During this same time, homeopathy was being developed. It was thought then that, because there is no substance in a homeopathic remedy, it must be composed of some sort of "light aether." Many people championed this notion, as well as homeopathy itself. Most hospitals in the 1880s used homeopathic medicines and allopathic medicine had not yet taken over. Homeopathy was seen as the new age, whereby healing would take place on the basis of light and spirit rather than substance. This was the late 1880s.

The fact that Michelson and Morley famously failed in their experiment makes it difficult for scientists today to hear biodynamic practitioners talk about the etheric. They firmly believe that the ether was discredited more than a hundred years ago. Moreover, there were at least fifteen experiments that followed that original one. Morley himself did another experiment in which the apparatus was placed on a turntable that turned on another turntable, intended to eliminate the motions of the earth and gravity. He had a special apparatus built with its legs embedded in glass, with the room made of a certain kind of canvas that would not allow the light to shift during the experiment, all in an attempt to remove any error from the experiment. These attempts always ended with approximately the same amount of "aether drag" as the original experiment. In classical physics, the experiments are considered failures, but the results are nonetheless reproducible.

Homeopathy and the synthesis of protein structures is linked to a strange phenomenon called "the chemist's beard." In chemistry, for instance, a lab in Buenos Aires was able to crystalize a particular protein substance on the exact same day that labs in Los Angeles and Moscow crystalized it. They had tried but could do it. Then, suddenly all three labs finally crystalized this protein. People have seen this phenomenon again and again, but no one knows why it happens. These are chemists; there must be an answer, so they came up with the chemist's beard theory. In those days, most chemists had beards. The theory says that chemists have beards, and chemists go to conferences and talk to one another. They also drag molecules with them from their experiments. After the conference, when they go and have a few drinks in the bar and talk to one another, their beards pick up molecules from what they are all doing and then return to their labs. This is the chemist's beard theory of why different labs all over the world would be able to crystallize proteins at the same time. This, of course, makes it difficult for them to patent their work.

In the case of the experiment with luminiferous aether, it was discredited, but people kept the experiment going because they believed there was something to it. Moreover, even though Einstein had said it was probably experimental error, later on, in 1920, Einstein said, "There may be a different concept of the aether that is not a ponderable medium." Thus, the great man himself kept the door open. The ether, as Steiner described it, may not be a ponderable medium but an imponderable medium. The question is this: How do we research an imponderable medium? Pondering the imponderable is meditation. To set up the experiments, we might need to change our consciousness. As we start to work into etheric realms, we work into life realms; as we work into life realms, the subtle interactions of light, energies, and nanosurfaces require a whole new vocabulary for experiments in the natural world. This kind of shift requires the development of imagination that

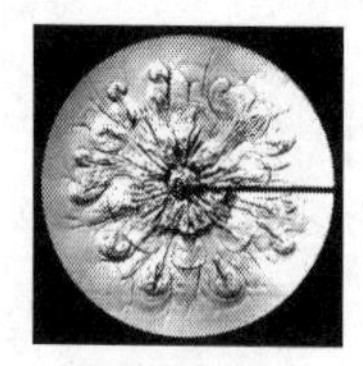
Figure 17

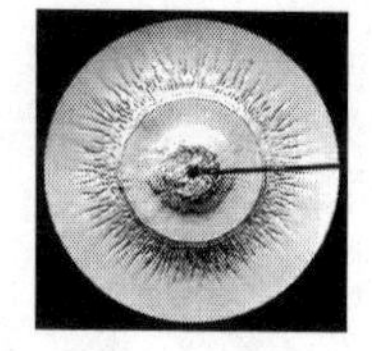
Figure 18

allows us to see the archetypes of the principles—the way those principles operate at the macro level of the cosmos.

Consider *figures 17* and *18*. They are "drop-pictures" as developed by flow research labs.[9] Good examples of these are Jennifer Greene's work in Bluehill, Maine, and Theodor Schwenk's work in Herrischried, Germany. The technique takes a sample of the liquid we want to test, and we mix it with a certain proportion of glycerin and place the mixture in a shallow dish. Then, at regular intervals, we drop a series of drops of distilled water into the liquid. Glycerin is a viscous alcohol and a known quality, and the distilled water dropping into the liquid works a force more deeply into the mixture with each drop. For this process to be reliable, the dishes must be completely polished and calibrated to an exacting protocol.

The little dark thing in the picture is the pipette that drops the distilled water. The drop of the distilled water falls, hits the liquid glycerin and sample mixture, and creates a form. Each subsequent drop pushes the drop further into the form. Each step is photographed, and the series of photos shows how the test liquid responds to the delicate formative forces generated by the drop activity. In this process, the formative (etheric) potential vitality of the liquid is revealed. This is the "drop-picture method." This was suggested by Rudolf Steiner as a way to study etheric forces, because the various liquids that have vitality each have a unique inner formative signature. The surfaces of vital water are much more capable of transmitting life and light than water in which the bonds of the inner surfaces that transmit energy are distorted by pollutants. Water has one of the highest surface tension coefficients of any substance, and anything in it immediately creates hindrances to the transmission of light and life forces. Thus,

9 See, for example, Schwenk, *Water: The Element of Life*.

the drop-picture method was developed to test the quality of water for its life-sustaining qualities.

In *figure 17*, we see spring water being tested by the drops of distilled water. We see that they form a whole rosette that looks like a flower of vortices, because the spring water has inner surfaces that interact and want to create the form of a sphere. The ideal form of water is a sphere, because in the spherical form all of the surfaces in that water droplet are coherent. The more spherical the form, the more potential it has, because it has minimal surface and, therefore, maximum energy. If the spherical form is an image of how light is gathered in water, when that spherical form goes in a plane, it turns into a vortex. The vortex is saying that the water would like to form a sphere, but it can form a sphere only in the air. When it comes in contact with some other substance, it tries to form a sphere by forming a vortex. An actual vortex would make a donut, circulating inward at the center and outward at the side. In an ideal sense, this would form a torus. That is *figure 17*—beautiful water that shows the action of how the spherical form is transmitted in the energetic planes in water.

Figure 18 shows the effluent from a sanitation facility. It is pure according to science, meaning that there are no bacteria in it. However, what they had to do to get the bacteria out has destroyed the ability of that water to take on the life, because they wanted to get all the bugs out. That water could have been made more alive by allowing it to move around stones and make little vortices to regain its spherical nature.

These are the principles behind biodynamics, and all those devices—flowforms and ways of stirring our preparations—are based on the same principle: we are creating inner surfaces that can be made coherent in order to intensify and concentrate the life and the light.[10]

Figure 19 (next page) is a photograph I took. Just get some water, stir it, drop some dye in it, and you see a vortex. The interesting thing about this picture is that it shows the planar form of the membranes, the sensitive

10 See the classic works on water flow and flowforms: Theodor Schwenk: *Sensitive Chaos: The Creation of Flowing Forms in Water and Air;* and John Wilkes: *Flowforms: The Rhythmic Power of Water.*

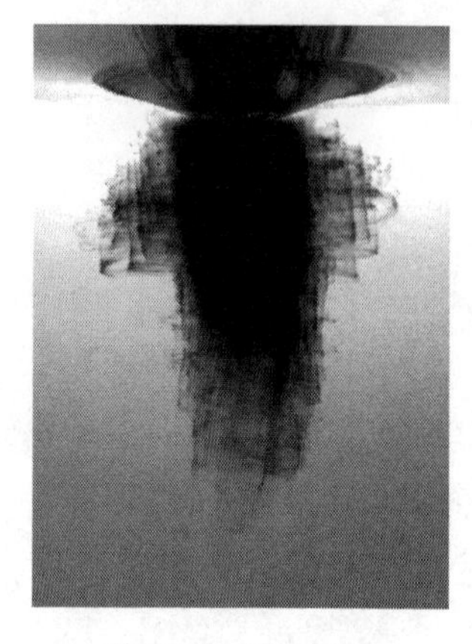

Figure 19: Ink vortex

membranes—one inside the other, inside the other, inside the other. This happens because, with each of those membranes, the dye enters that plane because it is a node; on one side, the water is moving faster or slower than it is on the other side of the membrane. The dye is entrained into the areas where the different speeds of the water are being worked out by the vortex. The center of the vortex is moving very rapidly, while the outside is moving more slowly, and as those speeds work out in the vortex, the vortex becomes organized, or actually becomes an organ that can receive the substance. The inner planes of the water in the vortex become sensitive to receiving whatever we place into the medium, because the action of the movements on either side creates a potential for the form to develop. This is an example of planes and sensitive surfaces as they arise from movements in nature. When you stir, you actually create an organ formation in the water. In stirring, you amplify the inner surface of whatever you are stirring, so that you create more energetic surfaces inwardly that take up whatever the living principle is in the substance you put into the water. This is the action behind stirring biodynamic preparations.

In *figure 20*, you see an action of airflow around an object, which creates an eddy. The eddy alternates side to side and creates a "vortex train," which is what happens in your vessel as you stir. Trains of vortices are generated that work out the energy of the various differentials of the movements of the sensitive membranes, and in the inner structure of the fluid, the amplification of the membranes creates a much richer inner surface. Within those surfaces, as they start to create laminar flow, as soon as you turn and break the vortex going in one direction, you create a condition of chaos and "microbubbles." These are nano-sized bubbles. In nature, we have the law of minimal surface, which says that the smaller the form, the more active the surface, as discussed earlier. If I make a sphere smaller and smaller and smaller, the surface becomes more intense, which is a major quality in nature. I want to intensify and amplify surfaces, because that is where the action is. It is where the macro meets the micro. The forces of the two movements meet in the formation of surfaces, inner surfaces.

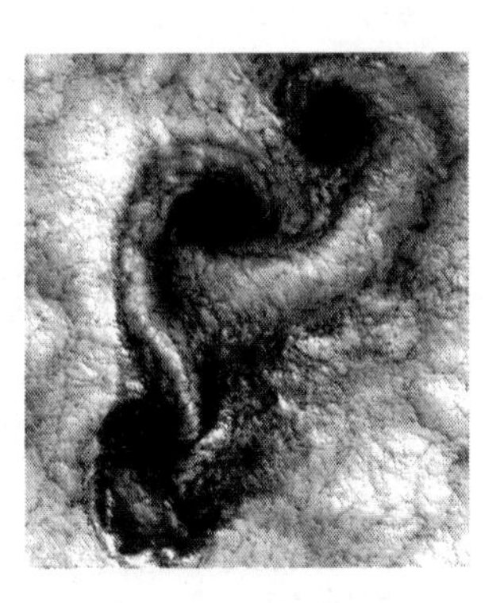

Figure 20: Air flow eddy

When I break the vortex, I create these ridiculously small bubbles. One experiment takes a solution of water and puts an ultrasound wave through it in one direction and another in another direction, and then focuses on a very tiny area where interference occurs between the two ultrasound frequencies. From that little area, bubbles will start to arise in the fluid. The experiment is intended to study the effects of micro-bubbles on the decay of alloys in ship propellers. When a ship propeller spins in the water, bubbles arise from the action of the propeller. The propeller is not pulling air from the surface; rather, the movement through the water creates a condition in which micro-bubbles form where the hydrogen and oxygen separate as a result of the rapid movement of the propeller. It was found that when a micro-bubble bursts, the decibel level is equivalent to a 747 going through your backyard. Moreover, when a micro-bubble bursts (there are photographs of these things), a vortex shoots out from the bottom of the bubble. In that tiny vortex, the local temperature goes to around 800 degrees and then stops. Therefore, engineers try to design propellers that make the least amount of micro-bubbles, and thus the least amount of sound, as they go through the water. Moreover, if enough micro-bubbles explode against the alloy of a propeller and the temperature spikes, it starts to pit the surface. Then, as micro-pits form on the surface, they create even more micro-bubbles, which make more micro-pits, and so on. This process is "cavitation."

This means that, when you break the vortex in the liquid you are stirring, you create an intense chemical, thermal reaction in the fluid. Whatever substance is suspended in the liquid gets blasted and atomized, and then the water that is being organized into a coherent set of streamlines absorbs that substance far more effectively. You then reverse the direction and the

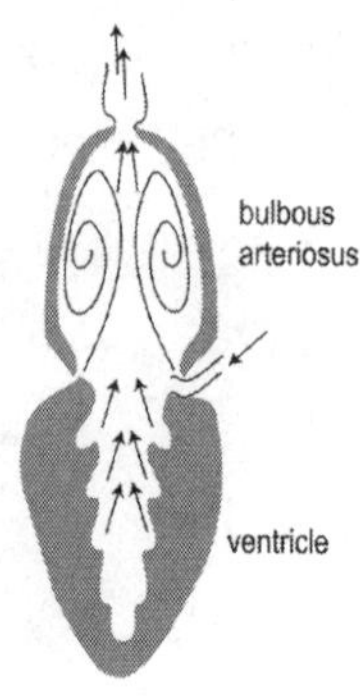

Figure 1: Fish heart

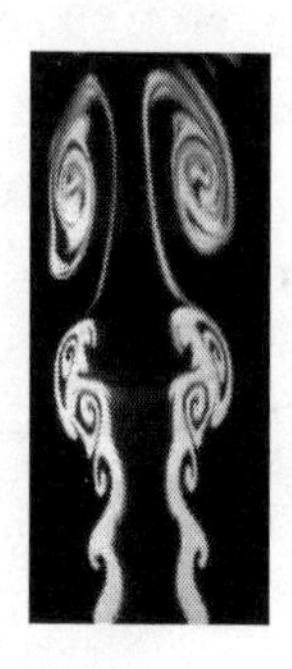

Figure 2: Vortex tripling

process starts again. Essentially, this makes the water into a digestive organ. The streamlines formed in the water are an analog of the way that the form of your intestines and mesentery allow your digestive system to operate. It is all about surfaces.

The more we pay attention to the creation of surfaces and inner surfaces, the more potent the digestive process. I am bringing these pictures because natural science has already discovered these concepts but lacks the cosmology that Rudolf Steiner provided so that ship propeller engineering can relate to spraying organic substances in a garden. If we take Rudolf Steiner's cosmology and cherry pick the conventional science, we can talk to scientists about our spiritual cosmological ideas in a way that is scientifically educated.

Looking at *figure 1* again, we see a fish heart, which provided one of the first inspirations for the development of flowforms. The ventricle below the vent is where the blood comes out, and the *bulbous arteriosus* receives what comes from the vent. The arrow on the right shows the inlet. As the ventricle expands, it draws blood through the inlet and contracts. You can see in the folds of the fish heart a whole series of little vortices that grow larger until they enter the *bulbus arteriosus* and create a separate vortex pattern. That is the action of the fish heart on the earthly, manifest side. *Figure 2* shows fluid exuded through a pipe into another viscous fluid, creating vortex trains. What you see is very similar to the action of the fish heart. The forces precede the substance. This is a fundamental law for this kind of work. Forces are the cosmic activity; substance is the earthly manifestation.

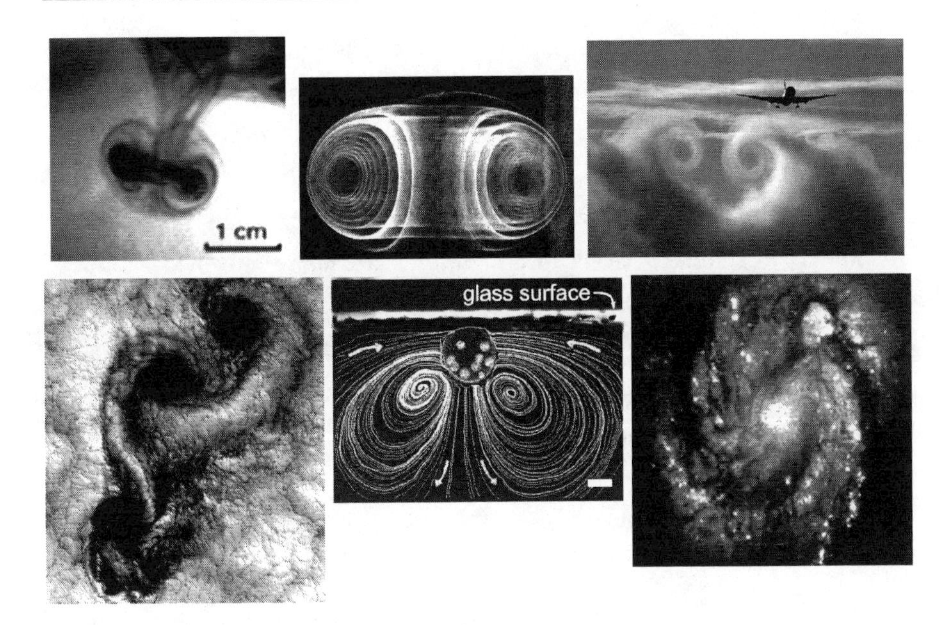

Figures 3–5 (above); figures 6–8 (below)

It is the same for the human heart, which is formed by the beating fluid before there is a vessel. In the embryonic sack, the fluid starts beating rhythmically. There are no veins, arteries, or heart yet. The beating fluid moves toward a small bag filled with cardiac jelly, and the movement of that flow punches into the cardiac jelly and creates the septa and valve structures of the heart. That is embryology. The form of the heart is created by the movement of the fluids, not vice versa. Therefore, we can say that the movement creates the form, and that the form is movement that has come to rest.

These concepts provide keys to understanding the agriculture course in a deeper way. Any organ from an animal is an organic process or movement that has come to rest. *Figure 3* depicts the situation in which a pulse of water is inserted into other water, making a vortex ring as it travels through the water. *Figure 4* is the same thing at the level of a one-centimeter vortex ring. *Figure 5* shows an airplane that has just flown through a cloud. The currents of air streaming off of the wing of the airplane have created a vortex ring in the wake. Now *figure 7;* the small round circle in the center is a "volvox," a microscopic microorganism with little hairs. It is a colonial animal, and the little hairs beat to create currents around it, as shown by the white lines around it. The currents go out and pull food particles into the volvox. This microscopic activity is similar to that of the airplane. In *figure 6* we see a "Von Karmen wave train." In the upper left of the image, in the center is an island in the Pacific, and all the white stuff is a bank of clouds moving across the volcano in the middle of the island. The clouds are being entrained into those incredible vortexual forms in the wake as the clouds get dragged across the mountaintop. The wind that goes around them creates Von Karmen waves. Then, *figure 8* is a picture of a spiral galaxy from the Hubble space telescope.

Therefore, the form principle in biodynamics and the study of form is really a key to imagination. Nevertheless, I have to study it in a precise way rather than going into fantasy.

III. Morphing Images • the Elemental World • Will in Images • Pictorial Consciousness • Crystal Forms • the Action of Minimal Surfaces in Nature

Now we will consider geometric forms. One key to agricultural imagination is being able to do a process with a picture called morphing. This means transforming one picture into another, and into another, while maintaining an inner feeling that the three pictures are coherent. When you can do this with a picture, it becomes a "tool" for entering the natural world by taking an image of something that presents a question of, say, what to do with a particular area of your garden or how you might like to work with a particular plant.

Say, for example, you have seen an anomaly—something unusual—and you would like to know how to research it. Now, use an image of the thing you see and try to morph that image in two ways—one back and one forward. If I see something in a plant I am working with and a question arises in me, I ask myself something like this: What did this plant look like just before it got to this spot, and what will it look like next? This is morphing. I take the present state as the middle and ask myself what it looked like just before it came to this present state. Can I tell what it will look like in its next state? When I morph something from the natural world, it is useful to learn then how I can take that change into silence. When I form the picture sequence and then imagine it in reverse, I am actually posing a question to the elemental world about that sequence. This is meditation in pictures. Rudolf Steiner says again and again that we must learn to meditate in pictures, not concepts. It is impossible for you to awake in another being with a concept. You can awaken in another being only through pictures that you control. I form a picture of what the entity, or thing, looked like just before and what I think it will look like next. It is not important whether the picture is correct. What matters is my intent in doing it. It is the movement of my will to harmonize with the will behind the life form that is important.

It is all about morality and one's intent behind the imagining. If my intent is only for my own benefit, then the answers will be connected to my intent to profit rather than to harmonize with the will of that entity. Suppose you wish to understand something about your cow. Esoteric wisdom says that it is best to try to contact the cow deva. The cow deva is the elementary being in charge of the cow species. In the ancient language, the being that guides the inner activities of a species is a spiritual being called a deva. *Deva* simply means a shining being. Native Americans call that being "Grandfather," and it is Grandfather in the natural world that we wish to address when we have an agricultural question.

Rudolf Steiner calls the Grandfathers *elementary beings*. They carry the consciousness of the species as spiritual entities. They are also called archetypes. Suppose I have a question about water. The modern Western concept for water is H_2O. The picture for water is a stream flowing around a rock or a solution dissolving a salt crystal. Therefore, I can talk about the bonding angles of the water molecule, or I can visualize a salt crystal dissolving into water and then the water evaporating and the salt crystal returning.

The key to this is found in the will, or intent, of my inner activity. I need to exert my will to learn a concept such as the bonding angles of water molecules. Once I have learned the concept, the concept is a kind of shorthand that substitutes for my will effort. I have already exerted my will to learn the concept, and now I just need to remember it. My memory fixes my will into the concept. I do not need the same intensity of will to recall the concept I needed to understand the concept in the first place. We all know this. Thus, when I am simply remembering an abstract fact, very little will is exerted in my memory. I just think of the concept "H_2O." In this way concepts are a kind of shorthand so I do not have to exert my will.

Words are the same thing. In conveying these ideas, I use words to take the place of the will I would need to use mime instead, for instance. Concepts are like placeholders for the will. Words are like corpses of the will in my thinking process. Everything in nature originates as the word-like intent of a creative being. It undergoes a transformation to become the corpse of the living will word. The living side of the will word is called *Logos;* it is the living part of the word that has become corpse-like. The word I speak is the corpse of the living, or creative, word. The concept I remember is the corpse of the activity with which I engaged the idea to learn it in the first place. To enter the processes of nature more deeply, I want to get my will engaged at another, more living level. This is imaginative consciousness; it is more living thinking. Imaginative cognition is thinking in pictures that transform dead thoughts into living, creative experience.

I have to take the dead thoughts that I want to remember or the recipe I want to remember and transform them into living thoughts. I take the recipe and the conceptual thoughts and lift them with my will into a realm where I do not have the same sense of entitlement about the answer. When I say H_2O, C_6H_2O, O6 + H_2O in the presence of sunlight, my will is not very active. However, if I actually follow in my mind the process of photosynthesis and the action of light through the various chlorophylls into magnesium and I have a picture of what magnesium is and what potassium does, not as a substance but as an activity in the plant, then the stage is set for me to meet some spiritually creative beings. If I know potassium as an activity, not just a symbol, then my will engages potash in a different way. To assimilate these inner activities of the creative nature beings, it is not useful to codify the pictures present in the natural world. In that case, I am just piling up the corpses. We have to think through the activities as a sequence seen in our inner eye.

Consider potash, for instance. Potassium builds ion channels. That is a concept, but if I imagine the potassium in the solution streaming from point to point within the plant, being pulled by phosphorus and carrying oxygen through to the carbon, then these ideas become living pictures. This is the way Steiner gave his chemistry lesson in the agriculture course. He described pictorially the activity of the substance rather than just saying, "Potassium forms a sulfate as a result of electron exchange." This is a concept. I want to take my concepts, especially about dead minerals, and revivify them as living pictures. The key each time is that I have a living picture; I have to invest will into it, because I must create it afresh in my mind's eye.

The issue of picture formation is a big one, especially for biodynamic practitioners. The simple key is to form pictures of a very simple sequence. For instance, this happened right before this; now this is happening, and this is how it looks; and this will happen later. I can take anything, from global warming to how many worms I have in my compost heap, and work with it in this pictorial way to build pictures that will allow my will to enter the problem in a moral rather than just a factual way.

I want to enter nature as a true human being. I wish to become a true human in the natural world, so that I can experience the natural world in a new, participatory consciousness. That new participatory consciousness is pictorial. Then, whatever my opinions are, they will be *my* opinions, and that is okay. Different opinions are okay. Nevertheless, I need to be able to substantiate for myself why I have my opinions. I need to shake hands with the elementals with whom my actions interact outside of my understanding. Pictorial consciousness is useful for our work in nature. I observe; I represent; I take it to silence; and I record something very simple.

Here is a little exercise to test how this process works. You see six images in *figure 1* (next page). The one on the left is a tetrahedron, an incredibly

Figure 1: Tetra inversion

important form in the natural world of agriculture because, on the molecular level, it is the basis for all of mineral structures. The surface planes of a tetrahedron are the building blocks for all silicates. Everything that comes from silica arises from a variation of the tetrahedral form. The tetrahedron is said to be self-reciprocal, meaning we can turn it inside out and it still is the same thing.

This series of images shows the self-reciprocating tetrahedron turning inside out. At the far left, all sides of the pyramid are equal. This means that the pyramid would fit perfectly in a sphere; the points of the tetrahedron are all equidistant from one another where they would touch the inner surface of the sphere, as well as equidistant from the center of the sphere. I have a sphere and a tetrahedron inscribed within it. The tetrahedron is a pyramid with four equal sides and four equidistant points. Each point touches the surface of the sphere; all the points are equidistant from one another on the surface; and this is the fundamental imagination.

In the second form, the image shows that the sphere is shrinking; it has moved through each point, and as the sphere moves through the point, it drags a plane with it. The plane is traveling inward along a line coming from the center of the form through the point. The plane represents the tangent plane created on the surface of the sphere where the line from the center touches (or is tangent) to it. I could draw a line from the center of each plane to the center of the sphere, and each plane would be equidistant from the center because each point is equidistant from the center.

In this imagination, the sphere is actually made up of an infinite number of tangential planes, all touching the surface of the sphere in one point. When the sphere shrinks through the form, it drags the tangent plane from the point of the tetrahedron into the tetrahedral form as a plane moving from the periphery to the center of the tetrahedron. This plane is seen as a triangle. Imagine a pyramid made of clay and cut off one of the tips. A triangle would replace the point. The triangle represents the plane that is moving. Now cut off all of the tips. That happens when the sphere starts shrinking through the form. All of the planes are moving toward the center at the same rate, meaning they are all the same size at the same time. All of the planes are actually on the shrinking sphere.

The center of each plane is equidistant from the center of all the other planes because they are in the sphere. That is the imagination, and this is known as a platonic solid, or form. Plato used this kind of imagination to train the very faculty I have been discussing. In Plato's mystery school, this geometric, imaginative practice was the basis for moral development. I need will to move this sphere through these points and watch the planes come through the form toward the center.

In the third image, the planes have pushed closer to the center, and then there is a box. Our exercise is to see if we can draw the form that would be in the box. I ask you to represent this series in your mind to see if you can discover what the form will look like that would be in the box. This is very strengthening for your imagination to do this. Take a moment before continuing and try to sketch what you think the form might look like.

In this process, the form in the box actually becomes another platonic form, the octahedron. The points come in touch with each other and a new form arises. These kinds of forms are often seen in textbooks on crystallography and mineralogy. The crystal forms are based on the molecular patterns of the mineral matrix. Moreover, a mineralogy book will describe that matrix as "a space lattice." This is the potential for particular planes in space to arrange within the solution to create that crystal form, and whether it goes one way or another has a great deal to do with temperature. Minute differences in temperature, pressure, and pH can cause very different crystal forms to arise from one matrix, and there you get an action of silica to become a myriad of different forms in an actual world of crystals and through the organic acids weathering of rocks to create humus.

The goal, or action, of the mineral realm is to form the crystal at a fundamental, molecular level. It involves the arrangement of tetrahedral molecules in particular patterns. They are invisible because they simply represent forces, but theories call them molecules. However, we could also call them tetrahedral planes in space, and all silicates arise from this form. When we work with this meditatively, it is very productive, and we begin to understand mineralogy in a whole new way.

As we wrestled with the previous exercise, a kind of feeling arises when a picture comes to mind that gives us a feeling of *knowing*. That feeling of knowing is our guide for working with the elemental world. It is not a mathematical feeling of thinking it is cool; rather, the mind follows a particular sequence of events, and we can feel our mind being guided by the lawfulness of the form. When that happens, the feeling of knowing is present. We get the feeling that this is the correct sequence, and such knowing is the key to working with elemental consciousness.

When we work imaginatively and a picture pops into focus for us, or when we can get it because someone was able to help us, this sense of "getting it" becomes our guide. That is what we look for when we are doing, observing, representing, going to silence, and recording. There is a feeling in the sequence that the picture coming to me has some connection with reality. I may not know what that connection is because I may not have formed the picture correctly. However, if I keep working with it, the archetypal beings behind the picture will work with me and participate with me in my dance of will. Eventually, I will be led to a person, a book, or source. Perhaps you will be surfing for something on the net and accidentally encounter a picture of a molecular structure or something tetrahedral and think, "That looks like the thing I was seeing." The sympathy of the forms is a language of forces that can lead to deep insights. You just have to check your sense of reality by finding links in conventional science to what you think you are seeing.

Angels can work in that way, too, to guide us to the issue being questioned pictorially in these exercises. If we have no question, we will get no guidance. If we have a question, there is help. The formation and dissolution of the picture is a kind of invitation to dialogue with the archetypal beings behind the forms of the natural world. They respond, but in their own way, and perhaps not the way we expect. Sometimes they have to send in a torpedo to soften us up a little, to get us to understand that this is not quite the way it is; it depends on how stubborn or clueless we are.

You may be wondering what this has to do with the theme of macro–micro. I included the exercise for two reasons; first, to point out the idea of a tetrahedron as an important object of study to understand soil science. Much in soil science has to do with what this little diagram represents. The second reason is the *feeling of knowing.*

Figure 2: Phantom quartz

Now we will shift to the image of the crystal. This one is known as the phantom quartz. Considering our discussion of the platonic forms and their structure, I would like us to look at the form of that crystal and represent it to ourselves as though it is growing in our mind's eye. Let us ask ourselves what it looked like just before the present moment and what will it look like next, although it is stopped for now because it is in that form. Then take the sequence to silence. Can I imagine what is taking place here in the formation of a phantom in the crystal? Observe it, represent it, take it into silence, and instead of forming an opinion just say, "I do not know. What do you think?" I think this may be the way it works.

You see these little forms in the crystal—the phantom. The question is this: How did it get there? Can you imagine a process whereby it got there? A phantom is very interesting because it reveals several sequences of its creation. Observe it, represent it, and take it into silence for a while.

According to crystallography, a crystal is a mass of inner-surfaces. Crystallographers research these things by shooting light into crystals and watching what the light does. If you shine a particular kind of light into a quartz crystal, it will not go directly through but rotate. This means that the interior of the crystal is coherent—it is not just stuff, and the surfaces in the crystal are built coherently and laid down according to laws of the way form operates. The most fundamental unit in a crystal is the tetrahedral form, and the molecular forms of the tetrahedrons are arranged in tiny spirals.

Figure 19

Tetrahedral molecules build onto one another and rotate like a twisted chain and, if I took a number of the tiny chains and repeated them again and again, the quartz crystal form would arise. There are passages between the spiral bundles of molecules that transmit energies through the crystal. The spaces in the crystal are arranged in rotating planes or surfaces, and it is the inner surfaces that rotate the light.

Now look at the crystal and think about inner rotating surfaces and chains of spaces in the crystal. The rock is a corpse of the living process that formed it; the living rock is a cosmic reality of light, and the soil is a child of the Sun. Rocks have their origin in starlight.

The inner surfaces of crystals carry charges, according to science, that allow light to move through them and create patterns in the crystal that allow it to store energy. If I solder two leads to a crystal and hold a match under it, I will get a current coming from the crystal. When I remove the match, the current moves in the opposite direction. These responses are called *piezoelectric* and *pyroelectric* effects. The crystal holds light, and the phantom (see *figure* 2) reveals the way the inner surfaces are formed and that the crystal is composed of an infinite number of inner surfaces.

Thus, when we form pictures we are first confronted with the world as it always appears to us. I look at a crystal; it is a rock. However, when I begin to do exercises in imagination, I can experience the crystal form as a kind of intelligence; that intelligence is of an extremely high order, one that does not make mistakes. It simply is. However, if that highly ordered intelligence were all there is, there would be no life. Life demands the space of freedom in highly ordered forms, and because there are spaces for life, there is room for change. That room for change is growth.

Looking again at *figure 19,* we see the idealized form of a self-reciprocal tetrahedron known as a stellate-tetrahedron. In esoteric circles, this is called a *merkabah*.[11] It is said to be the form of a human light body.

---

11 *Merkabah* is the divine light "chariot" of ascended masters used to connect with those in tune with the higher realms. *Mer* = Light; *ka* = spirit; *ba* = body (see

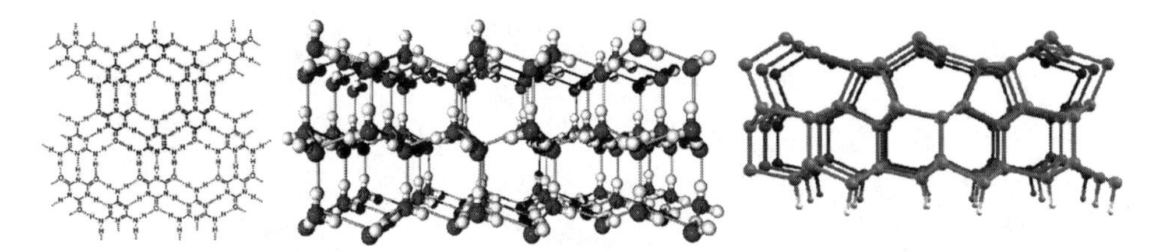

Figure 20: Crystal lattice Figure 35: Water structure Figure 36: Silica structure

If they grow this way and rotate, then you get a quartz crystal; just spirals of those tetrahedras in a different form. *Figure 20,* from a mineralogy book, shows a space, or a crystal, lattice. Compare to the Cladni plate images and you will see some where the water is vibrating (*figure 6*). Now look at *figures 35* and *36*. *Figure 35* shows the molecular structure of water, while *figure 36* shows molecular structure of silica.

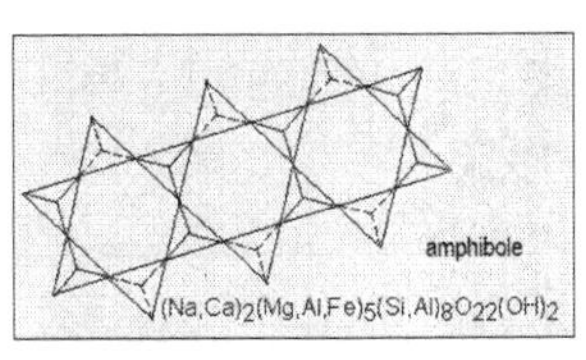

Figure 21: Amphibole

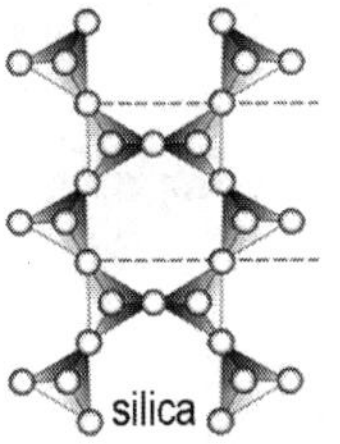

silica

Figure 22: Silica tetra

Basically, water is a living crystal with infinite inner surfaces that gather and transmit light. Looking at *figure 21*, we see an amphibole, the molecular star form of a volcanic silicate mineral. Mica can eventually arise from amphibole and produce the primary soil weathering cycle. *Figure 22* shows silica itself. We see chains of tetrahedral molecules rotating in the same direction, moving the light like channels of light streaming through the silica crystal.

Figure 23 (next page) shows an amethyst, known as a Brazil twin. One set of tetrahedras is going right, and as soon as that forms, another set goes

http://www.crystalinks.com/merkaba.html; accessed Feb. 2013).

Figure 23: Amethyst

left. They are twins. The gesture of the silica is to move light from one end to the other, spiraling along the crystal lattice, whereas the gesture of the amethyst is to go right and then left, right and then left. Looking at the way the silica crystal forms, we see that it shoots straight from the source as an image of light. The form of the amethyst, however, has a very different gesture. It is usually found in a geode. The crystal point is well formed, but the column is very short, with the clusters focused inward. The light does not stream through, but is held in the twin. Both the quartz and the amethyst arise from the tetrahedral molecule, but the differing molecular dynamics of the two can be seen in the overall growth structure of the finished crystals. The gesture of the inner structure of the molecular, tetrahedral forms gives us a picture of the dynamic that is present. There is a biodynamic preparation we can make using amethyst when we want the plants to grow one way rather than another.

A geode forms in magma. A gas bubble forms in the hot magma, the magma cools and contracts, and the gas is concentrated into a hollow space. The magma contracts and cools, and the gas creates a little blowhole to release the pressure. As the magma oscillates between cooling and heating, silicates streaming through the hot rock are injected in waves into the blowhole. The inside of the cavity eventually fills with a molten mass of very refined silicates in layers. The liquid or colloidal silicate in the old gas bubble solidifies, and eventually the molecular structure forms crystals on the inside. The metals in the silica colloid are influenced by the pressures and temperatures unique to this mineral form, creating the twinning process.

Esoterically, silica is considered a sense organ that enables the Earth to see into the cosmos. It is a sense organ for the Earth, looking toward the periphery and drawing in light wherever it is. It has an affinity for light. In *figure 24*, we have tourmaline, another silicate crystal form. Tourmaline is a massive action in the body of the Earth. For example, the whole mountain range from Sonoma Mountain through the basin of Santa Rosa was at one time offshore before the oceanic plates were sub-ducted under the continental plate.[12] When that happened, the minerals in the plate were pressurized. In pressurized rocks, water that seeps into cracks liquefies the silica, starting flows of what geologists call "rock milk." As these flows move through the rock mass, the solution encounters little particles that were broken up by the grinding plates. The solution flowing across the particles creates an eddy on one side of a particle. In that eddy, on the lee side of the flow, a long crystal forms. That is the dark mass in *figure 24* near the particle. A short rind of crystals forms on the upstream. Then that evolution settles, but the flow continues. The oscillation of the temperatures, velocities, and pressures of the flow leads to a reactive cycle in the crystal formation. In the presence of boron, there is a crystal form that grows into the upstream side, which balances the tourmaline crystal molecularly. In the figure, that is the gray area around the black area. The second evolution is then followed by a series of balanced flows that create the tourmaline crystal with its unique properties.

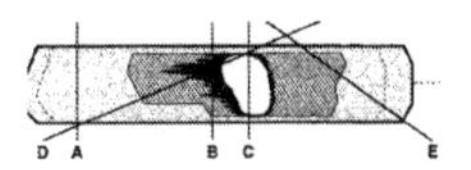

Figure 24: Tourmaline

This process is *tourmalinization,* which happens on a massive scale wherever rocks are under pressure. So sitting here we are in the midst of tremendous balancing forces of the tourmaline in the surrounding mountains. Tourmaline is a balancing silicate. Rock crystal is a streaming silicate. Amethyst is an enclosing or enfolding silicate. Three evolutions of silica as gem forms, but the form of the gem tells you the way it handles light. The forming process tells you something about the crystal and its potential for use as a tool for agriculture.

In *figure 25* (next page), we see a form known as a *phyllosilicate,* and now we are coming close to home. *Phyllo* means leaf. This is any silica that forms a leaf-like structure such as mica. The structure of mica draws in water between the surfaces of the layers. If the water is acidic from something that has rotted, it begins to leach potassium and other minerals out of the tetrahedra. The little tetrahedra above and below are the molecules of mica. The tetrahedra multiply the surface area inside the mineral. Because you have

12 The Sonoma Valley is part of the Coast Range Physiographic provence. See http://en.wikipedia.org/wiki/Sonoma_Valley; accessed Feb. 2013.

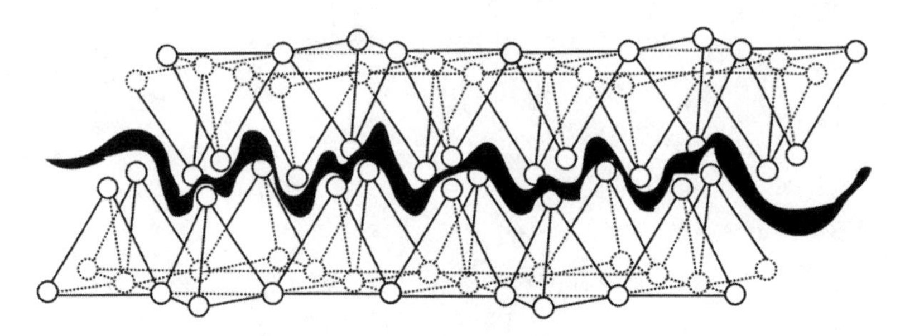

Figure 25: Phyllosilicate

tetrahedra hanging down and sticking up, the water circulates around much more surface area than it would over flat surfaces. Surfaces are the active energy areas in the world. Where surfaces interact with very weak acids, the acids leach out potassium, sodium, magnesium, phosphorus, all that plants need for growth in the soil solution. Thus, the tetrahedral form is the basis for silicates; it forms the active sections in layers and surfaces, which plants need to grow, and it is light that forms the structure.

Water gets in between the layers, drawn in by capillary action, and interacts with all these forms, and suddenly we have minerals in solution flowing through the soil and the plant itself, which, as we shall see, is made up of the same type of surface structures—xylem, phloem, cambium, pith rays, and so on. Those layers are the source of energy transfer in plants.

Figure 26 shows the form of capillaries in animal tissue; they form the hexagonal shape of a crystal because the body is designed to maximize the amount of blood that gets to each cell. These tissues have forms allowing for the maximum distribution and the most penetration of blood between cells. It is striking that, even in the animal body, the functional form is crystalline, a hexagonal form.

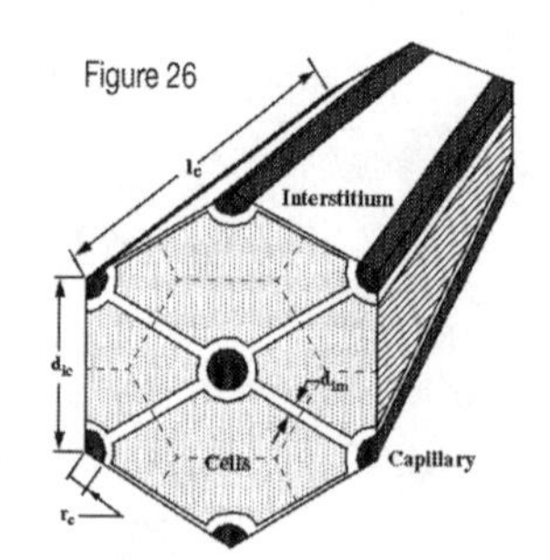

Figure 26

So we have gone from rocks through plants and minerals to animals, and it is the same picture—macro to micro. Where did that form come from? It comes from circles. Where did circles come from? They come from out there somewhere—the infinitely distant.

Figure 9 Soap bubble minimal surfaces

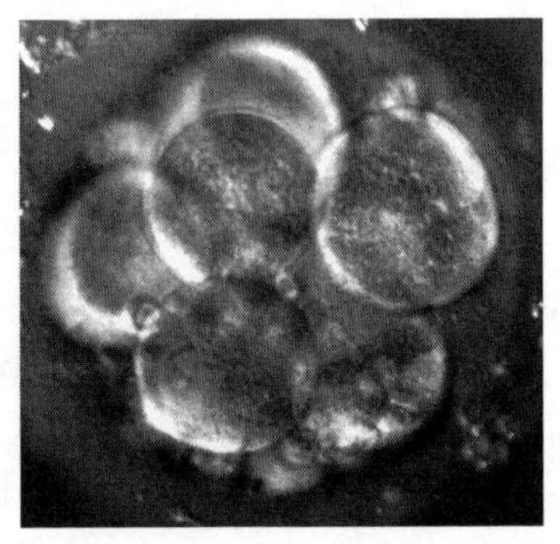

Figure 10: Morula

Considering again the law of minimal surface, look at *figure 9*. It shows a little cage made out of wire that somebody dipped in bubble solution. What we see is a cellular form infinitely repeated in plants. This form is common because of the law of minimal surface, as discussed earlier.

Think of a balloon. An inflated balloon has a certain amount of surface. As I keep blowing it up, the surface gets larger and larger and larger until the tension in the surface cannot be held together by the mass inside and the balloon ruptures. If I have a cell that begins to imbibe nutrients, it will swell larger and larger until it reaches a maximum size, about thirty times the normal size. Only one cell in the body does that—the egg cell. It can do this because the function of the egg cell, upon fertilization, is to break down inside into thirty little cells that will have the same volume as a body cell.

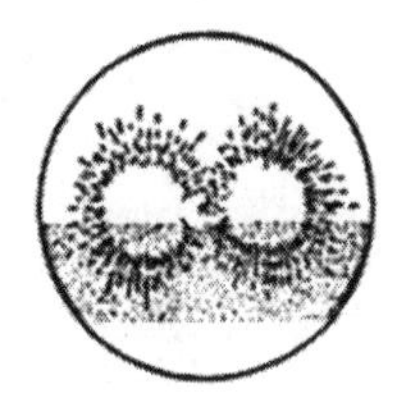

Figure 11: Cell division

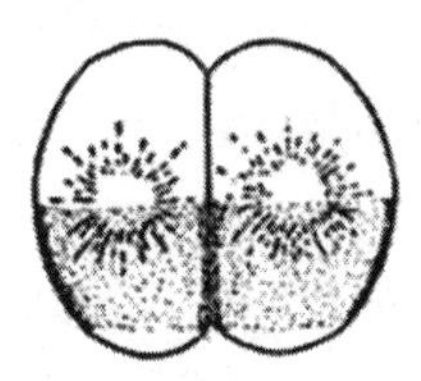

Figure 12: Cell with daughter cell

Figure 13: Daughter cell dividion

The only cell allowed to violate the law of minimal surface is the egg cell, and the reason is that the egg has to form something called the *morula* (*figure 10*). A cell has to maintain its integrity if we are to have life. A cell maintains its integrity by maintaining a certain size so that the cell surface does not siphon off too much energy from the cell mass. If the cell gets too big to maintain integrity, it splits. *Figure 11* shows a cell whose nucleus is dividing into two nuclei. *Figure 12* is an image of the cell that has created a "daughter" cell. *Figure 13* shows the next phase, in which the daughter cells also divide. Eventually, the egg divides into thirty or so cells. When the egg is fully divided it is called a *morula* (Latin for mulberry). The morula is pictured in *figure 10*. In the division of the egg, one cell divides and the protoplasm in that cell is still too large, so the cell divides again and again to establish integrity.

In regular cell division, a body cell becomes so large that the relationship between the surface and the mass is threatened. When the surface of a growing cell becomes too large in relation to the mass, the cell starts interacting with the environment too much and cannot maintain its integrity as a body cell. Therefore, the cell forms a membrane between the two parts that start to divide at the nuclear level (*figure 12*). The need for the cell to remain at a certain size is the law of minimal surface.[13]

The implications of the law of minimal surface go far beyond embryology, although those implications are very evident in embryology. *Figure 15* shows a human embryo at twenty-one days, in which you see a kind of saddle shape. *Figure 14*, from a website on geometric forms, is a picture of what's known as a "saddle." It is about the geometry of surfaces known as *topography*. The idea in terms of the law of minimal surface is that curved surfaces in nature have unique properties. An organism forming a curved surface can begin to control the exchange of energy between itself and the environment. The ideal form in nature is a sphere because every spot in the sphere is at the same elevation from every other spot on a sphere. This idea comes from a discipline called topology, which is used to design machinery and the most efficient surfaces for maintaining energy flow. In topography, I

13 While a sphere is a "minimal surface" in the sense that it minimizes the surface-area-to-volume ratio, it does not qualify as a minimal surface in the sense used by mathematicians. For a detailed explanation of this concept, see http://en.wikipedia.org/wiki/Minimal_surface (accessed Jan. 2013).

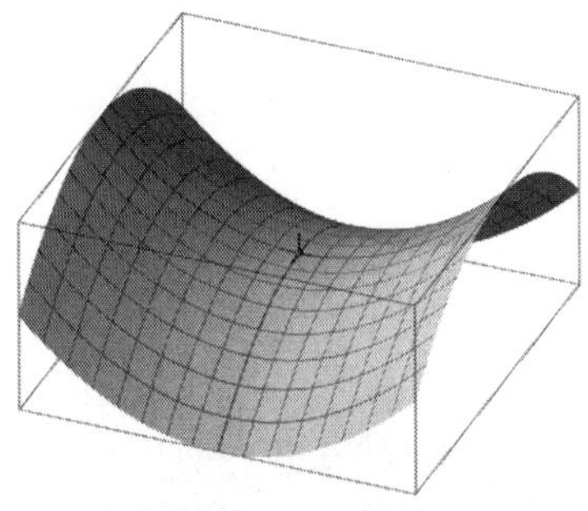

Figure 14: Saddle form geometry

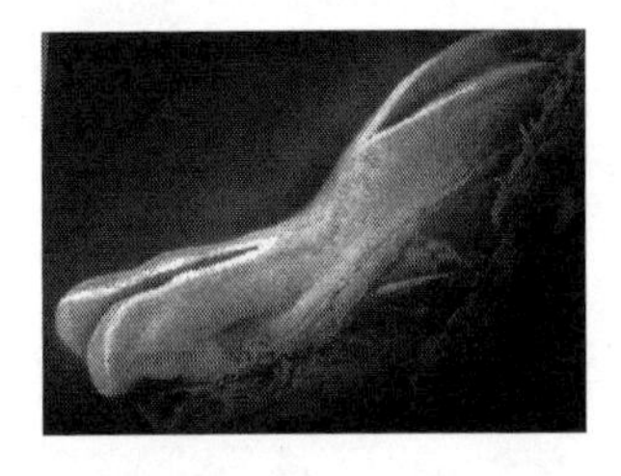

Figure 15: Human embryo, 21 days

can go all the way around a sphere and expend the same amount of energy as I would on a flat plane. In that world, a flat plane and the surface of a sphere are the same thing.

A growing organism begins to move away from the spherical shape; it is most efficient at holding energy, but it must adhere to the law of minimal surface. Topographically, a sphere that forms a membrane needs to create a transitional form that obeys the law of minimal surface. The saddle shape is the form in nature that can make such a transition from sphere to plane and still hold on to minimal surface. The saddle form allows an organism to maintain integrity while beginning to proliferate. If an organ has a saddle shape or series of saddle shapes, the law of minimal surface is involved. The organism will form saddles to hold on to a spherical form.

Figure 16 shows a highly evolved topographic form. A computer has generated many saddle forms to create a form that could become a construction girder. The repetition of many saddles leads to a form composed of perforations. Saddle perforations make structures light but strong. Saddles are a way

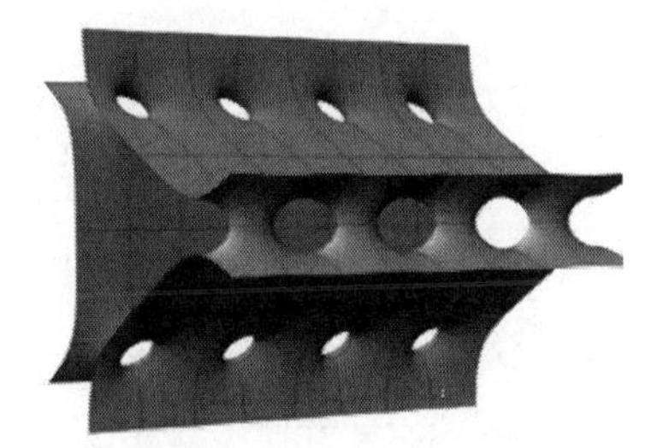

Figure 16: Computer-generated saddle form

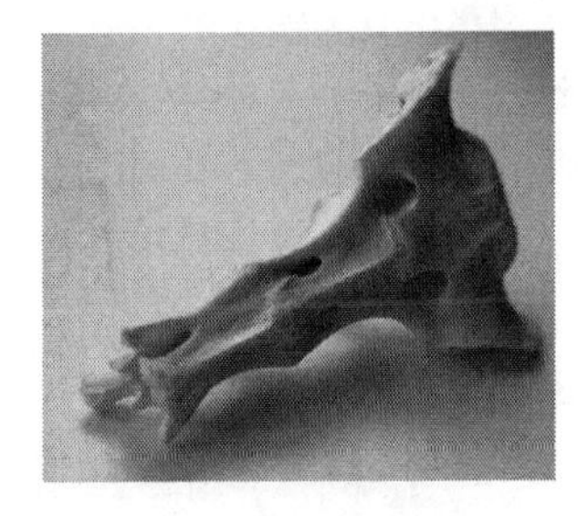

Figure 17: Rabbit pelvis

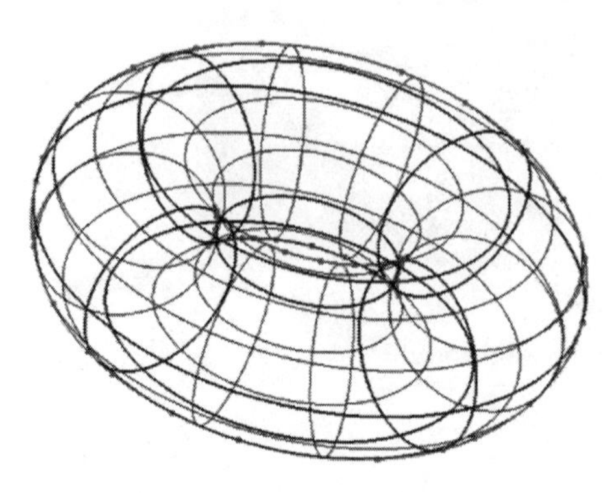
Figure 18: Torus

of building things that are complex but maintain the spherical archetype of the law of minimal surface. While *figure 16* could be an engineering plan for an airplane girder, *figure 17* shows a close-up of a rabbit pelvis. The correspondence is striking.

Finally, *figure 18* shows a *torus,* in which every part of the surface is the same altitude from every other part of the surface. A torus is a donut shape and the way nature makes holes in something (a pelvis, for example) that needs to be strong but light. I can perforate a bone with a torus to make a hole in it that still adheres to the law of minimal surface. In a toroidal donut shape, all the sides of the hole are circular in cross section. This can be seen in *figure 18.* In a torus, every space on the surface is still within the law of minimal surface; the form can become larger without sacrificing integrity. On a torus, we can go in any direction on any surface and, as on a sphere or flat plane, not change elevation. We can go in and out and around a torus and always move on an unchanging surface. This is the strange world of mathematical topography.

Therefore, a torus, along with a sphere and flat plane, is another form that exemplifies the law of minimal surface. This idea is fundamental in the formal language of nature. It also helps us understand the etheric, life forces in natural forms. Minimal surfaces dictate the way animals and plants build organs.

This brings us to *capillary action.* The law of minimal surface points to the fact that growing things must economize, since the fluids within them must also adhere to laws of the living world. In a sense, dead minerals must become living minerals. In a plant, the fluids are minerals that have been enlivened. Science recognizes them as living crystals. Proteins in nature are actually crystalline, as are acids. The border between the living and the dead is sometimes blurred. In this context, it is useful to recall Rudolf Steiner's comment that soil is a child of the Sun. This is why he felt that the silicates are so important; they support life because they have infinite inner surfaces.

We have seen that water has infinite inner surfaces. When a plant interacts with water in the soil, the surfaces within the plant react with the surfaces in the water, which react with the surfaces in the mineral, which in turn allow the exchange of nutrients in the plant. Functionally, root hairs, or *mycorrhizae*, and all the stuff of agricultural science are really about how many surfaces can exist in a very small area. The law of minimal surface has to do with capillary action, without which there could be no life.

As spaces in nature become smaller and smaller and smaller, strange things begin to happen. If I have a sphere the size of a lemon, the surface is two-dimensional and the mass is three-dimensional. However, if the sphere is made continually smaller, at a certain point the sphere becomes so small that the rate of surface reduction cannot keep pace with the volume's rate of reduction. There is a certain point at which the sphere becomes smaller in mass than it does in surface. At the beginning, the mass is much larger than the surface. Paradoxically, at this scale, the surface increases. This phenomenon describes the action of vapor, as well as molecular structures, lipids, and minerals in soil solutions. These things are so small that they live in another world, moving from earthly reality to the cosmic; a solution lives in another dimension. A living solution in a plant lives in the cosmos—it is almost not present here. These strange phenomena form the basis of homeopathic dilutions and the ways we stir biodynamic preparations in water. This explains the importance of the law of minimal surface; it is truly important to understand that the smaller the object, the larger the surface.

It may seem completely counterintuitive, but this is what makes life happen. Life itself is counterintuitive. From the view of physical laws on this side of the great divide, life does not make sense. Fundamentally, from Steiner's perspective in biodynamics, life is invisible and primary, while substance falls out of it into visibility and becomes secondary. This is just the opposite of how we would view this issue in terms of organic chemistry, in which substances create life. Steiner's picture shows that life is in the light and then falls out as substance as an image of the light process. We can see how the law of minimal surface is the basis for homeopathy, but it is also the basis for how plants can take up extremely diluted solutions to become alive in those plants. When surfaces come together, they create a kind of inner force by which the structure in the water is strongly attracted to the surfaces; this is capillary action.

Through capillary activity, the subtle attraction causing flow through these minimal surfaces exerts great activity in the plant world. Plant growth is characterized by the activity of life forces moving from the soil up toward the Sun. When life moves from the soil up and into the Sun, light can interact with it, and then the chemistry—the life chemistry that minerals have created—rises into the realm of the Sun. The Sun's energy acts through the plant organs, which are images of all that I have been describing.

IV. Ethers and Elements • Polarities of Potential and Manifestation • Humus–Clay Relationships • the Living Mineral–Plant • Chlorophyll Structures • Alchemy of Ethers Falling into Elements

I begin with an excerpt from *Extending Practical Medicine,* by Rudolf Steiner and Ita Wegman,[14] because it describes how etheric forces operate in life. He explains the form-giving process, or *astralization,* the first step of which involves the separation of the lower level from a higher level. He describes the process of separation of a being coming from the cosmos to Earth. Esoterically, this is the process of the separating the etheric life force from an element, the manifest form of a life force and the corpse of an ether.

14 Steiner and Wegman, *Extending Practical Medicine: Fundamental Principles Based on the Science of the Spirit* (CW 27).

People in biodynamic circles often confuse elements and ethers, speaking about them if they were the same thing; in fact, they are almost opposites. This excerpt describes the process of separating the ether from an element.

> In the plant we have material substance transformed by the forces raying inward to Earth. This is the living substance. It is continually interacting with the lifeless [matter]. We must conceive that, in the plant, living substance is perpetually being separated out of the lifeless. It is in the living substance that the plant form then appears as a product of the forces raying in toward the Earth. Thus, we have one stream of substance. Lifeless substance transforms itself into living; living transforms itself into lifeless. In this stream, the plant-like organs come into being. (p. 28)

He is speaking about the macrocosmic forces, of which levity is another force, along with other concepts such as potential, plane, periphery, process, and imponderable. Levity is another way to describe the macro side of the realm of forces.

The life of the plant retains the potential for sugars to form in a levity state. The carbon, hydrogen, and oxygen are kept in levity state within the sap through the action of the life of the plant and the capacity for that life to interact in chemical affinities. The life ether and the chemistry, or tone, ether are signatures of the highest ethers. The substances of the plant are embedded in the sap solution as the potential patterns by which sugars manifest in the plant. However, these substances are still energetically engaged in the life of the plant. They are present as sentient or sensitive potentialities for attraction and repulsion. It is not yet sugar, but in a sense it is designated sugar. It is in a life-light form that will eventually fall into what we call sugar. The ethers that animate the life and chemistry separate from the sap; the substances fall into manifestation. The sugar exists *in potentia* while still in the life of the plant. The life, the activity of light, and what he calls the chemical ether (or what I have described as the music of the spheres) is the potential for particular intervals; C_6-H_{12}-O_6 (the basic formula for carbohydrates) is a melody. The chemical structure of sugar is a melody and represents a kind of potential of intervals and nodes that, when the life leaves, it will fall into what we call sugar.

Nevertheless, it is already present in the solution as potential. That is the first part; that is separating out from the ether of an element. The element would be C_6-H_{12}-O_6, or carbon–hydrogen–oxygen, and the ether is the tone ether, or what Steiner calls the "mathematical ether." This is the mathematizing principle, or *mathesis,* of the cosmos. It is the geometry of potentialities, the unsounded melody present in potential that must pull away from the cosmic pole and fall into substance. Then we get something like fructose. This is the first part of the process: the separation of the higher from the lower.

The second part is building form from separated substances. Once the carbon, oxygen, and hydrogen have separated from the ethers, they are like notes of a melody that is no longer being played; the melody remains potential and has not yet manifest until the sugar falls into a certain ratio of C-H-O. That unsounded melody is still back on the periphery as the soundless potential of light, while the sounded, or manifest, pattern has resulted in sugar. In the plant, this process involves water and light. The formula for photosynthesis is 6 CO_2 + 6 H_2O, which in the presence of sunlight yields $C_6H_{12}O_6$ + 6 O_2 (carbon dioxide + water + light energy → glucose + oxygen). Light is needed for photosynthesis, but the light itself is also organized into ethers according to Rudolf Steiner: life ethers, tone ethers, light ethers, warmth ethers. The cosmic light is already organized into relationships, but there is not yet any manifestation. The valances, bonding angles, and molecular patterns of substances are fallen corpses of what was once a living, ordered, and unmanifest energetic entity, cosmic light.

Consequently, people say that Rudolf Steiner does not recognize atoms. In fact, he does use the word *molecules,* but for him the molecule is just an image of this ordering principle of the light. Today, the molecule is a kind of god, because many do not recognize the Great Creator. We can think of the forces animating a good friend in contrast to the substances of carbon, oxygen, and hydrogen in the tissues of our good friend's body. Which part of your friend elicits your friendship—the animating soul warmth and inner motion of the soul or the substances of the body? That is the dimension that Steiner addresses with these ideas of the etheric. It is in the etheric realm, the realm of potential, that we experience the primary order of things.

The warmth ether is the closest to its element of fire. The warmth can be thought of as a cosmic source of the enthusiasm that one thing has for another. We can imagine warmth as the enthusiasm that an animal has for a particular fodder, or a plant has for a particular substrate or a particular climatic niche. Perhaps a particular cultivar has an affinity for a particular acidity in soil structure and has to have a certain slope and exposure to light and warmth. You would call this enthusiasm of the cultivar by the term *terroir* if you are a wine grower. *Terroir* names a concept of the blending of time and space—a melody of time and space that eventually manifests as something. That thing then manifests sugars, but also fragrances, esters, oils, and color. Some people even try to do experiments in which they grind up some granite, drop it in a barrel of wine, and then say, Oh, I can taste the *terroir* of the granite. However, I do not think that this is the way it works.

Esoterically, *terroir* is an energetic entity, or being, with a particular kind of intelligence; an archangel. That archangel is composed of the enthusiasm for a time and place that says all these things go together in this particular way, and the plants living in that *terroir* accept the light in a certain way and move it in a certain direction, which becomes a signature of that place and time in the final substance. Some people say that *terroir* does not exist, because you cannot prove that a certain mineral is present in so many parts per million in the final product. Such an attitude is materialism squared. In reality it is the whole plant weaving through time and space that makes up *terroir,* which is not just minerals; it is the enthusiasm for life of a living entity embedded in a particular web of life. It is living force and enthusiasm for being—in a word, *warmth.*

Nevertheless, there are other ethers besides warmth. There is the light ether, which could be thought of as the sensitivity and intelligence that guides the enthusiasm. Then there is the tone, or chemical, ether, which reveals the sublime notes of the celestial wisdom of relationships that Rudolf Steiner has also called "number ether." Finally, there is the combination of warmth, light, and tone ethers into the highest ether of all—the ether of life itself. Life is in the ether, in the cosmos, in the potential. However, that does not mean the potential of the ether realm is random. It is highly ordered. For instance, tone ether is the ability in the cosmos for one thing to be related to another in a certain proportion and ratio. It could even be described as distance and scale, because it is the basis of our musical system. Distance between one note and another is a pitch or note in the cosmic choir. This means that the distance between Saturn and Jupiter is a pitch. It is an "unsounded sound." If I have the organ of a plant or animal, even if I take the organ out of the animal, the form of the organ is still connected to the related melody.

This is the idea behind the bladder and intestines and the mesenteries used in biodynamic preparations. The cosmic melody that manifests in the form of the organ still carries the potential of the particular ether combinations that were active in the life forces of that organ. If I take that organ and insert another substance in which there is a synergistic effect in the realm of form, I can enhance the activity of the substance. The task is to find forms of plants and forms of organs of animals or minerals that share the resonances of similar form motifs. To find those resonances, I have to ask myself this: How did that plant come into being? What process is active in the plant? What minerals does it mobilize? Can I find a harmonic of that mobilization force in what that animal organ does for the animal?

Is there a resonance between the becoming of the animal organ and the becoming of the plant substance itself in the ecosystem in which the plant lives? If I find that resonance, then the resonant structures of the shared formal motifs allow me to say there will be synergy, because the life forces that form both of them are harmonic. Then I get amplification of the potential forces. That is life ether in action. We can analyze the substance that comes out of there and say it is just still manure. However, its activity is different. This is the key. That activity shows up as qualities—keeping qualities or tasting qualities in the final product. This type of imagination of resonant formal motifs extends even to the ability over time of entering the plant at a different level rhythmically, so that we get the motion of the Moon through the star Moon and the sidereal, or star, Moon and the Moon phases, and all these rhythms begin to speak a formal language as though a piece of music. Even the rhythms of putting an organ into a compost heap and then removing it is a big deal in terms of harmonic, or resonant, form motifs.

When you pack a cow horn with manure, it is no matter of indifference how long you keep it buried. The burial rhythm has to do with the Sun. It is a forty-day solar rhythm that alchemists understood. They called it an alchemical month. All of these things that are fundamental for biodynamics seem pointless to materialistic science. In fact, behind these practices is very clear scientific thinking—not so much about the molecular action of substance, but about how those substances come into being from the cosmos. This is where we need the faculty of imagination to see this system of agriculture as a science. In biodynamics, we do not just ask what a substance is, but about how it came into being from a spiritual existence. The appearance of substances is generated, regulated, and harmonized by the energetic form, or what I have been calling "the form motif" in which the substance appears. The surfaces within forms are especially important, because the structure of surfaces in living forms is where the interactions between one side and the other side of the equation occur. In nature, forces always interact through some form of surface.

The great agent for surface in the mineral realm is silica. The great agent for surface in the plant realm is water interacting with minerals. In the animal realm, it is the formation of tissues and sheaths. In an animal or plant, it is the layers of various tissues that allow organs to develop. From the perspective of formal motifs, the layering and organizing of selected tissues in a plant organ is equivalent to mica formation in a granite face. The mineralogy, chemistry, and structure of the plant organ, as well as the form of the mineral process in the native rock, are resonant in their formal language.

Here, I am merely summarizing what has been discussed so far. The form is a picture of how a substance comes into being. When we train ourselves to perceive form, we are provided a doorway into a way of working with the natural order to enhance the metabolism of a plant or retard it to enhance ripening. You get imaginations from the resonant formal motifs on how to see more deeply into the activities of the natural world.

When we have to deal with this realm of nature imaginations, we encounter statements that Rudolf Steiner makes in the agricultural course: "The fragrance that spreads around out there in the aromatic meadow of

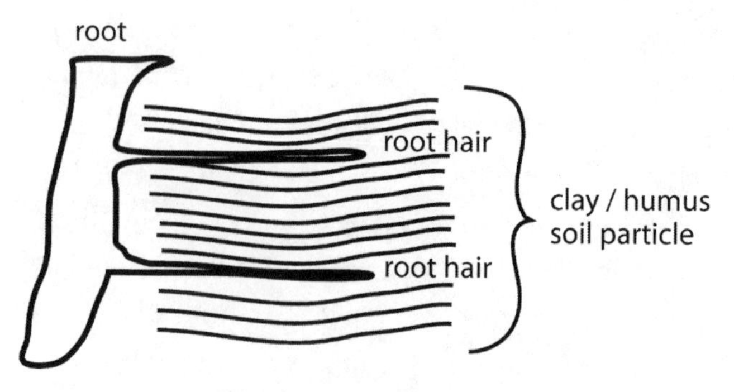

Figure 3: Cation exchange

wildflowers is what works on the plants from the outside. The fragrance in the meadow works on the plants from the outside" (p. 67). You may think that the fragrance comes *from* the flower, but Steiner would say that the fragrance *visits* the flower, that it visits in the light.

How should we understand this? Steiner seems to be describing the way a plant can take up the light because light is the means for the substances manifest in the plant. A fragrance actually comes to the flower as a process of the plant ingesting sunlight. A plant creates organs that scientists call antenna pigment complexes, which draw in particular frequencies of light and reject others. This form arises, for example, in the arrangement of pigments in the leaf. Combinations of the plant's colors, waxes, and oils allow it to isolate certain frequencies that it needs from the sunlight. Within the plant itself, the hierarchy of the way sunlight is moved through the plant is quite complex, because the sunlight creates a kind of energetic potential and substances move into the flow.

Figure 3 depicts cation exchange. On the left is a plant root with the minerals being absorbed through the plant's root hairs. On the right is a picture of a clay crystal and a bit of humus, which is a colloid. A colloid is a substance in nature from which life originates. In nature, secretions that support life have a colloidal nature. Protoplasm in a cell is colloidal. A colloid is hydrophilic, meaning it has a deep affinity to water. If you place a colloid near water, it will absorb the water because a colloid has infinite inner surfaces. Humus is a colloid and has a molecular structure that is layer after layer of acids in chains organized so that water is drawn between the colloidal layers and imbibed by the substances on either side. When colloids absorb water, they swell. Jell-O, for example, is a colloid made from the skin and bones of pigs. In a pig, the connective tissue starts out as blood and then forms a colloid as gristle, or connective tissue. It goes from a fluid to a solid, but colloids can also go from a solid to liquid. Anything that is gel in one form, then dries out, and becomes hard (such as a fingernail) is colloidal. Any connective tissue such as a tendon, ligament, or cartilage (such as in an ear or nose) is a colloid. Anything that starts out as a kind of mucus that then solidifies into a mineral structure is colloidal. The mineral form of the colloid is composed of inner structures that resemble the layers of a crystal.

Being a colloid, humus is the perfect form for plant roots that have fine hairs on them to imbibe water. Humus presents a perfect structure to the root hairs. The hairs grow along the microscopic, crystalline planes of the colloid, whereby they have access to the maximum surface area that interacts with the tetrahedral plane structures of the phyllosilicates that create soil solutions.[15] The inner planes of the phyllosilicates interact with the plane like structures of the clay/humus complex colloids and provide optimal surface area for the spreading of the root hair. Thus, the root hairs go along between the layers in the humus, which is formed in the context of the layers of the clay particles in the soil. The clay particles in the soil and the humus are resonant in their forms.

Figure 3 shows a diagram of a clay particle. It is a flat sheet. The humus particle is also a flat sheet. On the left are the root hairs seeking those flat sheets by which to spread the surface area of the root hair. The more surface area the root hairs have, the more drought tolerance the plant has. This is *surface in action,* because in nature the energy in a system is carried only at the surface, not in the mass. It is the same in electronics. If we have a wire and want to send a current along the wire, the current will flow only along the surface of that wire. Researchers have drilled holes into big thick wire, inserted probes into the center of the wire, and insulated the probe. When a

15 *Phyllosilicates* are sheet silicate minerals, formed by parallel sheets of silicate tetrahedra with Si_2O_5, or a 2:5 ratio.

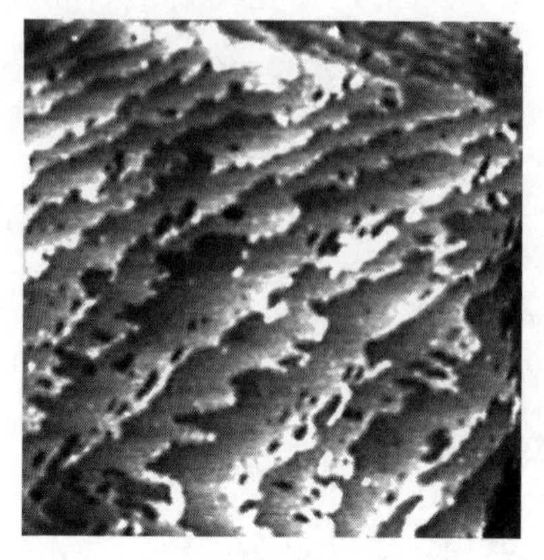

Figure 4: Microphotograph of mica

current is applied, no current is moving through the center of the wire. However, a probe on the surface of the wire will register electrical flow. Therefore, if we want to send a lot of current through a wire, we do not use a big, fat wire but make a cable woven out of many fine wires, each one insulated and carrying a current. In this way, we get strong voltage at the end. If we use a big, fat wire, the voltage will fall off halfway along the wire.

Surface is where the action is, whether electrical wires, the car battery plates, or a capacitor in a circuit. In the old days, mica was the capacitor used in radio circuitry. A capacitor regulates the current flowing through the circuit and can be used to accumulate voltage on the outside of the capacitor—a *charge*. This layered form is present in nature where charges need to be held and strengthened.

This is the function of humus, and the charge to be held and strengthened is a *life charge*. When humus is full of water, the soil is charged with life. A plant sticks root hairs into the capacitor of the humus, which discharges life into the plant. This is called growing. The flow of electricity is not life, of course, but only a shadow of life. This is cation exchange of what we call the clay humus complex.

Figure 4 is a microphotograph of the structure of mica. It looks like a layered Greek or French pastry. Mica results from a growth process in the mineral body of the Earth that Rudolf Steiner calls the remnants of a time when the whole Earth body was alive like a plant. In his view, the whole Earth body, the mineral body of the Earth was alive like a plant. The micaceous elements at that time held the potential for what later evolved as the leaf structure of plants. That was a time when the whole Earth body was alive according to Steiner's picture. The minerals themselves were alive. They were plastic and alive, because the plants had not yet formed into what we call plants today. They still existed as huge single-celled plant-like organisms. In the fossil records, the stems of these huge plants are called *stromatolites*. Rudolf Steiner calls that time in evolution the formation of the "world mineral plant." Subsequently, there was an articulation, or separation, of the minerals and plants. However, in those early times, all the forms that would eventually become plants were held in the mineral realm, which had life somewhat like the plants of today.

Then the plants separated; the life part of the plant separated, and the minerals fell as a kind of corpse. When the minerals fell into a corpse, they kept the structure of what they were in potential. Therefore, what we call the leaf-formative principle left an imprint in the mineral realm of what it used to be when it was alive in the mineral, or crystalline, realm. Today, we call that imprint mica or, more technically, the mica process. Mica is a constituent of granite. That imagination of when the Earth was in a more latent, or potential, condition is what we can now see when we look at the form of the mineral mica embedded in a rock face. The form of the mineral mica has the potential to generate the clay humus complex.

Mica has a tendency to generate the formation of root hairs. This has not gone unnoticed by industry. They mine mica, chop it into little bits, and put it in an oven to bake. After it puffs up like puffed wheat, they sell it in bags as vermiculite. The function of vermiculite in a planting mix is to increase the surface area so root hairs can have more action in the soil. Thus you can see that substance mica today is a remnant of the time when the plants were living in the minerals but now has fallen. That is a fundamental picture in thinking about biodynamics. The plant nature of this layering structure of mica eventually becomes the plate-like structure of clay, and clay is the agent of cation exchange.

Clay structure comes from the mica nature in rock, but it gets tumbled about and weathered in the water. It interacts with the acids of matter that has decayed and softened. The plate-like particles become finer. The finer the particles, the more their mass and surface begin to change relationship according to the law of minimal surface. The finer the clay particles, the more potential they have to work in the levity pole—that is, the more potential they have to be active in exchange, because the surface becomes huge

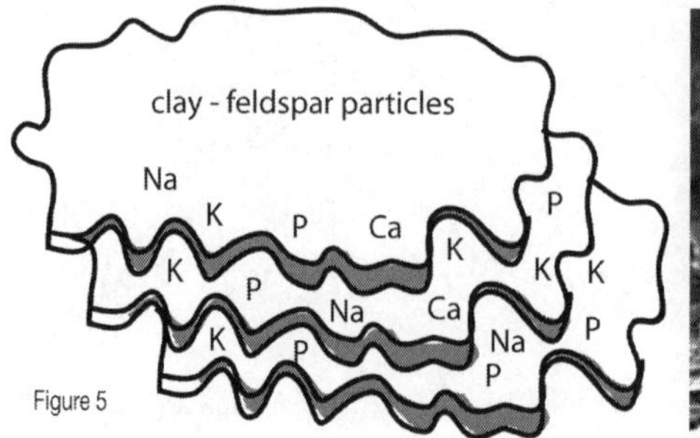

Figure 5: Clay and humus interaction

Figure 6: Clay particles

relative to the mass. Thus, there is a process you can use in working with clay to create a substance called *terra sigillata,* which is what the Romans used to make glazes for their signature black and red ware.[16] Those beautiful black and red glazes on Roman pottery are actually just clay.

Through a special process only the finest particles of clay are taken off. It is like activated clay. We do some experimental sprays with *terra sigillata.* It is a very interesting substance. Because of the refining process, it is clay to the nth degree. A solution of *terra sigillata* is filled with nanoparticles of clay with huge surfaces in relation to their mass. *Figure* 6 shows the clay particles. Look at them and then look at the "pastry" structure of the mica (*figure* 4). They are very close to each other, because they come from the same weathering process. Moreover, *figure 5* is a rendition of the way the clay particles (again, as in *figure* 4) and humus particles interact: through surfaces.

Plant roots eventually penetrate what lives in the soil and the soil solution flowing between the phyllosilicates. The plant roots take up the soil solution and lift it through the action of the clay, into the air and sunlight. Forms in the plants receive and assimilate the planes of light from the Sun. Moreover, it should not surprise us that the forms assimilating these planes of light are themselves planar. Antenna theory states that the antenna should be an image of the signal it is designed to receive.

In the plant depicted in *figure 7,* we see a stained microphotograph of chlorophyll cells. They appear similar to mica and clay. These stacked, layered planes represent the preferred forms for receiving and transmitting

16 For more details on this, see http://en.wikipedia.org/wiki/Terra_sigillata.

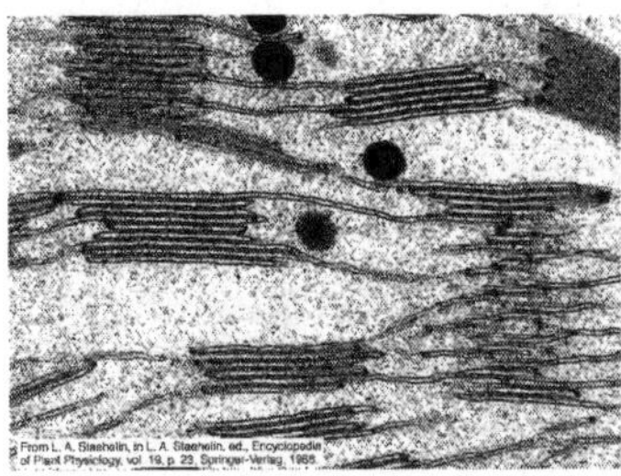

Figure 7: Chlorophyll cells

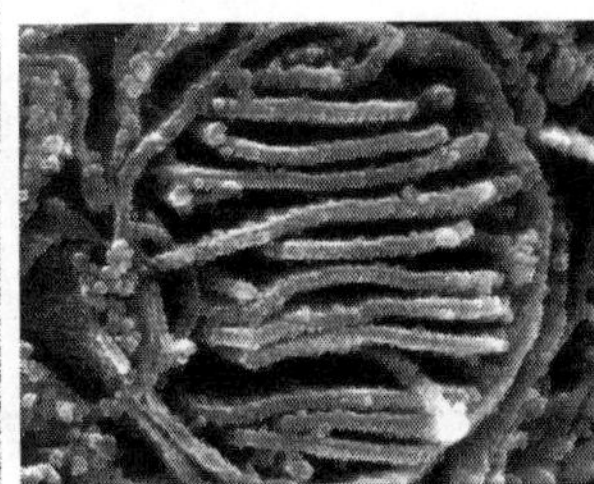

Figure 8: Grana

Figure 9: Starch grains

forces in the natural world. The fine layers and surfaces laid one upon the other allow great fields of force to build around the layered structures. Like a car battery, with its many small metal plates that interact with the electrolyte, the structural form motif of very thin plates interacting with a fluid solution is a fundamental pattern in physical science for structures designed to hold energy. This is also the form of the chlorophyll in leaf structures. The solutions carried in this flow through chlorophyll cells come from the weathered minerals and are the basis for photosynthesis in the presence of sunlight.

Chlorophyll cells are organized into larger structures, or units, in the leaf structure known as *grana,*[17] shown in *figure 8*. We see layer after layer and surface upon surface in the energetic formation of the plant. Sunlight enters and hits the chlorophyll, after which photosynthesis takes place and produces starch. *Figure 9* shows starch grains, which look like the chlorophyll cells, while grana look like clay particles. They are not little round balls but flat surfaces. All of this allows interaction, resonance, and harmony.

The starch grains are stored and moved through solutions in the plant to create the energy compounds needed by the plant for its life processes. This takes place primarily in the leaf. *Figure 10* (next page) shows a cross section of a leaf *primordium,* and the little black things are the primordia of the leaf around the *meristem.*[18] The meristem is made up of little layers of tissues,

17 *Grana* are the dense, green, chlorophyll-containing bodies in the chloroplasts of plant cells.

18 A *primordium* is an organ or tissue in its earliest recognizable stage of development. Cells of the primordium are called primordial cells. A *meristem* is the tissue in most plants consisting of undifferentiated cells (meristematic cells) found in zones of the plant where growth can take place.

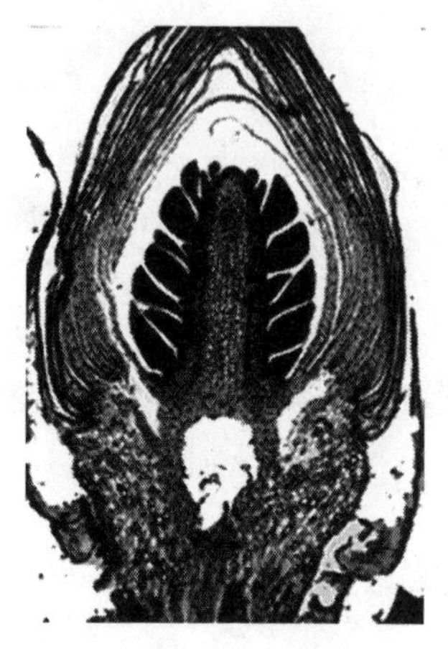
Figure 10: Leaf primordium

and the primordia come out of that. Surrounding it is the bud scale, which is just many more layers—many surfaces gathered around a center.

Now when I take surfaces and gather them around a center, those surfaces interact with the peripheral forces, which build up and create potentialities within the area within the surfaces. There was a time when people played what they called a kissing game. This was in the 1700s, when people began to play with static electricity. They would take a large blown-glass ball, put it on a spindle, and take it to the country fairs. They would find a really pretty young girl and have her sit in a kissing booth, and people would pay a quarter (a large amount in those days) to kiss the girl. Anyone who could kiss the girl's lips for more than five or ten seconds would win a prize. What those people did not know is that the girl had her hand on a silk cord, which went to the back of the little booth and someone with a glass ball and a piece of fur. A man would turn the ball on the spindle while pressing the fur against its spinning surface to build up a charge of static electricity. Like a braided transmission wire, the fur had many little surfaces that accentuated the charge picked up on the surface of the ball. The ball was attached to the silk cord that the girl was holding. As soon as a man got three inches from her lips, a spark would shoot off of her nose and knock him away. Everyone laughed and there went the quarter.

This illustrates what happens when we have numerous layers. Fur is made of many fine protein wires, and the glass ball acts as a perfect reservoir for static electricity. A silk cord woven from many silk threads is a perfect conductor of static electricity. It would not hurt the girl, but the person who came up to the girl got a shock. This is a picture of surfaces and their forces in nature. A sphere is the ideal form in nature for holding a charge, because there is no point to discharge the accumulated charge; think of the torus, which I described earlier. A sphere in nature could be anything from a plant bud to placenta or the layered hair of a cow horn.

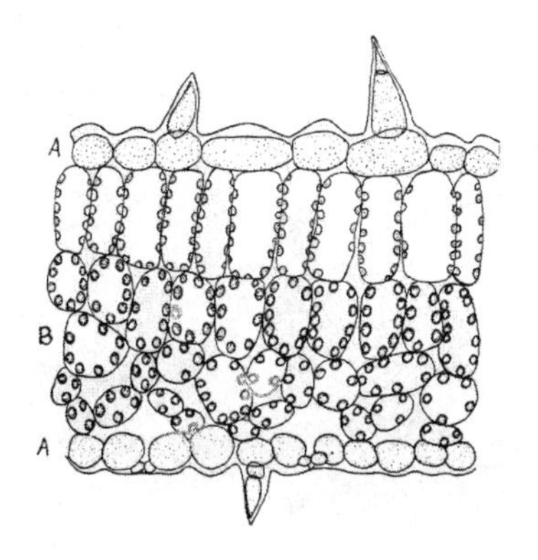

Figure 11: Leaf cross section

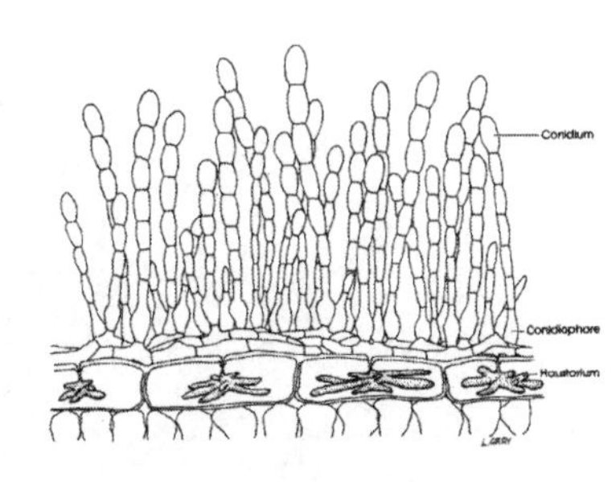

Figure 12: Powdery mildew

Fundamentally, the layering of surfaces in nature allows organisms to hold energy. That is what nature is trying to do with the plants by creating manifold surfaces within the structure of a plant. Looking at *figure 11,* we see a leaf cross section, a structure with the epidermis at the top and stoma at the bottom. The little round things are the grana. We can see that the leaf is layered with forms that either pack close together with one another or are round. The leaf consists of layers, and when light enters it is spread out through all the various structures. Each layer has its own way of absorbing the various wavelengths. When the proper amount of light hits a surface for that particular light form, it is drawn in; the chemistry of the plant is built around this economy. Note, too, that this organization does not go unnoticed by the parasites in the natural world.

Figure 12 shows a powdery mildew. The bottom structure shown is actually the epidermis of the leaf. The top structure in *figure 11* is the bottom structure shown in *figure 12.* Powdery mildew forms little spore capsules, a rod composed of the round balls coming out of the top of the epidermis. At the bottom of the rod is a little foot that penetrates the leaf along the surface of the epidermis, where the mildew draws vitality from the leaf. Many parasites have a similar plan for penetrating the surface and spreading sub-dermally just below the surface. For parasites know that surfaces are the places in nature where there are intense energy exchanges, so they adopt strategies to engage the surface energies as part

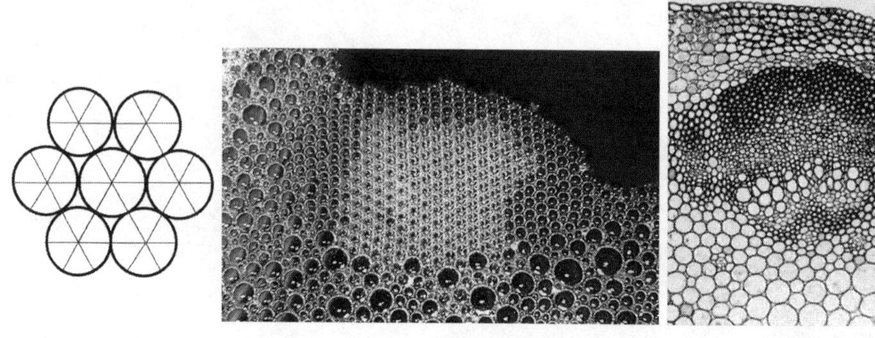

Figure 13: Six-circle hexagon Figure 14: Soap bubbles Figure 15: Plant vascular bundles

of their own life cycles. Thus, if I want a picture of why surfaces are so important, what can I examine?

Figure 13 shows the archetype of the action of surfaces. All of those circles are the same size. Now the great magic of the Sun and its circular form—in fact, all of the planets with their circular form—is that the circle contains a magical principle. The magic is the possibility of drawing a circle with a compass, and then I can draw that exact same sized circle around the perimeter of the original circle exactly six times. The compass will bring you back exactly to the same point where you started. This seems improbable, but it is the basis for everything I have talked about today. If I connect the centers of those six circles, the form of a hexagon appears. Furthermore, if I pull one of those hexagons up and out into three dimensions, the circles become spheres and I get a tetrahedron when I connect their centers. In two dimensions, the form looks like *figure 13,* but the same thing takes place in three dimensions with spheres, and we end up with close pack, or cellular forms. The magical principle of the circle describes the idea of surfaces, circles, and the geometry of the spheres.

If we want to see close pack in a form, look at soap foam in the bottom of a sink (*figure 14*) after draining the water. Close pack is the structure of foam. *Figure 15* is pretty much a picture of a close pack, showing a cross section of xylem and phloem in a plant cell. The circular form in close pack allows maximum interaction of surfaces. Fluids are thus distributed through the inner surfaces.

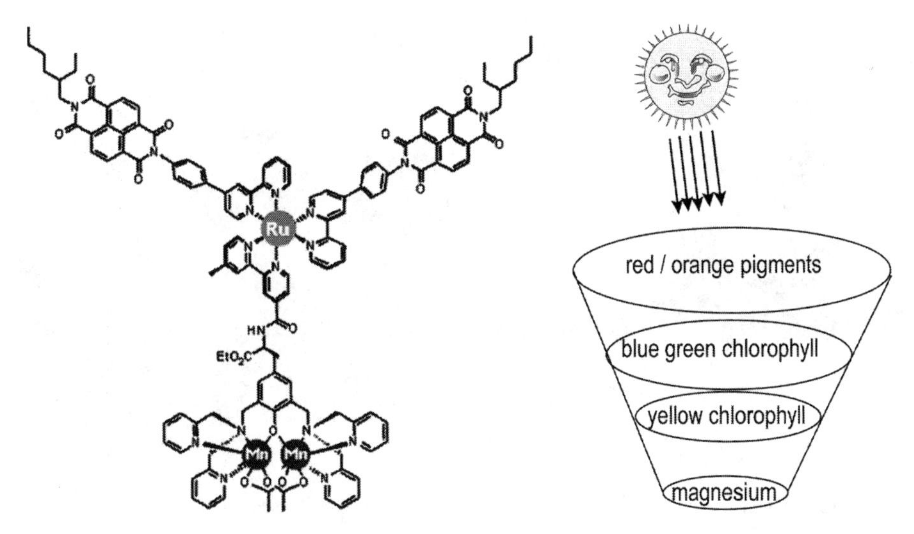

Figures 17 and 18: Light antennae

We will finish this section with a little summary. The major item in plants is chlorophyll, which has a particular structure called "light antenna." In *figure 17,* we see the molecular configuration of light antenna. The upper chain, shaped like a dish, is a type of chlorophyll. The nexus in the center is another type of chlorophyll. The branch going down from that little cluster is another type of chlorophyll. Then, right at the bottom, we see a molecule of magnesium. When sunlight comes in, it hits the "dish," is bounced around, and then is harmonized until it can be absorbed by the chlorophyll just below the center. This is chlorophyll A and chlorophyll B. Finally, the harmonized light is transmitted more deeply into the structure, where it activates the magnesium, which, as a metal, is filled with the properties of light. The transmitted light is then converted by the metal into energy for the plant. These chlorophylls are found in the plant's pigmentation, which the plant uses, based on sugars and alcohols, to draw the light into and through layers of different chlorophylls that filter it down until the light can touch a metal. Once it touches the metal, the light energy is regulated and harmonized by the pigments and eventually becomes the energy that makes the plant grow. For grape growers, this is what happens in the pigments of the grape skins. Those pigments are energy-transforming substances

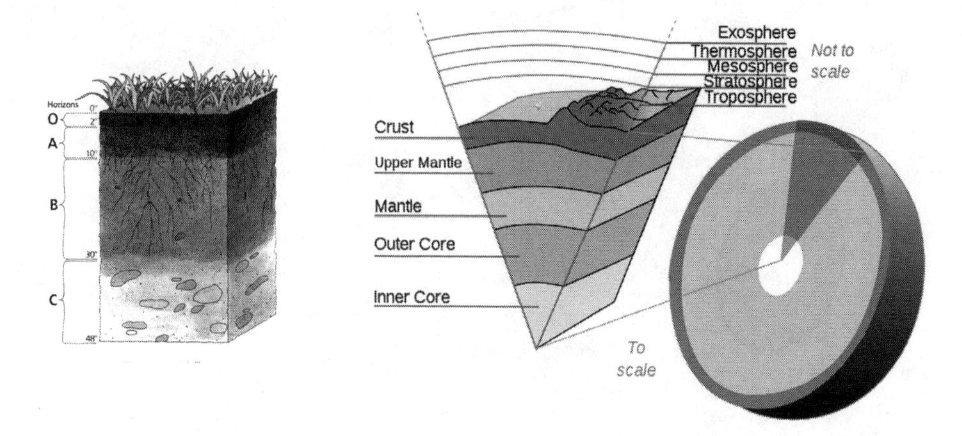

Figure 19: Soil profile

Figure 20: Layers of the Earth

known as phenolics. *Figure 17* shows those pigments as a molecular diagram. *Figure 18* shows this process schematically. We see that it looks like a dish antenna made of various substances that move the light down into the plant.

Figure 19 is an image of the soil profile in which the living layer is at the top. Below the topsoil (O), sub-soil (A), and the sub-sub-soil (B) is bedrock (C). We see the living part at the top and the silicates, pure and crystalline, below. Rudolf Steiner's picture tells us that silicates in the various layers take up the cosmic light. We could think of them as a light antenna for the earth. They take in the sunlight and reflect it back out.

The Earth's layers are pictured in *figure 20*. It is a cross section, with a mountain on top of the various geological layers. This structure appears similar to the light antenna in *figure 18*.

Whether we are speaking of the Hubble space telescope, geological cross sections, plants, silicates, or the molecular structures of various substances, the patterns are resonant images of the way the ethers descend into the elements. This structuring is the basis of energy transfer.

We can finish this section with a whole other dimension of Rudolf Steiner's work—that of the spiritual hierarchies. All of the structures discussed can be thought of as interrelating spiritual beings. The dynamic context of substances could also be discussed in terms of a spiritual context of forces of attention, soul forces, and moral forces of attentive will coming down via the light of consciousness, through all the hierarchies down to the angels. This is the macro, or super-macro, model of the micro model I have described. Whether macro or micro, it is the surfaces that provide the interface for energy transfer, the source of life forces in nature and in the soul.

V. Cow Horns and Silica • Form as Substance and Process • Ectoderm and Endoderm • Attraction and Repulsion • Electrical Devices • Capacitance and Life Charges • Preparations 500 and 501

There is a common myth in biodynamics—that a cow horn is made of silica. It is not silica but the protein substance keratin. Nevertheless, it is associated with silica and, to understand its relationship to silica, we need to distinguish between silica as a *substance* and silica as a *process.*

Silica as a substance refers to the thing itself, which I can hold in my hand. If we take it to a lab and analyze it, we would say it is made of silica, or silicon dioxide (SiO_2). It is a mineral. If we take a cow horn to a chemist to analyze it, the chemist would tell you it is made of keratin, a protein substance similar to hair or skin. The reason that we use cow horns is because they represent an intense silica process. Their substance is a crystalline protein, but they act the way silica does in the mineral realm of the earth body; cow horns draw in forces, or light, from the periphery. Their particular form causes this action.

The previous section discussed forms and surfaces in nature as agents in the natural world for energy transformation. Now we will look at form as substance and as process. If we look up the word *form* in the dictionary, we will find about thirty different definitions—everything from income tax forms to the verb. Such distinctions go back to Plato, Aristotle, and beyond. Those philosophers had unique words for the various qualities of form—as an activity and as a finished product, as an idea and as a process. They distinguished in their various uses of the word *form*. They recognized that form is a magical thing in the world. To illustrate this, here is a common definition of a quartz crystal structure, followed by the definition of the form of protein in a cow horn. You can be the judge of how they are related.

A basic structural unit of quartz is a group of three connected silicic acid tetrahedral molecules that do not form a closed triangular surface. The three molecular tetrahedra of the crystalline motif lie in three different planes rotating around one axis. This picture is of three pyramids touching one another at their tips in a spiral line around a common axis.

Now, the protein in a cow horn: Repetitions of that tetrahedral unit along the axis form the layers of the silica crystal. Each set of tetrahedra in each molecule is connected to two neighboring groups that lie above and below it. The motif is repeated vertically through the crystal to form columns that are linked in the form of spiral bundles. So the molecular structure of a quartz crystal is molecules that are formed as tetrahedra that are linked end to end that form columns that make bundles of molecular pathways within the crystal. The entire crystal can be seen as a spirally woven bundle of such chains. The bundles of chains are wound around the vertical axis of the silica crystal to form helixes. The helixes are formed of chemicals bonding and do not form a fibrous texture in the crystal. But the helices are a true geometrical feature of quartz that has important implications for the symmetry of all siliceous crystals. So these molecular forces make vertical helices—helices structured within the crystal at a molecular level with spaces in between the actual molecules where there is kind of a void space that draws in water. And this makes it possible for silica crystals to weather out into other mineral forms.[19]

19 See http://www.quartzpage.de/gen_struct.html; also, http://en.wikipedia.org/wiki/Tetrahedron.

This is from the wonderful website quartzpage.de, showing great animations of molecular structures. This site really helps in understanding the formation of soil.

Now a quotation about the protein structure of the cow horn. Chemically, we can take any protein and make a crystal from it if we make it pure enough. It is an interesting fact that proteins can be made into crystals. Proteins can be synthesized by crystallizing them. And the crystals have characteristics of minerals. This is a description of the substance of a cow horn, keratin.

> Keratins are the chief constituents of structures that grow from the skin of vertebrates. These structures include: Among mammals, the hair (including wool), horns, nails, claws, corns, and hooves, which are made primarily of keratins. For example, hair, a filamentous outgrowth from the skin that is found only on mammals, involves fibers comprising nonliving cells whose primary component is the protein keratin, a long chain (polymer) of amino acids that naturally forms a helix fiber and subsequently winds two of the helix fibers together to form a much stronger "coiled coil" fiber characteristic of keratin. The keratinized cells arise from cell division in the hair matrix at the base of a hair follicle and are tightly packed together.[20]

The crystalline structure of keratin sounds remarkably similar to the description of quartz does not it? Both are formed of regular spaced large chain helical molecules.

From this we can see that the substance of the cow horn is not silica but the form or formative process is like silica. This is a critical difference and Rudolf Steiner in the agricultural course is asking us to educate ourselves in the language of form. This is because the language of form links the realms of mineral, plant, and animal. The substances are different, but the processes are similar. This is worthy of further thought.

The dynamic forms of processes that carry substances though processes of transformation are a language that allows us to see more deeply into the dynamics of natural systems. In the examples of silica and keratin, they each go through a similar process of becoming. As a result, they have a mutual

20 See https://www.newworldencyclopedia.org/entry/Keratin (accessed March 2013).

resonance through their forms. They are formally harmonic with each other. This is why it is useful to place a particular kind of mineral or herb in a particular kind of animal sheath; the formative process of both substances is similar even though the substances are not identical. This is fundamentally an alchemical way of thinking, which makes it difficult to justify certain biodynamic procedures in a world that bases justification on the identity of substances and materials. People who live in a world where science considers the material as the identifier have trouble understanding the idea of life gesture or formal resonance as a basis for scientific work.

In the materialistic world, it is impossible to say that silica has any relationship to a cow horn. To think so is considered bad science, because it is true that *they have nothing to do with each other as substances*. However, when viewed as formative processes, silica and keratin are analogs to each other in different realms. The formal resonance means using a cow horn is synergistic. To do so amplifies the silica, because the two are linked by their formal motifs. However, for silica there is a step up, because the silica is moving from mineral to animal. The horn is like a transformer or capacitance condenser in its form. By putting silica in a horn, I am lifting the silica into a field of form that is siliceous but comes from life. The silica is resonant to the silica form of the horn, and that link is through the action of the form. Nevertheless, the form is a result of the similar process of the two things becoming. How a thing comes into being creates the potential for the substances to move in such a way that their form has coherence and significance. As stated in our quotation, the fact that it is a rotating tetrahedral chain has significance for all silicates. Eighty percent of minerals come from a silica matrix.

The tetrahedral form is really important because it is an archetypal form in the mineral world for eighty percent of the minerals. You can meditate inwardly by transforming a tetrahedron, and you will eventually be able to understand the way your bedrock becomes topsoil. Learn to do the tetrahedron exercise in your head, and elemental beings associated with the soil processes and your land will realize that you are interested in them. Maybe they will come and talk to you in your dream world. By holding a picture, you invite the beings who animage nature to engage your consciousness. If you repeatedly hold that picture, you will begin to make a doorway into their world. They, in turn, will become interested in the fact that you are forming your thoughts in honor of them. Think about it. If you knew that someone were holding your picture in their mind and sending you love every day, you would have a special kind of connection when you met that person. This is basic human capacity. You have elemental beings in the crystalline structures in your own body. When you hold these things in your consciousness, they connect to those outside of your body. Once you get to know them, they create in you what Rudolf Steiner calls "nature imaginations," which are the key to biodynamic agriculture. They are the key to working land in a new, imaginative way—a holy, or sacred, way. Biodynamics is a mystery school. It is not meant to be simply a recipe but a means to lift consciousness.

In the future, this meditative approach to understanding significant form will be a whole other branch of science as it was for the ancient Greeks. Even during the Middle Ages and, later, the Renaissance and Enlightenment, it was still a whole branch of science—*alchemy*. Then it had to go underground because of materialistic thinking, but now it is coming back because it is the balance to the rationalistic, reductionist side of research. Imaginative consciousness allows human beings to enter the natural world in a completely different way. It could be called *the language of form*—the discipline of imaginative, formative, cognition. The idea of using pictures to research, on the basis of their form, the relationships among minerals, plants, and animals was known in alchemy as *the method of analogy*. It was understood that analogy is a way of thinking that allows a person to see into substances or plants in a different, more living way.

Imaginative cognition gives a person's mind access to the invisible activities that create materials. This is the basis of alchemy. Classical alchemists worked at a time when nature still had spirit in it, and they sought to see the spirit in matter. If you are a Western person, your life body is configured in such a way that you will probably not see the spirit unless you have taken some spirituous substance. Then you may see some spirits, but then you cannot rely on whether you are really seeing a spirit or the spirituous substance. However, it is possible to see spiritual beings with your waking day consciousness fully intact; that way of seeing is imaginative consciousness, the basis for the method of analogy.

ectoderm – nerve
endoderm – digestion

Figure 1: Layers in an embryo

We will use the method of analogy here to compare the form of a cow horn to structures in the embryo and some principles of electrostatics. The goal is to show how the combination of the cow horn and manure makes sense. So please look at *figure 2*. This is an image of two layers in an embryo. The upper layer, ectoderm, is a proteinaceous membrane; as a result, it has a very strong silica process just as keratin does. It is not fibrous but has a quality of forming layers. It is siliceous in its activity. The actual substance of it is proteinaceous rather than siliceous. However, the way it forms is very much the way silica forms in sheets and surfaces and layers. The ectodermal layer in an embryo is the outer layer. The prefix *ecto* means outer; it is an outer *derm* (skin)—ectoderm.

Embryologists have followed the evolution of the ectoderm all the way to term in the fetus; the ectoderm is the source of all the nerve apparatus—nerves, axons, dendrites, and brain tissue—wherever there is nerve tissue, which evolves from the ectoderm. Thus, we could say that the ectoderm represents the nerve layer in the embryo. The nerve layer is where the sense organs are embedded. In the structure of the sense organs, we see the same type of proteinaceous sheath formation—the eye, the ear, and even the skin. Basically, our brain is an island of skin that has been taken inside and folded to create maximum surfaces.

Research has shown that the brain becomes folded under the impact of sensory impulses that come into the fetus. The more sensory impulses, the more the outer ectoderm, or sheath, expands and becomes active, and the more it needs to fold to accommodate the growth. Therefore, wherever there is a nerve, layer after layer of tissues are formed. The formative action of the sensory impulses acting on the developing nervous system gives us a picture of what we could call the action of peripheral forces, the forces of the cosmos, the forces of light, and what we could call warmth. The nerve transmits these forces of life, but because it transmits life it cannot be alive. It has to be simply a transmitter, not an editor.

The nerves sacrifice life in the organism to allow light, warmth, and sound to enter the organism from the periphery on the outside. That is the function of the ectoderm. The ectoderm and the nervous tissue eventually become transmitters or transceivers. That is they receive energies from the periphery and then transmit them to the center. In a general way we could say that is the prime function of the ectoderm. On the other side, we have the endoderm, the inner skin. In the very beginning of development, the embryo has these two layers: ectoderm and endoderm. The endoderm gives rise to the digestive system—the intestines, colon, stomach, and so on.

The function of the digestive aspect of the organism is to secrete. In the beginning of life, we have one layer, the nerve layer, which receives external impulses and is intimately connected to a layer designated to secrete when stimulated by impulses from the nerves. The little embryo is maybe one-thirty-second the size of the head of a pin and it already has these two layers. The formative activity consists of impulses from the outside that hit the outer layer of the disc and are transmitted to the inner layer. The inner layer then begins to secrete in response to what has come in from the outside. The secretions go into the general flow of life in the organism to promote stimulus for growth and development; in this way they impact the outer layer. After a short while, the outer layer receives impressions from both the outside and the inside. This is the law of the organism.

ectoderm – nerve

mesoderm – blood

endoderm – digestion

Figure 2: Layers of embryo a little later

The flow of impressions in embryonic development rapidly creates a third layer, the mesoderm (see *figure 2*). In the early stages of development, the mesoderm gives rise to the blood and eventually becomes muscles, bones, organs, and anything that is not a part of the digestive or sensory system. Now we have a three-stage embryo, with blood in the center, the nerve impacting the blood, and the blood receiving the forces from the nerve and interacting with the endoderm, which secretes substances into the blood. From this point through the whole life of the organism, the blood becomes a mediator between the digestion and the nerve. This is the function of the three layers in the embryo. Rudolf Steiner calls the blood a tablet written on from both sides.

The blood then carries the reactions of what comes from the outside and what comes from the inside through the body as a go-between. In a human being, the blood is a being acting as a counselor or problem solver. In the organism, the blood carries the reactions to sensory impressions that fire off secretions in the glandular system. The blood deals with the hormonal cascade of impulses from the glands. That is a picture from physiology.

The three-layer embryo of nerve, blood, and digestion is a meditative image we can work on for years. Using this image to make pictures, small wax models, and so on, helps us understand what Steiner is talking about, especially regarding the way biodynamic preparations are made and the organs we use for them.

Regarding the preparations and the sheaths of those preparations, Steiner used a very precise language of form. To understand that language, it helps to keep in mind this idea about the three layers of the embryo. If you work with this image every once in a while when you are reading or looking at something, the light will go on for you about a passage in Steiner's *Agriculture Course*. Working with this image will eventually provide insight into what Steiner calls the threefold organism, which comes right out of embryological development of sheaths, or layers.

In response to a sensory experience the ectoderm stimulates the glands to begin secreting. The secretions trigger responses in the life organs that stimulate the endoderm to secrete. Suddenly, the life of the animal is being carried through its body by secretions. Secretion is a truly important element in biology, because it is the pathway of communication for impulses of life in organisms that are not hardwired. Most people think that communication among the body's cells is based on one cell touching another, sort of like an electric circuit. However, you can take a little piece of garlic and put it in your shoe and, in a few minutes, people will ask if you ate garlic for lunch. The garlic does not know you have a bunch of cells between your foot and your mouth, but the substances in the garlic create such interactions in your blood that your blood and secretions just take off so that you have suddenly aromas in your mouth that originated with your foot. There is no way blood could get there that quickly.

This is the reality of us as physical bodies; we are more subtle than generally believed. In nature, this is an example of the subtlety with which things can move in the natural world. It is the key to how I can take something from a plant world, put it into an animal organ, and know that it will have a specific action. The activity of secretions reveals a subtle level of life, or what Steiner would call etheric forces. The etheric realm includes enzymes, hormones, pheromones, fragrances, esters, flavors, oils, and so on, all secreted by living things when stimulated by energies outside the organism. Organisms contain sensory organs—even plants. One sense organ in plant is chlorophyll; it is a sensory organ for light. It is like an eyeball. That sense organ draws in light and, out of that, creates secretions that become part of the plant's photosynthesis process.

The plant responds to sensory impressions by creating dynamic substances. That is just as animals do, except that animals have a lot more organs to create dynamic substances, glands, and so on. There are also glands in plants—oil glands, for example. Just squeeze an orange rind with a candle like the kids do. The fireball that comes out of that skin is from a gland in the orange skin that secretes aromatic oil. The idea of secretion is important for understanding how an organ transforms a substance with which it comes in contact. There is an analog here. The life realm has analogs in the realms of physics and technology, which can be useful for understanding some of the more delicate relationships in nature.

Now let us shift to something called electrostatics. If you are interested in this work, I suggest a book titled *Man or Matter* by Ernst Lehrs. It is a can opener for rigid thinking. It is not an easy book. It is an important book for understanding Rudolf Steiner's alchemy. It is one of those books that you may need to read several times before getting it.

In his book, Lehrs describes what he calls the *two electricities*—one type is the derivative electricity with which we light our world and use in cars and that we are actually stealing from the Earth. The other is original, or primary, electricity, which science calls static electricity and whose application is electrostatics. Electrostatic devices do not have the power to drive an engine. Consequently, there was not a lot of development in electrostatics until the computer. Now electrostatic devices are everywhere. Your ink jet printer is an electrostatic device, as well as the gigantic Van de Graaff generator used to accelerate particles to the speed of light in particle accelerators. Static electricity is used in filters that pull fly ash out of smokestacks in industry and to spray paint on assembly lines.

Static electricity is a prevalent form of electricity in the natural world. It simply indicates a potential to build a charge that can hold life. To maintain a charge, we do not want a discharge. Forms that carry and hold on to life without a discharge are round. In nature, round forms can hold the maximum amount of charge, life, and potential. Static electricity is also held without discharge by forms composed of many layers. Natural forms linked to life are layered because the static charge is carried on surfaces. Charges are carried on surfaces, and if I have many, many surfaces I get many, many charges.

If we look at these ideas alchemically, as Ernst Lehrs does, the charges can be seen as polarities of fundamental forces of gravity and levity, the two most fundamental forces in the natural world according to alchemists. Gravity moves down and leads to manifestation. Levity moves up and leads to rarified states. These two forces are found in infinite combinations and permutations in nature. Lehrs sees the positive charge as fundamentally an earth, or gravity, quality of manifestation undergoing a process of being influenced by levity, the force of the cosmos. A positive charge is an earth condition induced to lift into levity.

To understand this, here is a little scenario. Let us say that I am feeling a little down and someone comes and says, "You look like you need a hug." Okay I get a hug, and there is a little exchange there. Now I am feeling a bit better, and my earth is lifting into levity. What I am feeling for you is sympathy. I just received a charge from you. We used to call it a "contact high" in the 1960s. It is a charge. Something has been transferred, and a new condition is unfolding.

The opposite of a positive charge is a negative charge. In a negative charge the levity of the cosmos becomes imbued or influenced by the gravity pole of the earth. This time you are feeling pretty good in the morning, and then you hear a story about some bad thing that happened somewhere in the world while you were sleeping. This is asking your levity to come down into a little box. You feel, "Do I have to?" We could call that feeling antipathy, or a negative charge. Antipathy does not mean that we hate something; it is simply a feeling that we cannot get any satisfaction in a world where nothing is the way we want it to be. Sympathy, on the other hand, means that the world is going according to our wishes.

Earth lifted to cosmos	+ charge = sympathy
cosmos drawn to Earth	– charge = antipathy

Figure 3

Therefore, those two charges could be linked to feelings. The feelings we have in our soul when projected out into the phenomena of the world become positive and negative attractions, charges, repelling forces, attracting forces, atomic valences, bonding angles of molecules, and so on. In the view of alchemists, all of these physical forces are felt as the feelings of the world soul. In our abstract sciences we have forgotten that these forces actually are the equivalent of feelings in the life of the Earth; they have become mere calculations of force. However, when they were discovered during the time of Benjamin Franklin, alchemists were trying to understand these forces. The language they used to describe them was a *soul language* for the properties of attraction and repulsion. It was like a magical, magnetic kind of thing. They were searching with an awareness of who they were and finding certain forces in the natural world that, to them, seemed to behave like an ensouled being.

Today, they are just abstract forces, and we have forgotten that they actually have something to do with us. We just have powered devices that use so many ohms, or the resistance of this is a certain amount, and so on. Our power devices do not run on the primary static electricity but on the secondary, or derivative, electricity. This is because we have learned to deplete the static electric charge that is the basis of the natural world and turn it into derivative electricity for the purpose of powering our devices. We create secondary electricity when we run a wire through a static field. It causes a discharge in the static field. Using a generator, I can then pull current from the discharge and use it to power a device, and I can use a battery to store it. When I take a wire, a little loop, and run it through the field line of a magnet (a magnet is basically a little Earth), then I can put a little meter on it and lead it over to my light bulb, which suddenly starts to go on because I am pulling, or deriving, a current of secondary electricity from the static field of the Earth. I am deriving power from the static field of life. I use it in the form of a secondary electrical force that runs my device.

This is why our electricity depletes life, and yet people who study these things equate electricity in a light bulb to the electricity in life, simply because it has positive and negative charges. We live in this kind of funny world in which natural forces are trying to hold onto life energy by creating forms that can build up and hold charges of life. Rudolf Steiner calls those forms *sheaths*. He chose certain sheaths that have these kinds of patterns in them. The function of the sheaths is to enhance the forces in the substances. In that way they function in a way similar to capacitors. Alternatively, we could call them batteries, but that is not quite accurate. The function of the sheath is more like a capacitor.

From this, you can see how our devices are little shadow images of the way the natural world operates. They are shadows of life, because they draw energy from the natural world rather than enhancing it within the natural world. The principles of the forms of devices can be used as analogs, but not the forces used in the devices themselves. The advantage of understanding the formal motifs of the devices is that we can imagine new forms in the natural world that would allow us to concentrate the natural energy to build a charge. Consequently, the Earth regains the energy we have been stealing from her for our devices. For me, this is biodynamics. Moreover, we are taking from it massively. So when we begin to give back to the land, this is sometimes called the virgin hymen. Viktor Schauberger[21] said that this membrane, the virgin membrane of the Earth, is depleted by our use of these kinds of devices and by the misuse of iron implements. We can revivify it by understanding the forms and forces active in the natural world. The forms of electrostatic devices are more useful than the forms of electric devices in this regard.

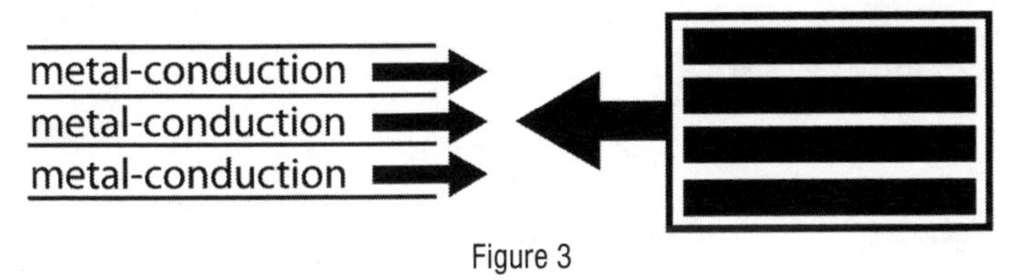

Figure 3

Now we can return to the cow horn. There is a principle in electrostatics. Metal plates conduct charges, so I obtain two metal plates and hook them up to a battery, a source of stored charges. Once the plates are hooked up to the battery, the stored charge flows from the battery to the metal plates and creates a charge on the metal plates. As long as I keep the battery connected to the metal plates, the charge will flow from the battery to the plates. The plates absorb and conduct the charge until they are *charged,* or saturated, so to speak, with the charge. Once they are charged, the plates create a field, or potential, between them. That field will register the amount of charge from the battery that the plates are able to hold. However, if I were to take the wire off the battery, the charge would flow back off the plates trying to go back to the battery. The plates will hold a charge while I have a constant

21 Viktor Schauberger (1885–1958) was an Austrian forester/forest warden, naturalist, philosopher, inventor and Biomimicry experimenter. See, for example, Cobbald, *Viktor Schauberger: A Life of Learning from Nature;* and Bartholomew, *Hidden Nature: The Startling Insights of Viktor Schauberger.*

power source going into them. This is what we do when we charge a battery in a car or tractor. When I take the charge away, however, the power wants to flow back out. To understand this, visualize blowing up a balloon. If you blow up a balloon, you get a charge, a potential or a force in the balloon that wants to flow back out when you stop blowing into it. If you open the end of the balloon the charge runs back out into the room. This is a useful picture for understanding capacitance.

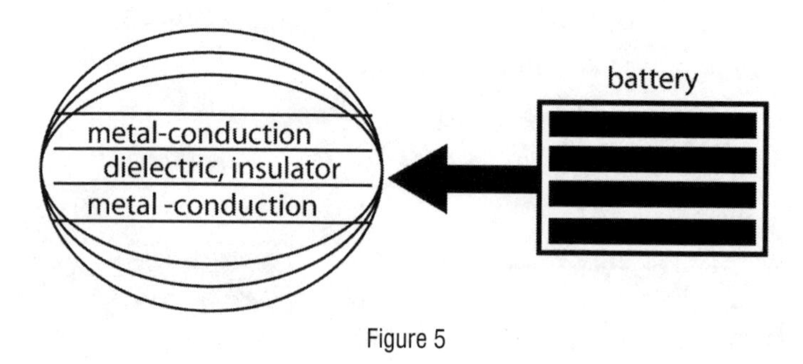

Figure 5

Now, if I put something between the two plates to resist that flow, it is called a dielectric. Silica is a good dielectric. Then, when I put a charge into the capacitor again, the dielectric allows a greater charge to build between the plates (this happens for technical reasons that do not concern us here). The resistance of the dielectric has changed the capacity of the two plates to hold a charge. Now when I take the battery off of the capacitor, the charge will start to come off the plates, but the dielectric has caused a field to grow between the two plates that stores a greater charge. Therefore, if I take a plate and dielectric, multiply them, and pile them together, I get a device called a capacitor. A capacitor allows me to put a certain amount of current into it, but it'll hold a larger amount of current. If I make a capacitor with very thin plates of metal and very thin insulators, I can stack them up and get a strong potential field.

The thinner I make the plates, the more capacity the capacitor will have to hold a charge after I shut off the battery. It forms a capacity to store extra charge. A battery is fairly weak, but if I connect it to a capacitor, the battery creates in the capacitor a much higher capacity to develop power. The key here is that a capacitor is made of inner surfaces layered to accept and repel, accept and repel charges entering from the outside. Thus it builds up stronger charges from weak charges.

As an analog we can consider the protein structure of a cow horn. It has a structure similar to silica, which is a dielectric. The protein layers in the horn continually receive very weak charges from the sunlight. Moreover, there are metals within the protein structure that strengthen the capacitance. When the cow walks out to pasture in the sunlight, her horn acts, in effect, as a capacitor on her head, building fields of energies that flow into her body.

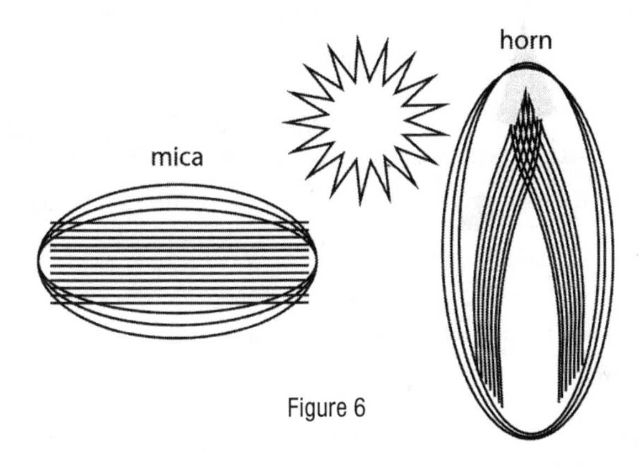

Figure 6

This may seem farfetched, but the principle of the how all these charges are formed and built into our devices is also active in the organic world. Again, we get back to surfaces. In *figure 6,* we have a diagram of the mineral mica. Mica is structured so that it is composed of silica plates with very fine metals dissolved in them. Mica, a natural capacitor, is composed of huge amounts of inner surfaces; dielectrics and conductors are sandwiched together in a pile. If I take that pile of surfaces and place it in sunlight, the light creates charges in the mica. When the light and warmth hit the rock face, a change takes place in the way the charges operate in that substance.

In a piece of mica, the actual energy is not in the device but in a field around it. That is where the capacitor stores the extra charge that is created.

This is the principle behind many devices used in the world today. In the beginning, when people did not know how to manufacture capacitors for radio parts, they used mica as a capacitor. They needed ways of controlling and regulating electrical flows, and they turned to mica to do that.

I heard about this man in England who was studying how bats communicate with sound. He would go out in the evenings with a sound truck, park out in the boonies, turn on the antenna, and then listen to and record the high-frequency bat communications. One morning he was sitting having his dinner in the sound truck after working all night, but he forgot that he had left the antenna on. He was sitting there and suddenly heard a high-pitched squeal that sounded like a bat. He looked out the window of the truck and there were no bats around, and then he saw that the sun was rising. He thought that was pretty interesting, so he decided the next day to leave the antenna on at the same time in the morning and try to find out what made the sound. The next day he left the antenna on again, and he noticed that, as soon as the Sun touched the horizon, the squeal came through his antenna. He noticed, too, that as soon as the bottom of the Sun left the horizon, the squealing stopped. This was very interesting. Then he noticed that he had parked his truck near a stone circle. Eventually, he saw that when the rising sun hit the stone, the stone gave out a high frequency signal. He stopped studying bats and started studying stones.

There are countless flakes of mica in the stone, each of which is a little capacitor. If I take a beam of sunlight and shine it on a cold stone, there is suddenly tension because some of the stone is getting juice and some is not. According to that man's research, if you have a big dish antenna, an audible sound arises from that interaction, which we could call a discharge. This is an activity simply from the sunlight interacting with the mineral. The warmth of the sunlight striking mica in a chunk of granite causes a reaction and activity.

Figure 6 shows a mica capacitor made of layers of mineral (potassium, iron, and silica) receiving light from the Sun, reacting, and creating a field around it. Now we come back to the cow horn, which forms as a result of skin processes in the cow that form sheaths of keratin as proteinaceous substance. The substance of keratin is similar to hair and fingernails. It gets compressed around a bony core, making layer upon layer after of proteinaceous crystalline structure that resembles the helical structure of a silicate. It has a very smooth round form, and the good ones take the form of a helix, or vortex.

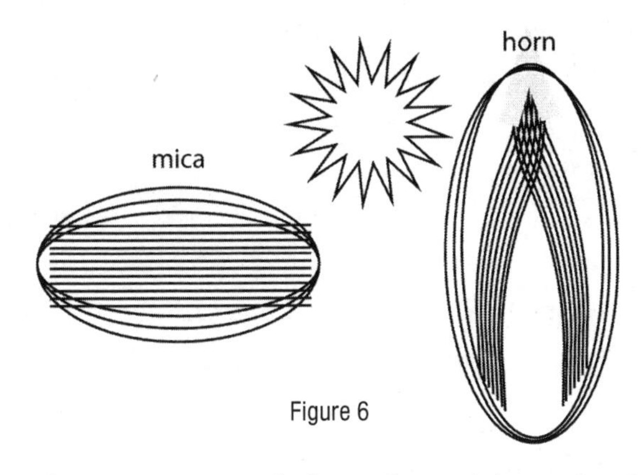

Figure 6

The horn's form uses materials from the periphery, the tip of the horn, and moves them toward the animal. The form itself is an amplifier and a capacitor (see field properties of the cow horn on the right in *figure* 6) This means that charges can be built up within the layered structure of the horn. However, the form of the horn directs the charges one way, toward the cow. In terms of electricity, such an arrangement would be called a diode, which allows a charge to pass in one direction but not in the opposite direction. Such a device is used in a trickle charger, which can charge a car battery overnight without allowing the charge leak back out. The device that does this is called a diode. A diode is a very special kind of capacitor. It allows the charge to flow in only one direction.

Returning to the cow, her antenna is made of keratin, which is a helical, siliceous-like material that receives sunlight and transmits it into the skin of the cow. The energy moves into layer after layer of tissues and is transmitted through the organism down toward the sea of blood in the digestive area. If we look at the structure of the intestines, we see they are layers of skin inside skin inside skin. The digestive apparatus of a cow is a big, organic capacitor that receives a "trickle charge" from the cosmos and streams it into the mass of organic, enzymatic food moving through the layers of the cow's digestive capacitor.

What happens to the cattle whose horns have been cut off? Here we can talk about fevers and infections. Typically, cows do not have the calcium forces that they need in their bodies. Those forces are used up in the formation of a fetus; we can see this in the rings around the horn. When cows lack horns, the forces they need to regenerate tissues are not available. In such a case, when the muscles of the uterus expand, the calcium that cows need to give birth is compromised in her immune system. Then the cow can get an infection and die when she starts to give milk to her calf. This was rare in the past, but now you call a veterinarian, who gives the cow a calcium injection.

There are great mysteries here. As I understand it, the cow's horns are not for fighting but for sustaining life energies. Their purpose is to induce and transmit life and light into the mass of the cow through the membranes of the inner skin, down into the endodermal area so that the blood can receive light. Basically, cow stomachs are tremendous capacitors, constantly fed by forces from the periphery.

At the other end of the cow, you have another set of capacitors in the hoofs, which lock the energy of the cow in at that end. Between these two, you have a thermal nuclear chamber called a cow stomach, which is a fountain of life. A cow takes in grass, by volume some of the least nutritious food on Earth, runs it through the mill, and charges it with life. The microorganisms living in the cow's gut bathe in those life forces and proliferate on the fermenting grass. The microorganisms change the fodder and permeate it with their own life forces. Then the cow siphons off those microorganisms, which are a rich source of protein for building the cow's body. Basically, the digestive juice of a cow is like the plankton that whales feed on.

Cows do not take the forces of life from the manure, because there is so much life that they give to their food. We take that sacred cow patty and put it into the capacitor, the cow horn, and we do what alchemy would call a *spagyric* process, the joining of the alpha and omega. That horn capacitor still works because it has the correct form. It is tuned to that product because together they are part of the whole process of the blood sandwiched between the digestive system and the nerve-sensory system. Manure is permeated with life. Then we put it back into the very organ that drew in life from the periphery. That is synergy, and that is then preparation 500.

Nevertheless, the genius of Rudolf Steiner says do not stop there. Let's take a silica crystal whose form has a very similar pattern to this one. We grind it ever finer, until we get it so fine that the law of minimal surface comes into effect. Recall that, according to the law of minimal surface, the smaller the particle the greater the surface. Now we have a siliceous material that is increasingly open to the infinitely distant plane to receive forces from the cosmos of light, and we put that into a capacitor that has the same pattern. Thus we get, once again, synergy for the cosmic light side. In preparation 500, we have synergy for the earthly life side. That is the genius of Steiner—to see that these things are analogs of each other and that they are connected to each other in scientifically reasonable, not fantastical, ways. Not being able to think about the reasons for this kind of work is okay in the beginning. However, over time, you will develop your imagination so that you begin to see the genius that Steiner brings, showing us how to take the work in biodynamics forward.

CHAPTER 4

ALCHEMY AND CHEMISTRY
BASIC PRINCIPLES

I. Alchemical Rules • Phlogiston Theory • Solve/Coagula • Colloids; Levity/Gravity

I really need to begin with a kind of apology to winemakers, because my master's degree is in fine arts, not chemistry, and what I discuss in this chapter involves a lot of chemistry. Therefore, some of this may not jive with your own view of chemistry, but please keep an open mind. Over the years, I have struggled with chemistry, especially the chemistry of living things. For thirty-five years I have on my own been using the alchemical worldview as a filter, because it has been extremely difficult for me to understand biochemistry through its own huge and complex nomenclature and language. I do not have the training needed to go into my garden and bring soil amendments or sprays to a plant with the feeling that a specific substance is connected to a specific plant.

My goal has been to develop a kind of inner sensibility about the soil and a relationship to plants that could be called an artistic experience—that is, the ability to recognize significant patterns. There is an idea brought by Daniel Pink called the "new MBA," which is the MFA, or master of fine arts degree. Corporations are hiring MFAs to look at spreadsheets, but not to be the bean counters. The CFO, or chief financial officer, counts the money, and the MFA talks about the patterns that are present. Companies get more overall pattern recognition out of the MFA than they do from an MBA. Seeing the bigger patterns takes an artistic eye and disciplined imagination. It takes a sense of pattern recognition to say whether there will be a crunch someplace in a spreadsheet. Once we get a sense of pattern recognition, artistic training in disciplined imagination allows a person to see into the future when things will either gel or fall apart.

Pattern recognition is a way of understanding deeper or broader archetypal relationships and not just specific or detailed data points. The more data points you have, the more you need a highly specialized language to make sense of what you want to understand. However, it takes a different kind of filter to look at data to see how the whole field is moving, rather than only a specific, very tight event. Because we know that everything in terms of data has to do with probabilities, nothing is ever really certain. Everything is always in flux and we need to make judgment calls. Years ago, when I was a chemistry know-nothing, I decided to embark on self-education to understand how a plant and the soil speak to each other. My research doorway was the study of weather and climate in the alchemical context of planetary motion. I eventually found out that, alchemically, similar patterns are present in both the atmosphere and the relationship between the soil and the plant, and even in just the soil itself.

Now we can consider some pictures that show patterns that might help us see how, for example, a particular root stock would perform on a particular *terroir.*[1] How can we encompass all the variables needed for all the little data points to come together in a peculiar way. It might be so many variables that they would obscure the bigger picture. Some growers even suggest that the idea of terroir is almost outmoded because of the way we study what happens in the field—that there are too many variables to use the term *terroir.* However, those who know how to look at all the variables know that something like terroir contributes to the whole of a given harvest year. It is an intangible, and that intangibility means all of the variables in the field of activity involving climate, soil, temperature, shade and light, drainage, bedrock, varietal, and whatever. How do the variables perform in a given niche? Terroir is a kind of palette of those variables.

The ability to "see" terroir is really a kind of clairvoyance. It is one's intuitive perception of variables, so many variables that you could not actually

1 *Terroir* is a term that refers to the relationships among the grapevine, soil, slope, climate, micro-climate, and genetic variety.

say that this causes that. Nevertheless, in the mix, you can begin to synthesize and be able to say whether you think there is a probability of something going in a particular direction. This kind of thinking is needed in the life sciences. It is very difficult to say something causes something else, because there are so many variables. This problem of seemingly random variables leads to the stochastic science of probability. The stochastic approach is the science of investors, because it is all about probabilities rather than facts of cause and effect. Stochastic science predicts trends.

In the soil and the plant, the plant's ability to pick up a particular nutrient is called availability. It does not matter how much you have in the soil because, in the great magic and dynamic of what a plant is, it can sometimes compensate and transform things that are not available. What matters is the availability in the relationship between the plant and the soil solution. In chemistry, that availability, we could say, has its own breathing process. Some of the minerals in the soil solution allow for availability, and others do not. There is something called a deficiency phenomenon. Is it in the tip or is it in the older leaves where you see the yellowing, browning, or some other symptom? Those are symptoms that you cannot put your finger on and say what exactly is happening. How much clay is in the soil? Are you on lime? Are you on volcanics? Those things make pinpointing a problem and solution very difficult. Nonetheless, we can teach ourselves to read the trends.

Let us consider some images of how an alchemist would look at calcium, sulfur, silica, and carbon. What is nitrogen? What is phosphorus? What is potassium? In an alchemical worldview, the substances have a kind of persona—even a consciousness. They have volition. They have a will. They are predictable; what calcium does is very different from what silica does. If the neighbor on one side of you is calcium and the neighbor on the other side is silica, you could have a clear picture of their dynamic. This is how alchemists looked at things. A substance is an activity that has come to rest. For alchemists, a substance is not only an activity that has come to rest, but also the consciousness of a being—the being of the archetype of a substance. That entity had a particular way of being and interacting that is called *elective affinity* in chemistry.

We could also say *valences*,[2] a more abstract term, while *elective affinity* has an alchemical ring to it. So those valences of whether we have a cation or an anion (negatively charged ion), or whether we have an acidic cation or a basic cation—all of this chemistry is essentially an affinity for combining or repelling. The great poet Goethe wrote a novel called *Elective Affinities,* which was about a *ménage à trois,* and each person in the drama was, mentally, a different chemical. The family drama of when Sally Silica met Carl Carbon or whatever. Goethe was pointing at the fundamental method of alchemy, analogy. The whole issue in alchemy is to find an analog that gives an inner picture with a feeling connected to it, a picture we can actually take into scientific consciousness, but also connects to us as a feeling.

For example, calcium in the soil has a fixing nature. This is coagula in alchemy. It is a quality that says this is mine, that is mine, and this is mine, too. That is what calcium does in the soil, and as a result it makes crumb in clay soil because it binds things together; but that also means other parts are separated. That binding quality is an analog used by Steiner in his *Agriculture Course* when he described lime in soil as "greedy." What was he talking about? It sounds unscientific or at least anthropomorphic. Sure, but he is speaking in alchemical language. The reason he speaks that language is to give a picture that has a feeling, so that when you go out and walk your land you get a feeling for the action of the lime in your soil. Moreover, if you practice it, the feeling comes up as a picture. I think I need to lime, or I think I have too much lime. Nevertheless, with the method of analogy we do not want to throw chemistry out the window. It is okay to use the language of polyphenols, flavonoids, and so on; such terms name families of substances that are active in particular ways.

The historical shift from alchemy to chemistry is very telling. It is connected with being able to determine the weight of air. In 1703, a chemist by the name of Georg Ernst Stahl (1659–1734) developed his theory of phlogiston. He saw phlogiston as a kind of hidden fire, a substance alchemists would call levity. When alchemists burned a piece of wood, they believed that there was phlogiston mixed with the original substance, and when they

2 *Valence* refers to the relative capacity to unite, react, or interact.

burned the substance the phlogiston was liberated by the action of the fire, whereby the substance fell into what they called a *calx*. This is a fundamental idea in alchemy. *Calx* is the source of our term *calcium* or *calcining*, which means to burn into a calx, or ash. Stahl saw when wood burns, the ash is much lighter than the wood, so he reasoned that there is a *vital force* in the formation of the wood that is released by the fire and that the vital force, or phlogiston, returns to "the crystal heavens," or periphery, the origin of everything.

Entities come from the periphery into a center, and when those entities leave their bodies, the vital force goes away to the periphery, from whence it came. When we see something die, it is quite apparent that whatever animated the corpse is now somewhere else. This is their picture, and thus Stahl was a follower of so-called vitalism. Calx was considered to be left over when wood is burned, because the phlogiston is taken away by burning. Some phlogiston proponents explained that the calx, or ash, is lighter than the wood because the phlogiston has a negative weight. This is a very interesting idea; negative weight means buoyancy, or what alchemists call levity.

Others, such as Louis-Bernard Guyton de Morveau (1737–1816), offered the more conventional argument that phlogiston is lighter than air. However, a more detailed analysis, based on the densities of magnesium and its combustion residues, shows that just being lighter than air cannot account for the increase in mass. When metals are burned, the calx is often heavier than the original metal. Stahl explained this phenomenon by claiming that phlogiston would leave and be replaced by air. He was severely criticized for that idea. Critics wanted to know where phlogiston goes. Can we weigh phlogiston? Exactly where will we find this phlogiston? When metal burns, it may become heavier; how does this happen? Stahl claimed that it combines with air, and they laughed at the notion.

Nevertheless, some hundred years later, the phlogiston theory was seriously challenged in 1779 by Antoine Lavoisier (1743–1794), when he proved quantitatively that burning metals in air to form a calx increased the weight by combining with what he called oxygen. The metal calx was subsequently put in water, which created an acid. Lavoisier called the element from air that added the weight "the acid former," or *oxy-gen,* the generator of *oxy,* or acid. Lavoisier showed that combustion requires gas that has weight, which was oxygen. It could be measured weighing the burning metal in closed vessels.

The use of closed vessels also negated the buoyancy, or negative pressure, that had presumably disguised the weight of the gases of combustion. Lavoisier burned things in a sealed bell jar on a scale. He showed that, as you burn metal in a bell jar, it actually becomes heavier by combining with oxygen, not because air replaces phlogiston, as Stahl had claimed. These observations solved the weight paradox and set the stage for a new caloric theory of combustion. Lavoisier's discovery showed that air has mass and disproved the phlogiston theory. The era of modern chemistry began when Lavoisier demonstrated the importance of quantitatively accurate weights. In the twentieth century, the atomic theory showed that oxidation reactions result in a loss of electrons. No one could figure out what was happening when things burn, but by the time chemistry had gotten to the idea of the electron shell, they proved that it loses electrons.

By convention, the element that loses the electrons undergoes oxidation, and the element or compound that receives the lost electrons undergoes reduction owing to the negative charges of the electrons. Modern chemistry is based on this idea of the electron exchange. Not many people have actually seen electrons, but we know through convention that they have weights and charges, and if you spend enough time in a lab, it becomes your whole world. Nonetheless, chemists eventually realized that these reduction/oxidation (*redox*) reactions do not always involve the transfer of electrons. By convention—and this is the second convention in this issue—the concept of oxidation number was developed to cover the instances when electrons are not really gained or lost in an ionic reaction.[3] Suddenly, we have a convention of a convention, and even though it is an ionic reaction, maybe there are no ions or electrons that are actually changing, but something is happening and we just call it *redox,* or oxidation-reduction. By using the oxidation

3 The *oxidation number* is usually equal numerically to the oxidation state, and so the two terms are often used interchangeably. However, oxidation number is used in coordination chemistry with a slightly different meaning.

number to describe a reaction, it does not matter if a compound actually contains ions. So where are we in the logic structure of this?

Now we will take a turn into the cosmic side of the issue—the total amount of matter (this is what we are speaking about when we describe atomic theory and electrons) in the universe, or the total amount of matter as measured by the cosmic microwave background. Unfortunately for theorists, that measurement accounts for only about thirty percent of the critical density of matter in the universe. In other words, they have calculated everything they can see, weigh, count, and measure, but it adds up to only about thirty percent of the weight that should be present. Therefore, their measurement is missing about seventy percent of the matter that should be in the universe. Seventy percent is not *nothing,* which implies the existence of an additional form of energy that accounts for the remaining seventy percent.

The nature of dark energy is a matter of speculation.[4] It is thought to be very homogeneous, not very dense, and not known to interact through any of the fundamental forces other than gravity. Since it is not very dense, it is unlikely to be detectible in laboratory experiments. Could this be phlogiston? Dark energy can have such a profound effect on the universe, making up seventy percent of universal density, only because it uniformly shows that something occupies what would otherwise be empty space. The two leading models to explain dark energy are the *cosmological constant* and *quintessence,* which also happens to be the fifth element for alchemists. Both models include the common characteristic that dark energy must have a negative pressure—that is it must be lighter than air. To explain the observed acceleration in the expansion rate of the universe, dark energy must have a strong negative force acting repulsively.

Now where are we in terms of weighing reactions in bell jars and measuring and counting things to better understand light—which is, if we go back to Stahl, something in living things that cannot really be weighed and measured. Dark energy is a source of energy in matter, and when you burn matter it is

4 According to physical cosmology and astronomy, *dark energy* is a hypothetical form of energy that permeates all of space and tends to accelerate the expansion of the universe.

released. Maybe it actually could go through glass, which does not sound so farfetched, given the things I am describing. If seventy percent of the known universe is made of dark energy, then glass is really not an obstacle to its movement. What we capture in the glass is the stuff we can weigh.

I am discussing this little scientific cul-de-sac as an alchemical disclaimer, because alchemists have said that there are the four elements of earth, water, air, fire, and a fifth, *quintessence*. For them, quintessence was what we call consciousness, the great force of transformation, the thing that turns one thing into another, the thing that turns a piece of wood into a table, the thing that turns a section of dirt into a vineyard. This is consciousness. Thus, the four elements always had the fifth element in the center. Sometimes it was called *akasha,* Sanskrit for ether, space, or "seed of thought." The fifth element carries consciousness of how the other four interact. According to alchemists, it is possible for human beings to harmonize their consciousness with akasha to get living things to do what they would not normally do, and so they looked for the rules for how akasha, or quintessence, moves—how the energy of transformation, quintessence, moves from one state to another so that they can rhythmically enter those activities and amplify the transformation from earth to water, from water to air, from air to fire, from fire to ash, and then from ash as the seed to a new Earth. They understood that everything is linked in transformation and that the greatest example of this is plant growth.

In the alchemical picture of reality, a consciousness is hidden in life that rules the transformation of energy from one state to another, from one phase to another, from solid to liquid, from liquid to gas, and from gas to warmth, and that these movements are akin to consciousness that human consciousness can work with as the birthright of what it means to be a human working in the natural world. They understood that, to do this, one must understand how a substance becomes part of a whole nexus of activities or dynamics, how biology becomes dynamic, or biodynamic. One must understand how the substances and active forces, as well as the parameters of those forces, interact for instance in a wine-growing terroir. *Terroir* is consciousness of place, time, and the movements of cold, wet, and dry things moving in time. Moreover, once I can learn to form pictures of how those things move in time, it becomes easier for me to form inner pictures, such as seeing that the soil has some mineral substrate in it and that when it rains certain substances come out of there that have to be taken into account. I can reason that maybe there is more potash than there is calcium, and because we are on volcanic substances there is more sulfur than phosphorus. Thus, any deficiencies can arise in me as a picture of why something shows up as a deficiency in the crop. That kind of dynamic, pictorial thinking is human consciousness entering akashic relationships that comprise the archetypical elements of earth, water, air, and fire.

Keep in mind that there are two fundamental ideas in alchemy; things go up and things go down. Going up is levity. Elements under the influence of levity are expanding and moving into a condition of *solve. Sol* is sun in Latin, and in chemistry a sol is a colloid. In working with minerals, there are three states of levity: a solution, a sol, and a suspension. In a solution, the substance unites with water and does not settle out. With a suspension, a mineral goes into water, but because of the particle size, the mineral eventually settles out. In a sol, the particular size of the mineral particle causes it to remain in the water because of the charge on the particle; the smaller the particle, the greater the surface area that can collect a charge. The power of transformation in nature depends on the size and quality of the surfaces that interact. The goal in biodynamics is to amplify surfaces, because in nature surfaces are where gravity and levity meet, and the smaller the particle the bigger the surface. With levity, we are moving toward sol, or solve. Steiner would call this realm of minimal surface the etheric, or life, realm. Key words in this area are *life, sun, force, sol,* and *levity.*

If I drop a calcium rock it will fall to the floor, but if I grind it to a fine powder and blow on it, that calcium rock will just float around the room for a while. The rock is now in a state of levity. To understand this, think of blowing up a balloon. The surface of the balloon has height and width but no depth. It exists in two dimensions. The mass of the balloon has height, width, and depth and exists in three dimensions. If I blow the balloon up the mass grows larger quicker than the surface and, eventually, mass overtakes the ability of the surface to contain it. When that happens the balloon's surface ruptures. Now think of the opposite. If the balloon could shrink, its mass would shrink more rapidly than its surface. Eventually the mass would become smaller than the surface, making the surface larger than the mass. This sounds crazy but it is a scientific fact.

A particle whose surface is larger than its mass is very small, in the nano range. Nano-size particles enable our blood to affect diffusion across cell walls. Nano particles create a sol. In chemistry, a sol is associated with colloidal suspensions. There is a whole other realm of interactions of charges in colloids, and those kinds of charges interact in the cell sap and determine whether particles are available for plant roots. From an alchemical perspective, this ionic exchange is closely related to the principle of levity, or ether. Alchemists would call that levity force *fire*. We burn something, and it moves away to the periphery. However, the fire element in alchemy is not just the fire in our fireplace; it is also the great process of lifting toward the peripheral forces, toward colloids and solutions. Things happen in solutions that defy the law of gravity.

If you drop a whale off a three-story building it makes a big hole. If you drop a horse, it makes a hole. If you drop a dog, it hits and bounces. If you drop a mouse, it will probably hit and run away. If you drop a spider, it floats away. The smaller the object, the less it is influenced by the counter-levity force—the forces of gravity, deposition, precipitation, or coagulation. Because there is a whole other set of forces that counter levity, a crystal can fall out of solution into gravity and weight. It is the tartar we have to scrape off the bottom of a wine vat. It is the calcium that lodges in your elbow. It is the urea crystals that lodge in the feet of a person with gout. Gout results when blood lacks the force of levity that can hold urea nitrogen in solution, so it enters a state of suspension and eventually becomes dense enough to form little crystals that gravitate to the nerves of one's feet. Gout means there is not enough levity in a person's life forces, food, and emotions. Gravity is also consciousness.

Of the four elements—earth, water, air, and fire—earth and water are connected to coagula. One representative of this is mud. Earth has the most gravity, or coagula, and water has is able to lift, even when earth is coagulated or precipitated. Water has a latent levity force called buoyancy. Go down to the beach and drag a log to the edge of the lake. Moving the heavy log against gravity requires much sweat and strain. However, when you get that heavy log into the water, you can even float around on it. People do not think this is strange. That buoyancy is a kind of latent levity in the water. When these forces interact, a subtle play of forces takes place between earth, which wants to pull down into gravity, and water, which buoys. That is the dynamic of suspensions and colloids and solutions.

Water can lift earth into a state of levity, but when water is lifted into an even higher levity state as vapor, earth must then fall. In alchemical language, that falling is called *sal,* or salt. It is basically a process of precipitation. When water goes up into air, it enters a new space. What we call water, H_2O, is actually two gases. This is a paradoxical thing to think about. Water is oxygen and hydrogen, which miraculously come together and give us an incredible gift by falling out of levity. Water is the universal solvent. Through gravity, air enters a gravity state and becomes rain—this is the way alchemists would put it. They would say that rain is the salt of air. Of course, they do not mean an actual salt, but a *sal* process of precipitation. Moreover, if it is cold enough, water forms a crystal. All rain comes from ice, and all ice forms around a nucleus. Ice nuclei are pieces of salt or carbon at 30,000 feet that become super-cooled, attract cold water, and form crystals. Crystalized water is overcome by gravity; it falls, enters a warmer air layer, melts, and becomes rain. This is just basic meteorology.

These dynamics, moving up into levity and going down into gravity, are the two fundamental polarities in agriculture. If you can get a feeling of the thinking behind these processes, it is possible to realize that some minerals are more gravity laden in the processes they support. Those minerals favor deposition. Other minerals are more on the solve side and favor dissolution. Some minerals and metals create very strong action, while others hold back, stop strong action, and lead to settling out. This is soil science. As a general principle, roots and stems are the organs of plants that relate to the gravity forces. Flowers and fruits are the organs of the plant that relate to the levity

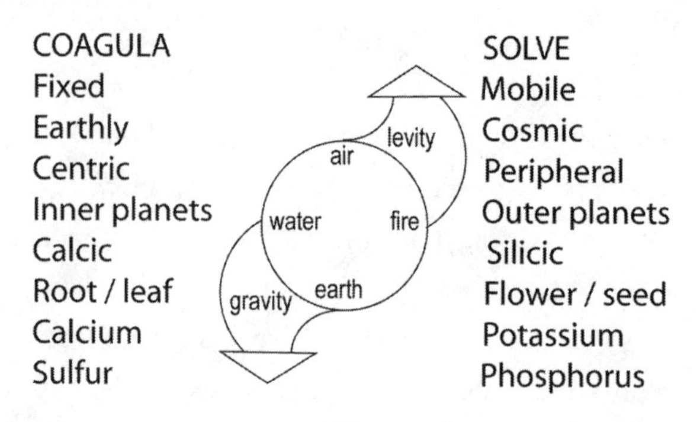

Figure 1

forces. The various minerals and metals tend to work in particular areas because of their function as either a gravity source or a levity source in the soil sap and within the actual electrolytes of the plant itself. These relationships of gravity and levity play out continually, just as they do in the human body. That is what I described as gout, calcium deposits, and so on. Plaque and atherosclerosis are both diseases alchemically related to gravity.

Basic alchemical language always involves these two polarities of levity and gravity. Is the process in the organ moving toward a solve or a gravity condition? This alchemical polarity makes a wonderful meditation. You can actually watch a plant grow in your inner eye, and then ask yourself which part of the plant is in a dissolving stage and pushing toward the meristem. Where is that development happening? What part of the plant is involved in the forces of deposition? Where are substances being deposited at the expense of growth? Burn a few leaves, burn a bunch of flower petals, and burn a root to see which one leaves more ash. The part with more ash is linked more to gravity. You can see a picture of the dynamic of gravity and levity in the particular plant. This kind of visualization is a way of training oneself to think more pictorially and analytically. The idea is to retain analysis while improving it with imagery.

In the center of *figure 1* we see a circle; earth, water, air, and fire. One arrow shows earth and water pointing down in gravity. The other shows air and fire pointing up in levity. On the left we have the coagula column. That is the gravity side of the ledger; coagula has a fixed nature. In his

Agriculture Course, Steiner calls coagula an earthly force and says it is useful to know the difference between earthly and cosmic forces. He also refers to the earthly forces as centric, meaning that they work from centers. They move out from a center and deposit things in space.

When I have something that starts from a point, explodes outward, and then builds mass around it, that mass is building out molecularly to conquer space. Much of our technology is centric, even anti-centric. We are very good at blowing things up and then forming things from the spaces created. We love to blow things up, which is how we get around in our cars; we are blowing up dinosaurs. We explode things and then take the ash and form it. That is on the left side of the ledger in *figure 1.* The earthly quality and the centric quality are linked to the element of earth. Centric means it rises from the center. Earthly is earth and water in the gravity pole. Earth describes the tendency to fix things, to make particles fall out of solutions and settle into substances or forms. An act of forming the thing is what alchemists call *sal.*

The other qualities in the coagula pole of *figure 1* are descriptions that Steiner also uses to describe coagula. The inner planets—Mercury, the Moon, and Venus—contribute a fixed nature to the patterns of growth. He connects their action to lime, the root, and the leaf. It is interesting that we have calcium and sulfur at the bottom of the coagula side. In the next chapter, we will go into detail about the action of calcium and sulfur according to modern chemistry and plant science and the fact that whatever makes the plant move forward needs another substance to hold it back. This is the principle of checks and balances in the economy of nature. Every substance that moves a plant forward needs to be paired with a substance that holds it back. Every substance that holds a plant back needs to be paired to a substance that helps it move forward. By moving forward I mean it moves from a center of growth toward the periphery, where growth is diminished in favor of flower and fruit production. The ultimate peripheral plant forms are pollens and seeds.

Rudolf Steiner describes pollen as little "fire ships." He is speaking in alchemical language about how a flowering plant is going through a combustion process, and when it goes through combustion the fruit retains "ash" products such as essential oils and acids that help it move forward. However, it cannot hold on to all of them, so they have to be dealt with in the ripening process. In the formation of grapes there are two key acids—tartaric acid, which holds back ripening, and malic acid, which moves ripening forward. The tartar is a more coagula kind of salt and the malic a more solve type of salt. The malic is part of the Krebs cycle of transforming energy to make new tissue. Malic acid is great, but if it goes too far the final product of ripening starts to go into oxidation. Thus, if a plant has abundant coagula substances it also needs solve substances to keep it moving, but not too much. This may be important for the grower. If your soil is deficient in the coagula substances, you have a lot of solve; or if your climate is deficient in solve, such as sunlight and warmth, it brings a lot of coagula through the cold and wet, so you need to provide solve. You need something to solve the fact that there is too much coagula. Paradoxically, that solution in nature is mildew. From the perspective of nature, mildew is a combustion process, because there is not enough fire in the plant's life process; there is too much earth and water, so mildew brings a fire process. Of course, mildew will ruin your whole operation, but it will solve your fire issue. You can see from this that alchemical language is concerned with the checks and balances of processes and how to operate in a given sector of nature.

Now look at the right side of *figure 1*. We see that solve is the action of mobility in contrast to the fixedness of coagula. A plant with an abundance of solve makes a lot of protoplasm. It runs toward the periphery. It may be a ten-foot cane on an overly vigorous vine, indicating too much nitrogen and vigor. Solve is the impulse to keep expanding toward the periphery. That is what the meristem does on the tip of a branch. Basically, the growth of a meristem is the effort of the plant being to work back to the Sun. As a spiritual entity, every plant is trying to return to the Sun. We human beings call this effort the "growth process." The spiritual plant is striving toward the periphery. This is the solve quality in a plant as it reaches out to become mobile. The solve pole involves the cosmic qualities, such as light and warmth, which are difficult to measure physically. You can have your photoperiod, the amount of light exposure. We have all sorts of information

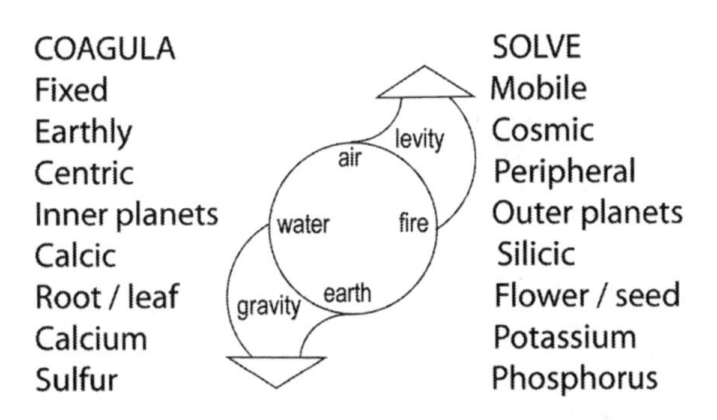

Figure 1

and know the temperature curves and the difference between night ripening and day ripening, but what exactly is a photoperiod in reality? Besides being a mathematical abstraction, what is $C_6H_{12}O_6 + H_2O$ in the presence of sunlight? What do we actually mean by "in the presence of sunlight"? What is actually active in sunlight? We can measure forces and we can measure time, but we cannot easily answer with a formula for what light is *as a being*. In the same vein, we could ask what warmth is as a being. How can we characterize it? These are not easy questions.

So the cosmic side of this polarity is very difficult to measure. There has been a lot of discussion over what is going on with light and with warmth. When science discovered that air has weight, and that oxygen has weight when metals are burned, it gave some answers and models, but we still do not know where the warmth goes when wood burns or where it goes when we digest a sandwich. We still do not quite know how the warmth process happens. Steiner calls these forces the cosmic, the mobile, and peripheral forces. He calls the solve forces *planar,* or surface, forces.

Science has a lot of pictures around planes of light. Our Earth is bathed in energetic planes of light. Science tell us that light does not strike us as particles but as planes, and that light energies propagate through wave planes, or plane waves, which are perpendicular to the wave of the light direction of movement. This is the picture from conventional science. Light planes rotate around an axis that is perpendicular to the motion of the wave. Each and every one of the infinitude of stars generates planes of light energies. Consequently, the Earth is bathed every moment and from all directions by energetic planes moving from stars and streaming to Earth, creating crystalline interfaces. Every time three light planes meet, there is a point; in Steiner's work, that point is at the tip of the meristem of a plant. That is his picture. There is something special about those meristem points; Steiner calls them "star points." The meristem aims at those star points. It is a representative of the infinitely distant plane.

The peripheral, or planar, forces are characterized by Steiner as suctional. According to astrophysicists, the ubiquitous energy of the universe, so-called dark energy, is sectional in its effects. Steiner said it was suctional in 1920, and people thought he was crazy. Now we are finding out that, indeed, the most common cosmic forces in the universe are suctional. They are drawing out things here on Earth and making them subtler and subtler until they exist somewhere else on the periphery. Steiner connects those forces to the outer planets of Jupiter and Saturn and to the spaces beyond the planets. In the plant, the peripheral forces support the formation of flowers and seeds. By forming the flower and the seed, the plant is going out into the Sun. It is going into the periphery, and as it does so it creates much finer substances. There is root beer and there is cabernet sauvignon. They are both beverages, but what are the phenolic overtones of root beer? The periphery is where the finer substances are articulated and where ripening and protein formation happens if we allow the plant to go there.

If we allow the plant to go into ripening and hold it there, it bathes in those finest peripheral forces, because the periphery, the skin of the grape, is your bank account as a grape grower. Why? Because that is where all the flavonoids are. That is where we find all of the things that are valuable to growers. The most refined parts of the ripening happen in the periphery of the periphery. It is not the seeds and the stems; they give you something but not the main thing. The peripheral forces are solve, or refined, forces. They are highly developed forces and, depending on very minute changes in temperature or pH, they are incredibly flexible, so that they can become a raspberry, licorice, or other things.

So the flower and seed can be seen in the mineral realm as potassium and phosphorus. However, even within a set of polarities there are polarities alchemically. For example, in the mineral realm potassium is a gravity (sal) representative in the fire pole. Also in the mineral realm sulfur is a levity (sulf) representative in the gravity pole. Alchemical polarities within polarities require a very different type of consciousness when we are doing research. Once I think I can say for certain what is real, someone makes a new rule such as the oxidation number. Then the case for ions needs to be seen as a polarity within a polarity, because there actually are no ions moving in a chemical exchange, but science needs explanations to maintain appearances.

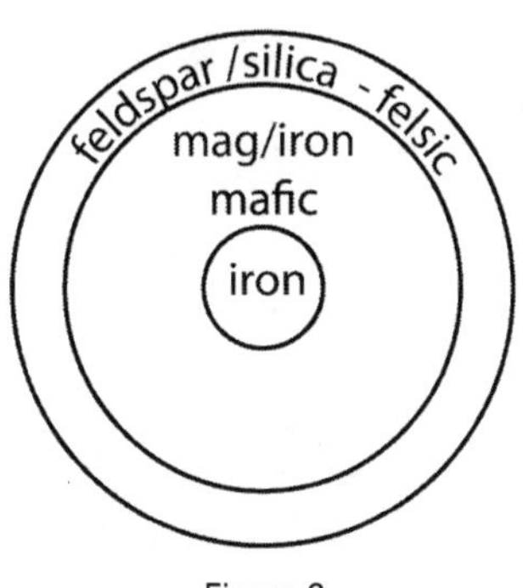

Figure 2

That is the crazy-making part of trying to pin down alchemical reasoning. Alchemists had little problem with such polarities, since they understood that we need to expect polarities within polarities. To be able to remain mentally intact in such a system, it helps to see problems pictorially, allowing us hold the problem in several different ways without feeling conflicted.

Now look at *figure 2*. This diagram represents the way the Earth is built. It has an iron core surrounded by a mantle of magnesium and iron. Magnesium and iron compose what are known as mafic rocks. The term *mafic* combines magnesium (MA) and iron (FE). Mafic rocks are minerals such as olivine and gabbro. Those are the dense rocks that surround the mantle as a kind of centric force around the iron core. Some ocean beds are made of mafic rocks. Most volcanic materials have a tendency to be *felsic*. This term combines feldspar (*fel*) and silica (*sic*). Those rocks represent the solve, or cosmic, pole. From among the feldspars, potassium, calcium, and sodium move from a mineral form to an active form that plants can assimilate. They weather out of the general silicate rocks of the Earth's crust. Granite (composed of quartz, feldspar, and mica) weathers out into many gem forms and most surface rocks of the continents. Most of the rocks on the surface of the continents are silicates unless they are of oceanic origin.

The weathering process happens among the silicates as derivatives of granite. Even in this basic division of two types of rocks, you have one that is coagula, or mafic, and another that is felsic and more the solve type that are dissolving to make everything else.

This is basic geology. In a typical school, you get the hair on the bump on the log on the frog in the hole in the middle of the sea approach to learning the sciences. You are given a list of 400 terms to memorize before anyone tells you anything. It may take years to get to the picture and see that most of the mafic rocks are in the center and most of the felsic rocks are in the periphery. Nevertheless, it helps to know that and it helps to know that the felsic rocks are in the solve pole and the mafic rocks are in the coagula pole. It helps to know these differences, especially if you have a lot of mafic rocks that you are trying to grow things in. With mafic soils you have a whole other problem than you would have working sandy soil or feldspathic soils.

I will give you an example of how thinking in pictures rather than in fixed concepts relates these ideas to various aspects of life. *Figure 2* is a diagram of the Earth body divided into the two most fundamental rocks in geology. In the center, surrounding the iron core, we see the alkaline, or mafic, minerals. On the periphery we see the acid formers among the minerals. This diagram of the Earth is similar to your digestive system, with its alkaline forces on the inside and all your senses composed of proteinaceous membranes on the periphery that resemble the formative processes found in the silicates. Alchemically speaking, the analog for your body in this model is the Earth. This kind of pictorial thinking is called analogical, and if you want to take the time to develop pictorial thinking and you have some science background, you can find the most amazing links between the way the pictures move alchemically and the actual nuts and bolts of what you would find in science in terms of analytics. You can find that the two poles of analysis and imaginative cognition dovetail beautifully into each other. Pictorial imaginations help you understand analytical data by grasping things from the outside. This enables you to look at wholes in nature rather than trying to build wholeness out of little pieces.

We are being driven into a way of doing science that will be unable to grasp the subtlety of life forces. Computers and the memory they use force us to see the world increasingly in pieces. The more pieces we get, the more difficult it is to actually reach a conclusion. Has science determined whether chocolate is good or not good for you? Both. Is wine good or not good for you, and what does moderation mean? Our way of gathering information makes it likely that as we get more information we are less able to hold the big picture of these dynamics. This is because, alchemically, everything out in nature is actually a picture of the human being. If you can understand this and work inwardly with pictures and even take those pictures of your work in nature into sleep then your meditative life as an alchemical practitioner helps you build analogs that allow you to see action of medicines, whether pharmacology, homeopathy, psychology, neurology, plant physiology, geology, or climatology. How many "ologies" do we need to understand that we are deeply connected to what is out there in nature.

II. Mineral Cross • Lime • Feldspars • Silicates • Amethyst • Inner Surfaces • NPK • Salt • Sulfur • Mercury

As a supreme alchemist, Steiner brought together plant substances, mineral substances, and rhythmic gestures in animal organs and linked them to planetary movement patterns. When these different realms are brought together in the proper protocols, the result is synergy. My purpose here is to bring what he did and why he did it. Of course, this is from my perspective as an aspiring alchemist and should be taken with a grain of salt. When you see these realms organized this way, the mineral, plant, animal and planetary signatures appear to be very similar. They may be in different realms but the gestures are similar.

For years I had a website, DocWeather.com, and I used take pictures of a cloud, a wave, the American River, and a mountain configuration in northern Colorado and then compare their energetic signatures in an article on my site. I would get some nasty responses from people: "I am kind of interested in what you are doing, but please stop putting this bullshit on the Web. These are not connected to each other, so stop putting out this garbage." I would say, "Well, I pay for the website, so you can have your own website and do that."

Those comparisons of what we call gestural motifs were used in ancient times, when they were known as the doctrine of signatures. The doctrine of signatures as it stood in those times has been rightfully debunked, but the idea behind it is still very useful. What they did with it was on the mystical side of science—what people today call pseudo-science. The danger of alchemy is that it can lead to mysticism and pseudo-science. On the other hand, pure science can lead to abstract delusion. Somewhere in bringing these two polarities together is a middle way. In physics, the middle way is not so important, since abstractions work well in the physical realm. However, in agreement with Goethe, it is in the life sciences where the real issues are to be found. In the life sciences, it is in the play of things that reality is found. It is not so much *what* the substance is, but the real play of forces. To a physical chemist, the formula is the determining factor, but in the life realm it is always the fact that one substance is buffering another.

The issue in alchemical work is that formative activity becomes substance, and then the substance behaves as it does because of the formative activity that gave rise to it. In the realm of life, the activity is primary, since much of the activity is in the reciprocal actions of buffers. The activity supersedes even what the substance does, because the substance in a typical reaction is going through so many changes between, say, an acid and a base, or one temperature and another that, in a laboratory dish, that same reaction would neutralize and die. In life the most delicate reduction and oxidation reactions are happening simultaneously at the nano level and there is little neutralization to distort the reaction. In nature there are all sorts of complex ionic changes that flow into one another without confusion. Out of the chaos a substance emerges. I analyze it and say this is what it is, but as soon as I analyze it, I take it out of life, and it becomes that thing right there without all of the magical ability to sustain polarities in a living way.

In the ancient world, that inability of substance to stay in the levity field of life was the pole of coagula—things that had become fixed. This thing right here is this thing right here to the fixed mind. When the ancients said *earth* they did not mean this thing right here forever. They meant something that had fallen out of potential. In that system, a thought would be earth. The thought, once it appeared, was no longer in contact with the potential for forming it. They considered the formation of an embryo from a solution in the womb to be an example of earth. That was an earth; an embryo to them was like a crystal, a crystal like a thought, and a thought like an embryo. They were all coagula, or earth.

In fact, in alchemical language, there are *embryonates* in the mineral realm. This is a condition in which there is an ore concentration in a pocket of the mass of native, or "country," rock. Embryonate ore goes from a state of less concentration than ore through more concentrated stages to finally become actual ore. Paracelsus, the great alchemist, called these formations embryonate. The language that alchemists use for nature has a lot to do with seeing the human being as an archetype for the natural world. This is the method of analogy. Analogs are pictures of material states going in and out of manifestation. The pictures of "inanimate" things resonate with processes in the formation of human life. Consequently, they are considered to have continuity with human life. This was the source of the imaginations that created medicines from natural substances. The substance, although the corpse of an activity, is animated by the process that formed it. That potential for activity remains in the form of the substance, even when that activity has ceased to form it. The life was considered still present in potential, even after the corpse has fallen. Then, the form of the corpse had potentialities for activities in it.

Today we call the recycling of dead forms of life *composting*. We take dead substances and, suddenly, they become alive again. It is a great miracle. Nature says to give her all of your dead things and she will put them back

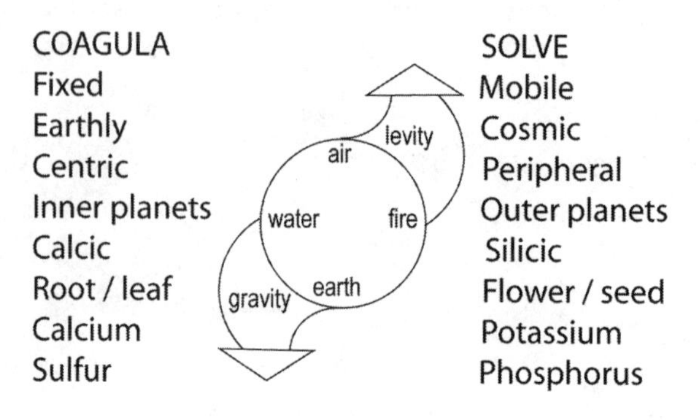

Figure 1

together; thus the dross becomes black gold. Death becomes the great miracle of life. The fact that we can stand up is a miracle. It defies the great law of gravity. Where does that force that allows us to stand up come from? An alchemist would say that is the human inheritance of the force of levity. With levity we overcome death.

In *figure 1,* again, Earth and water are in the gravity pole and tend to be connected to coagulation, the force of the sal pole. It is a process whereby potentialities fall into the realm of the fixed, or coagula. When there is a condition in which I am trying to understand an entity such as a plant I am tending, alchemical imagination allows me to picture how the plant manifests. Manifesting life forms are going in the direction of coagula. They come from a totally non-physical warmth state. Although we can measure temperature, we do not really know what it is. With temperature, one thing is always relative to something else, but on a scale of what? Absolute zero to who knows what? Absolute zero is the ultimate coagula, and who knows what is the ultimate solve. Somewhere in between these two is the remarkably narrow temperature range that can support life as we know it. The coagula side is earth and water; the solve side is light and warmth.

To alchemists, warmth has more levity than fire. They would say that the warmth has a corpse in the form of the fire. Warmth is living potential; fire is the fixed condition of warmth. That may be an odd way to put it, but this is how alchemy would look at it: fire is coagulated warmth, the corpse of warmth. Fire that you see and burns things is a corpse of the potential for warmth to extend throughout the universe in ever-finer states of existence. Standing around a fire, we can feel that warmth is cast off the fire as it returns to the solve condition at the periphery. We can stand ten feet away from a fire and feel the warmth—that is, the unmanifest form of the fire. Technically, I should say that fire is the slightly manifest corpse of warmth. Fire is tricky because it is the least manifest of the four elements but is still more manifest than warmth.

Therefore, alchemists would distinguish between warmth and fire. Fire is going toward coagula, warmth toward solve. They are held in a delicate condition that can cook your food or burn your house down. Fire as an element coagulates further into the air element, which further coagulates into the water element. The water element in turn coagulates into the earth element. The two great coagula forms are water and earth. Water has levity in it, but it also has gravity, which allows it to coagulate into ice. Elemental water also coagulates by combining with earths. We can see that in the formation of carbohydrates—*carbo* (earth) *hydrate* (water). A carbohydrate is earthed water. This is how alchemists would see it.

Earth and water are the gravity, or coagula, side, and air and fire are the levity–solve side of the four elements. Between water and air is a kind of membrane, which I have described from the point of view of physics as phase shift. From the solid–earth to water is a shift in phase. It takes a certain amount of energy to lift something from earth into a fluid condition. Suppose it takes one caloric unit to go from a fixed condition to a fluid condition. In physics, it takes seven times the amount of caloric energy to go from a fluid condition to a gaseous condition.

Each time there is a phase shift, it takes a huge input of additional energy to shift to the next phase. As the reaction comes back down through phases, that energy is released. A huge energy input is required to go from water to air through the boiling point. However, there are these magical places—like the top of a lake—where the temperature never gets to a boiling temperature, and yet there is a huge amount of evaporation because other laws come into play, such as the law of minimal surface, which we have discussed.

In the law of minimal surface, a reversal takes place as substances start to move away from coagula. Substances at the level of minimal surfaces move into a solve condition and start to go into volatile states. Substances at the minimal surface scale do not need the same temperatures required at a larger scale for a phase shift. In this realm, then, we get to the domain of life, and what Steiner calls *attenuated substances* becomes the activating principle—the action of the smallest particles rules, the action of the extremely small.

When Steiner talked about attenuated substances back in 1924, people were skeptical, but the concept of attenuated substances supports a huge industry today, semi-conductors. Attenuated substances at dilutions of parts per billion can make a big difference in electronics technology. In today's world, alchemical air and fire can be found as substances so small that they begin to exist in a whole other universe, where there become huge forces because the particles are so small. This sounds kind of crazy, but it is the world we live in. In Steiner's work, the whole purpose of stirring silica in water is to multiply the surfaces in the water through the formation of countless micro-bubbles, the action of which is extremely powerful. This realm of infinitely small particles and fluid membranes moving at different rates next to each other is where life reaches its maximum. Alchemists would call it the interface between the alchemical elements air and water.

Again, this is the method of analogy. People look at the analogical method today and think it is a bit stupid. Nevertheless, the imaginative consciousness behind the method is useful when you wish to see more deeply into complex subjects such as mineral transport in rock faces or in soil solutions.

Figure 1 had coagula and solve with a list of terms for the polarities Steiner uses in his work. If you read Steiner's *Agriculture Course*, whenever you encounter the terms in *figure 1,* that chart will help you orient. *Figure 2* is a picture of the way science views the Earth's organization. In the center is an iron core surrounded by magnesium and iron, or mafic, rock mass as the mantle, in turn surrounded by the felsic crust of feldspar and silica. The mafic layer tends to be alkaline and basic, and the felsic layer tends to be acidic. In your body you have an alkaline core with an acidic periphery—an

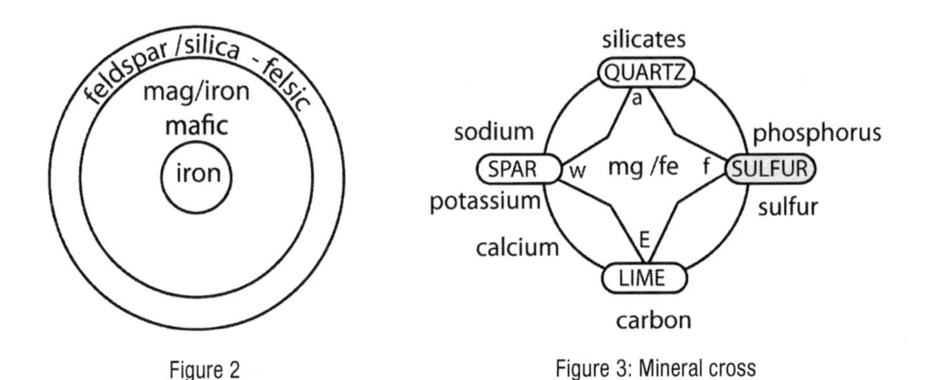

Figure 2

Figure 3: Mineral cross

acidic crust in the sensory organs. This is a good picture of the way organisms are organized.

The picture of the Earth is also a picture of what we could call a basic organization of an organism, within which are the fundamental polarities of life. *Figure 3* breaks down the basic minerals that we must engage when we are working with the land. The diagram shows the minerals and divides them into the elemental categories of earth, water, air, and fire. At the bottom is earth, the lime pole of the minerals in which the important substance is carbon. We think of calcium carbonate, but the lime pole includes all of the carbonates and carbon compounds. All earthly life is based on the formation of carbon compounds, which Steiner calls the carbon scaffolding in his *Agriculture Course*. Carbon is the foundation of life as we know it on Earth. Life is built on carbon. That is in the pole of earth among the minerals. You have lime as the mineral representative of carbon. In the realm of life all the carbonates—sugars and alcohols and all of those substances, generally anything that has to do with life is based on carbon. In the mineral body, carbonates, as representative of the earth element, provide a foundation for organic life but represent the end process of what we could call the sacrifice of the silica.

The opposite pole to the lime pole is the silica pole. In the Earth body, the silica pole includes all of the silicates. Seventy-five percent of rocks are silicates; they start out as silica and break down through weathering, which is a very interesting phenomenon. Weathering builds soils.

The quartz crystal, when it grows molecularly, contains channels and is not quite as solid as we generally think. If I take a quartz crystal, attach a lead to it through a galvanometer (a voltmeter), and put pressure on the crystal, a charge will come out of it. If I heat a silica crystal, a charge also comes out of it because of the space within the crystal.[5] In silicates, the molecular structure is a kind of energetic imprint of an archetypal being behind the formative process of the gem in the rock mass. I am told that, late in his life, Steiner became interested in using gems for healing. There was a lot of interest in using them for their healing forces. When I heard that, I realized that this is a door.

Years ago, I was in the Rosicrucian library in San Jose and found a little book written by an alchemist about fermenting amethyst in red wine. The idea was to pulverize and make amethyst small enough to ferment it in red wine to give it a healing effect. I have done this for many years and sprayed the diluted gem ferment on crops with good results. As a silicate, amethyst has a unique formative process known as "Brazil twinning."[6] Amethyst is a silicate and, like all silicates, is based on the geometric form of a tetrahedron—a pyramid that has four equal sides and can be contained in a sphere. The way tetrahedral pyramids link to one another makes sheets, coils, or some such form.

Molecularly, the tetrahedron is the basic silicate form (SiO_4), four oxygen atoms and a silicon atom (a chemical analog of carbon) that form a pyramid. The way that these pyramids link to one another determines the form of the gem. Amethyst is called a Brazil twin because of the tetrahedral arrangement. In quartz the tetrahedra link up in spirals. One is here, and then another one attaches to that one at a point but the second one is slightly skewed from the first one. This skewing of tetrahedrals keeps happening, finally resulting in columns of spiraling tetrahedral molecules that rotate light within the crystal. As the tetrahedra rotate, energetic channels are formed within the crystal. It is those energetic channels within the crystal that produce the charge when you squeeze the crystal or heat it. The

5 See http://quartzpage.de/, which describes various silicates, their molecular structures, and how the inner structure of silica or quartz conducts charges.

6 See, for example, http://www.quartzpage.de/crs_twins.html.

stress causes the crystal to change its spatial configuration, and there is a discharge of energy called piezoelectricity through pressure, or pyroelectricity, through heating.

Looking at the shape of the whole quartz crystal, we see that it grows along the central axis. It actually spirals along the central axis. Steiner says that if you spray pulverized quartz on crops, the quartz will acts as a prophylactic against mildew because, alchemically, the long spiral gesture is a light gesture—it is an image of a plant flowering. Quartz is a kind of mineral flower growing up and out into the light realm. When you grind the quartz very fine and then stir it in water, the surfaces in the water pick up the formative imprint of the way that quartz grows. The water takes on the molecular formative gesture of the mineral streaming toward the light. Mildew loves water and earth, which are found elementally in the leaf of a plant in a cold, wet season. Mildew is a fungus, and the silica crystal is a prophylactic against such invasions because it has the molecular formative gesture of a growing plant stalk streaming away from earth toward the light pole of its flowering cycle.

Although amethyst is a silicate, it has the Brazil twin molecular formative pattern in which one tetrahedral group goes right, the next one goes left, and the next right, and so on through the crystal. If you look at the overall formative gesture of the amethyst, it does not grow in long spikes but within geodes as short, prismatic, held-back points. It is hard to find a three-inch-long amethyst crystal, but it is easy to find a geode with many of them in a cluster. Therefore, we spray amethyst on plants such as cabbages and lettuces that you want to hold back from flowering. You do not want a cabbage to emulate the streaming light formative gesture of the quartz crystal. It would turn into a shepherd's purse (*Capsella bursa-pastoris*).

This is an example of the analogy method based on molecular science; the molecular formative gesture of an amethyst becomes an analog for the type of growth process you wish to induce in a cultivated plant. The red wine? I do not know what it does, but I think the plants love it. I just put that in there, but the Rosicrucian guideline said red wine; the amethyst has an affinity for that.

The mineral has to be ground to an extremely fine powder. You grind it on a slab until you no longer hear it grind; this brings it close to the nano level. Then, make a dough with the powder and wine, and put the dough into a horn, and bury it for a summer and autumn. After it comes out, put that in the red wine and let it sit and ferment there. I spray this preparation on my cabbages when the moon is moving through water signs. I call it the moist prep 501. It brings light, but a moist light rather than the summer light of the Sun. I need moist light in Northern California. We have enough sunlight, and you can fry plants here by spraying the quartz prep on them. The dilution of the finished wine prep is about one tablespoon per gallon.

Ferment is an alchemical code word used to describe anything undergoing a hermetic transformation. *Hermetic* means sealed off from the atmosphere. A cow horn is a hermetic vessel sealed off from the atmosphere. Amethyst powder in a gallon jug of wine is a hermetic operation. The sealed vessel creates a condition in which the transformation takes place. The wine is an analog of the weathering process found in the country rocks where weak acids from dead animals and plants percolate through the rock to draw out feldspars and other mineral compounds. In nature, silicates are attacked when acidic rainwater moves in the channels and interacts with the mineral micro-surfaces there. If the water is pure in there, there would be no action. When things rot and rain falls, the rain carries those acids from the rotting things up into the crystal spaces and starts to pull potassium, sodium, and calcium out of the interfaces through chemical ionic reactions.

Depending on what is in the solution that goes into the quartz, you get different minerals coming out of the country rock and moving through the soil solution. The different solutions start to act on the various formations, whether volcanic, sedimentary, or granitic. The soils that result from this weathering process are analogous to the divisions in the large masses of country rock known as *plutons*. These are large masses, hundreds of miles long, composed of molten rock that has been buried in the earth. Gradually, the various minerals in the pluton settle out like raisins in pudding, and the various levels of the pluton give rise to different soil types when that section of pluton makes its way to the surface to be acted on by the weather.

Dark rocks (olivines, pyroxenes, and gabbros) are mafic and settle to the bottom of the pluton. The clear material at the top is felsic minerals. Even in a little pluton, in the area under where you are living, that same type of acid/basic organization is present. The mafic rocks are rich in iron and magnesium and settle out to the bottom. The felsic layers are where we find gems.

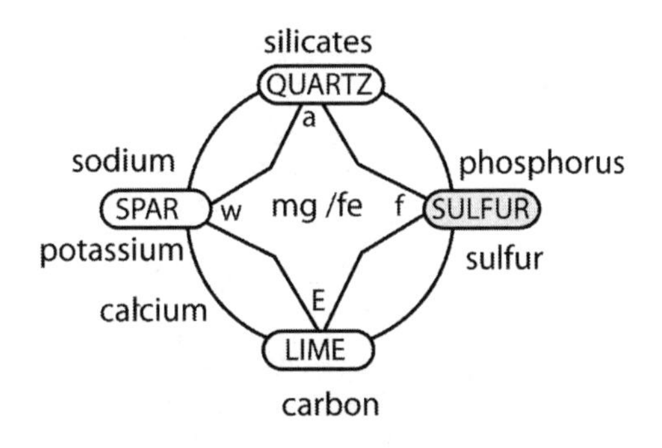

Figure 3: Mineral cross

I am describing to you the alchemical method of analogy. I just made an analogy between amethyst crystals and cabbage. To empirical science, such an analogy is absurd, but when we use pictorial thinking we can picture formative gestures from different realms and connect them. We are looking for analogous patterns of movement of Goethe's "becomings." It is a type of pictorial thinking. Analogy helps us understand that we can take an amethyst and spray in on cabbage or lettuce and get results. I can spray it on my lettuce and have it in January. Lettuce loves amethyst. It is as though they are making a little geode with their leaves. I cut the lettuce, put it in a bag, and when someone opens the bag it will smell like lettuce and have a certain quality and bouquet.

Taking something from one realm and applying it to another is what Rudolf Steiner did. He took plants and put them into animal organs or took minerals and put them in animal organs. He was using the method of analogy in his *Agriculture Course* when he taught that two things can travel the same path of becoming, even though they are in completely different realms. They have the same kind of becoming process as they move from unmanifest into manifestation, even though they are in completely different realms of life. They have a similar, we could say, formative gesture. They have a similar becoming. If we can find the way that the becoming evolves, we can take things from different realms that have similar evolutions, put them together, and get synergy.

In *figure 3*, the mineral cross diagram, the water element (*w*) is to the left with the mineral spar, representing the feldspars. These are part of the action of water on the silicates. Whether sodium, potassium, or calcium, the minerals that come out are based on the acid that goes into the silicates. Different portions of minerals come out, determined by interaction of the liquid acid on that mineral. There is a kind of flowing quality to the formative gesture of the feldspars that is different from the fixing quality of the carbon.

Carbon is the building block or foundation, whereas the feldspars always interact with the solutions moving through the rock. Geologists call these solutions that flow continuously through rock masses "rock milk." In weathering soil or in substances deeper in the magma, water seeps in and is pushed through the rock mass. The various pH ranges of that water create conditions whereby minerals are continually pulled from the rocks as the weathering process. The weathering of the soil from silicates into all the feldspars makes it possible for plants to pick those minerals up to make their own bodies. That is the spar pole in *figure 3*. I put the spars with water because they are acted on by water; they have a flowing quality in the soil solution. Potassium, especially, creates the ionic channels allowing plant nutrients to flow within the sap. It works that way in the mineral body, as well as in the plant itself. Potassium is a kind of a shunt for ions to move from one place to another in the forming of a plant, so that is why it is very active in the beginning of the year. Potassium uptake in plants is very active because it is the active ionic transfer agent in soil solutions.

I put quartz and all the silicates with the air pole of the mineral cross. I already described the molecular structure of quartz. Its inner molecular structure has an energetic, tube-like form of inner openness. In a quartz crystal, once the first layer of molecules is laid down, those rotational planes within follow the winding pathways, as though the quartz is made of countless paper-roll-like windings with multiple inner layers—highly organized surfaces within surfaces.

That is why we use it for computers. When a synthetic crystal is made for a chip, they put a seed crystal in a vat of molten silica. Then the vat slowly turns in one direction while the seed crystal slowly turns in the other direction. After this winding, a crystal called a boule will have formed. It is about the size of a small water bottle. In that boule, every surface is identical, wound to be molecularly perfect. As the crystal moves in one direction, the solution in the vat moves in another direction. This allows the silica to wind, molecule by molecule, around the seed. Finally, they slice the synthetic crystal and etch it for use in a computer. Silica is an infinity of inner surfaces.

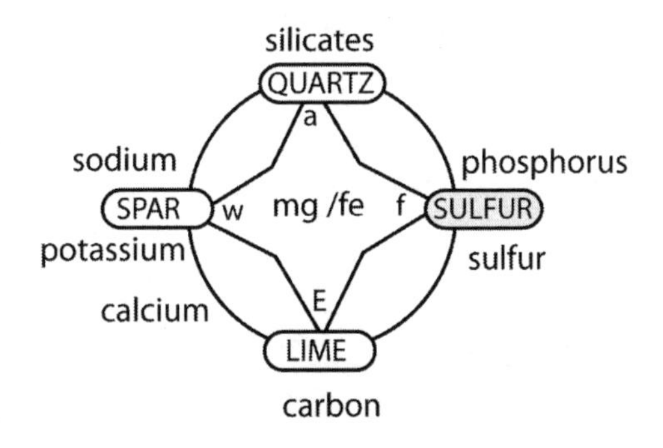

Figure 3: Mineral cross

In the language of alchemy, surface is air. If we want to understand this, we can go out and look at a lake. Where water and air meet, a surface appears. Chemically, air also includes the principle of "inner surfaces." Chemistry is the science of having one substance come into a more intimate relationship to another substance to form a compound. The ultimate agent for this is water. However, in water there is the great mystery of how two "airs" (oxygen and hydrogen) sacrifice to meet, fall in love, and have a child. That child is water, the corpse of air. Simple experiments can reveal the inner surface of a drop of water filled with minute streamlines of forces. We get an energetic discontinuity wherever fluids or gasses are moving in opposite directions, or in the same direction at different velocities, or even in the same direction at the same velocity but at different temperatures. The discontinuity is a surface boundary where there is the maximum of energy transfer. In nature, energy is transferred along surfaces, lines, or boundary gradients of discontinuities. It is the surface characteristics of substances that create the huge differentials that result in such phenomena as lightning.[7]

7 From Wikipedia on electrical voltages and surfaces: Most authorities agree that whatever may be the origin of free electricity in the atmosphere the electricity of enormous voltages that disrupts the air and produces the phenomena of lightning is due to the condensation of the watery vapor forming the clouds. Each minute drop, as it moves through the air, collects upon its surface a certain amount

In this explanation, the smaller droplets have a larger surface-to-mass ratio and can hold a greater charge. As the minute droplets coalesce into larger drops, the surface of the larger drops is reduced in relation to the mass of the drop. This creates an overload of charge that must be discharged. The overload builds until the drops become so large that they cannot hold on to the charge and a discharge is stimulated, resulting in lightning.

The law of minimal surface is why a multi-stranded electrical cable is not made of a single solid wire. In electrical devices, current is conducted only along the surface of a wire. To accommodate the conduction of large volumes of current, the electrical cable must be made of many small wires, which give more conducting surfaces in the cable.

In nature, *surface* is the real workhorse. It is where the charges are carried both in organisms and in elemental phase shifts. Surface allows weathering, the exchange of gasses, and membranes. In the formative gesture of a surface, first there is an energetic membrane caused by some type of discontinuity in temperature, salinity, velocity, movement, or direction. Where that energetic membrane forms, matter has a tendency to fall out of potential and make a physically manifested membrane.

A good image for meditating on this is the formation of a sand bar in a river. The energy in a river is carried by the inner surfaces of the water. Where the water slows to make a turn or encounters some form of resistance, the diminished forces in the inner membranes of the water cannot carry the load of silt anymore. Where the water slows down, the energetic-flow membranes in the water will lack sufficient energy to hold the silt against gravity. The silt falls out because the energetic flow is impeded. The water has to drop the silt to conserve momentum. A sandbar is the resulting form gesture of the gravity-laden silt. Human bones are analogous to sandbars caught in the fluids of your muscles and blood flowing around them. Alchemically,

of free electricity. Then, as these tiny drops coalesce into larger drops, with a corresponding decrease in the relative surface exposed, (the law of minimal surfaces) the electric potential for discharge rises until it overcomes the resisting power of the air. This remark will be more clearly understood when it is considered that, with a given charge of electricity, an object's potential rises as the electrical capacity of the object holding the charge is decreased, which is the case when the minute drops coalesce into larger drops.

the surface function is air. Therefore, silica is the air mineral that represents inner surfaces.

The fire pole is sulfur, but alchemy calls it *sulf*, which refers to any agent in which the levity force and the gravity force are mingled so intimately that gravity is almost overcome. If we get sulfur rock from a volcano and hold a match next to it, the rock catches on fire. This is not your usual rock. Geologists call sulfur "rock grease." It actually is a kind of grease, but as a rock. In rock grease, the levity force (which wants to go to the periphery) and the gravity force are in a very delicate equilibrium.

The sulf pole includes minerals that represent very high forces that are not quite manifest. One is sulfur itself, as a rock, and the other is phosphorus, which is not actually a rock but can be formed elementally through a complex chemical process. In nature, phosphorus is always in compound, either moving away from or coming into being with these other minerals, the feldspars or carbonates. Therefore, phosphorus is the great agent of change among the minerals needed for plant nutrition. It is the fuse that ignites the movement of minerals in the plant's sap. Because it is so volatile, phosphorus is susceptible to becoming fixed. This is because, in nature, everything that reaches a peak in one pole is in danger of becoming its exact opposite. It is a great alchemical principle of polarity that, once something becomes so intense in one pole, it calls into existence the exact opposite to maintain balance. The farther it goes in one direction of development, the more it calls the exact opposite to balance the one-sided development. Phosphorus is a very volatile metal and the great transformer among the metals needed for plant development.

This is why it is in the center of NPK—nitrogen, phosphorus, and potassium. N, nitrogen, is another volatile and an air. Most of the air we breathe is nitrogen, and not much is oxygen. Nitrogen is in the air pole and is very quickening to life. P, is phosphorus, also very volatile. It combines with many things, and it, too, is constantly in danger of becoming fixed because it is so volatile. K, potassium, is potash. It is actually a sal mineral from feldspar, but it reacts with the water to get ions channeling through things. The three major nutrients that you supply to a plant—nitrogen, phosphorus, and potassium—are the most mobile. This is why they are substances sold in a bag as synthetic fertilizers. Plants use them up and they need to be replaced.

Sulfur and calcium are the least mobile, and are considered secondary in conventional agronomy and soil science. Being mobile means the substances are used up quickly; we have to just keep supplying them. The secondary substances slow things down. The primary substances are solve. The secondary ones that hold back and settle out into the spaces created by the primary metals are coagula. We could say that the volatile substances are levity, and those that hold back are gravity. When we consider deficiencies in plants, phosphorus and potassium deficiency are found along the veins in the older leaves on a plant; it is the yellows and purples. Calcium and sulfur deficiencies, by contrast, are found in the terminal parts of plants. There we see browning and such.

Why do some deficiencies show up in the terminal parts and the others in the veins? Using solve and coagula in a pictorial way of thinking, solve forces move to the periphery, and nutrients flow to the periphery. Phosphorus and potassium cause nutrients to flow toward the growing tips. Potassium builds ion channels to keep the nutrients flowing toward the tips of the plants. When we have a deficiency of these metals, everything will be moving toward the tip away from the veins and we see a color change that reveals a deficiency in the veins of the older leaves. It is because everything is moving away from the older leaves, trying to become solve on the periphery.

However, in the opposite situation, nutrients are fixed in the older leaves through calcium and sulfur. In this case, deficiencies will be found in the terminal areas, because the nutrients are being fixed in the older leaves, meaning if I have a calcium or sulfur deficiency, it is difficult for the plant to go into fruiting. Sulfur and calcium support the fruiting process, but they have to be dragged up to the tip of the plant to support flowering and fruiting. When we have a deficiency in sulfur or calcium, we cannot get good fruiting because the calcium and the sulfur are fixed in the older parts of the plant and there is not enough of these minerals to go to the periphery, so the periphery of the leaf tip changes color or the fruit does not set properly.

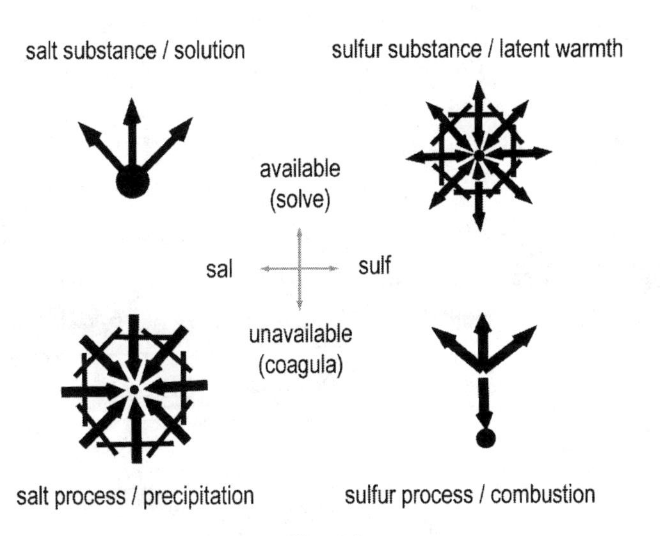

Figure 4

These symptoms are standard plant physiology, but when we look at them in terms of solve and coagula, we see the dynamics of phosphorus and potassium in one area, and active calcium and sulfur in another. That is what I am trying to express here; the alchemy that gives this kind of solve/coagula picture allows us to do a process of phenomenology, but we can place it within a larger context in which the dynamic becomes perceptible.

Look at *figure 4*. In alchemy, a polarity always has another polarity within it. If I look at solve as a polarity, I have phosphorus, a non-metal connected to fire, and potassium, a metal connected to water. Within the solve pole, I have a mini-polarity of levity and gravity. Phosphorus is very solve in its action in a plant, and potassium is a little less solve. This is why alchemy came into disrepute; the idea of having a polarity within another polarity seemed like snake oil. Modern scientific consciousness likes to organize things so that everything gets figured out. We have the periodic table, ion exchange, and all those things, which are great but there is also another picture.

The solve pole is also called *sulf.* The complement to solve is coagula, and in that pole we have the term *sal,* or *salt*. However, in the sal pole I have sulfur that supports combustion. In the opposite situation I have placed potassium, which is linked to water in the sulf pole. This does not seem to make sense, but in nature everything has to balance. In the sulfur pole, I also have an agent that is not quite so much volatile but is very active in moving toward the periphery—that is, potash. The sal pole is the pole of gravity, in which I have an agent that is not quite gravity—sulfur. Thus, I have a little sal/sulf in coagula and a little sal/sulf in solve. The picture shows that potassium is a lot more active than calcium in moving to the periphery, even though they both come out of the feldspars. Potassium is a lot more active than calcium is in creating the movement of nutrients toward the growing tip in a plant.

Speaking as an alchemist, Steiner approved of using feldspar in the cow horn instead of silica. Some purists believe it must be mountain crystal, but he would not say it that way because he was wiser than that. In the cow horn itself, the keratin forms a relationship between a more siliceous or proteinaceous elements in the horn and a more calcic gesture in the setting of the protein. In the forming of the horn, there is an oscillation between those two formal gestures, which can be seen in the rings of the horn that represent the calcium demands made on the female's system during lactation. In laying down the horn, there is an alternation of the proteinaceous or peripheral keratin formation and the calcic or centric inability of the system to produce keratin owing to lactation. This alternation results in the rings of the horn. This is just to show that everything in nature is an alternation. The principle here is that there is a formal gesture in the horn that shows a tension between two polarities of protein (solve) and calcium (coagula). The form of the horn is a result of this tension. This leads to some interesting ideas about the formative gesture and form.

Research was done in India, where it can be difficult to obtain cow horns because cows are considered sacred. Therefore, they used clay horns in their experiments.[8] They were able to get a nice large horn and used it to make a form, cast in clay. Then, they made their prep 501 in that. They tested a cow horn, a clay horn, and a control and were unable to see any difference in the action. I shared this information once in a talk, and I got a strong reaction from a dairy farmer, who questioned whether this was really biodynamics.

8 See: http://www.biodynamics.in/Perumalcowman.htm.

I cannot answer that, but I think the experiment was well done and points to biodynamic research in action. It is biodynamics done according to the unique requirements of the place. They were unable to get cow horns, so what could they do? The research is called "mud horn." They make the horns out of the mud where they are working. We are considering the possibility of casting horns using clay slips, trying one with high mica content and another with high silica content. This may be considered heresy, but why not try it. Can we make a capacitor? Basically, that is what the horn is—a capacitor. We are taking a weak charge and trickle charging to build a stronger capacitance.

So, each polarity contains another polarity. That gets a little complex, but it is not a matter of recipes but thinking. Recipes are only the stimuli for future research questions. Readymade answers are the death of research. Recipes are only the beginning and help build an organ of perception. Once we build an organ of perception, we see that there are, within polarities, other polarities. These polarities are sal/sulf, and in the middle is mercury.

Look at *figure 4*. An alchemist would say that sal is salt, but the salt has two modes; one is a process and one is a substance. If I take a salt crystal as a substance and put it in water, the salt as a substance wants to move toward the levity pole and go into solution. Salt as a substance represents gravity and coagulation. However, when I put the salt crystal into a condition of levity in the water, the levity pulls through mercury to move the salt into the polar condition of levity. Mercury is transformation—rhythmic transformation. The mercury in the fluid condition pulls the salt into a solution. Where did my crystals go? As salt, they are now in a levity condition. This is the issue of tartaric acid crystallization in the grape industry. Tartar is an acidic salt, but it is in solution in the grape juice. As an acid, it has the potential to form a crystal. It has potential to be a crystal.

That salt, as substance, wants to go into solution, so alchemically we could call that condition *substancia*. In old Latin, this means "the essence of being," not "stuff" as we know it today. In the sal pole, I have a condition in which the substance moves to solution. The opposite in the sal pole is when the salt moves out of the solution and becomes manifest. That is the process

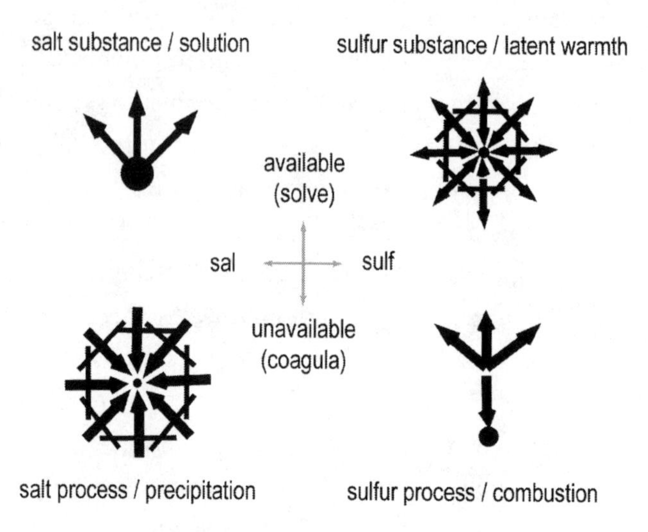

Figure 4

in which a salt moves from solution into manifestation. We can call that a "salt process." I have two polarities in the sal pole, the substance and the process. The process gives rise to a substance. The substance disappears into a process, and these two states alternate. Tartar, as a process in the potash economy of a growing vine, eventually becomes a crystallized substance after a long cycle of development. The solution and crystallization of the tartar in the fluids of the grapevine has a lot to do with the energy economy in the plant. Grape tartar, or potassium bitartrate, in the growth process is a foil to malic acid, which is important in the ripening process. Malic acid is the key to the Krebs cycle of energy uptake and transformation. It is buffered by tartaric acid. These acids and buffers within solutions govern the going in and out of manifestation in growth and ripening of vegetables and fruits and whatever. Sal/sulf, sulf/sal, sal/sulf—they alternate, depending on the temperature, the pH of the soil solution, the volatile acid content of the juice, and so on.

We have sulfur as a substance on the other side in *figure 4*. Looking to the left, in the sal substance, the diagram shows a fixed substance going to the periphery. That is the upper left of *figure 4*. The lower left shows something coming from the periphery into a center, where it forms a manifestation.

That little dot in the middle at the lower left is the cube of salt as a crystal. These diagrams are meant to show the polarity between sal as a substance and a process and sulf as a substance and a process. They are reciprocal. You can actually meditate these diagrams and ripen yourself to perceive something as you scrape tartar out of a vat.

The other side, in the upper right, shows sulfur as a substance. We see arrows going both ways with a membrane in between. Sulfur as a substance is the rock I can light on fire with a match. In sulfur as a substance, levity and gravity are united in an intimate compound. Sulfur as a substance is almost not even here on Earth. It is like oil, grease, or gasoline—just about to volatize. This is why grape and fruit growers use it as a prophylactic against mildew; mildew likes earth and water but does not like fire. The sulfur brings in some fire as a substance. Sulfur as a substance contains latent warmth. It is a mineral that is very warm because it volatizes.

The lower right is the sulfur process, and the sulfur process is what happens when I bring a flame to the mineral sulfur and the flame liberates the sulfur substance into a sulfur process. When I bring fire and liberate the levity bound to the mineral substance, the fire, warmth, and levity, goes off, and what's left is an ash that falls down. This is a combustion process that leads so far into the periphery that a salt is formed as the polar coagula. Thus I have a coagula ash that comes from the sulfur process.

Now, the great secret is that, traditionally, grapevines were used by alchemists to create the ash used in the initial stages of the transformation of metals. The grapevine has a great secret salt hidden in it, which is tartar. Alchemists would burn grapevines and take the lye that comes from slaking the ash, and they would use that lye to form a solution that forms a crystal, which is tartar. They would then add that to the tartaric acid crystals from a wine vat and boil the two to form a solution that they would then decant and let settle. That solution forms very pure crystals of tartaric acid called cream of tartar. The alchemists would use that cream of tartar as the beginning for rendering lead into gold.

That was their standard beginning of the process of leaching the impurities from a mineral called *stibium,* or antimony (Sb). Stibium contains a great deal of sulfur as an impurity, and alchemists need to pull out the sulfur to obtain stibium that can start a process of moving toward transformation. Stibium is the great mercury among metals, a kind of chameleon. It can become many different things and form many compounds. It is like a tourmaline in that regard. The ability to form numerous compounds was known as *mercury*. Stibium is one of the fundamental mercuries, and grape tartar was used to liberate it.

So the salt hidden in the ash is actually the substance that attracts the life (water) in the plant at the beginning. The salt attracts the water into the plant to make it grow. To an alchemist, the burning and combustion to form an ash was an analog to a particular process found in the plant world. The alchemical code word for ash is *seed*. So to alchemists, an ash is a seed of the whole life cycle of the growth process of a plant. They saw it as a kind of seed because, out of the ash, they could take a salt and knew that the salt attracts the water in the first place. This kind of thinking is known as salt, sulfur, or mercury, or a sal/sulf/merc, and is another level of solve and coagula. When we start to work with life, and we have a wet spring and slightly sandy soil on a ridge top, those things become the variables that growers need to work with in imagination.

If you wish to think in this alchemical way, make a list and determine whether one variable is on the sal side and another is on the sulf side. You can ask whether a variable is a warmth-giver or another is a condensing or depositing variable.

III. Nutrient Availability • Potassium • Sulfur • Calcium • Phosphorus • Mineral Mobility

In the soil science language of today, the principles of sulf, sal, and mercury refer to the availability or unavailability of plant nutrients. You can have all the nutrients in the world in the soil, but if they are locked up and unavailable, you need soil amendments to make chemical or structural changes or possibly change a particular cultivar to something that will grow in a particular soil. In general, things that are solve, or sulf, tend to be available because they are active. In *figures 5 and 6*, I indicated "available" and "unavailable" for potassium, calcium, phosphorus, and sulfur. *Available* generally means that the solutions in the soil or in the plant are active in making exchanges. In this case, a soil scientist uses the term *availability*.

When elements in the soil are not actively exchanging, it is usually because a constituent in the soil has already gone through a coagula relationship and

SALT POLE: FORMING AND DISSOLVING

Potassium (mobile)
• Creates potassium channels for ionic transfer. It is active in clay complexes in mineral transport.
• Unavailable in feldspars and micas.
• Available in clay colloid humus complex, composed of weathered clay interstices

Calcium (selectively mobile)
• Strongest fixing agent for phosphorus and potassium. It needs nitrogen to become active.
• Unavailable in feldspars and lime.
• Available in calc phosphate process. Phosphorus and nitrogen activate calcium. Calcium fixes phosphorus over time as the calcium phosphate process develops in the earth through weathering.

Figure 5: Salt pole

SULFUR POLE: COMBUSTION AND ASSIMILATION

Phosphorus (mobile, very available)
• Adenine triphosphate - active in energy transfer for fruit and seed production. Phosphorus circulates freely in entire plant during growth pushing it towards flowering.
• Unavailable in phosphate rocks; it combines readily with calcium becoming fixed.
• Available in animal and plant residues. The largest source of phosphorus is in calcium phosphate formation. It is an important pH buffer.

Sulfur (Selectively mobile in young tissues. Mobility is dependent upon nitrogen.)
• Calcium locks sulfur into mature organs (leaves). Nitrogen shunts sulfur in the plant from mature tissues to rapidly growing areas.
• Conditionally available: Availability is determined by environment. It is important in ribulose light antenna of photosynthesis (ribulose bi-phosphate) when carbon enters the biosphere.

Figure 6: Sulfur pole

fallen into a fixed nature, rendering it unavailable. This kind of dynamic in soil solutions usually involves phosphorous and calcium. Calcium is very fixed and phosphorous is very active; they have a kind of love/hate relationship in the soil. Calcium represents the tendency to hold on, while phosphorous tends to push fluids out to the farthest reaches of the cells. Calcium is mostly unavailable and phosphorus is mostly available. I say "mostly" because together they buffer each other and give dynamism to the two poles.

The relationship between available and unavailable is a big issue for soil, but it also has deep implications within the plant itself in terms of whether the plant can go all the way through the growth process, flowering, harmonious ripening, and form a seed. Obviously the ultimate goal for the plant is to form seeds. Regardless of what you take from a plant, its goal is to assure the next generation. Wherever happens, the plant has to adapt and make that happen. Nut grass, for example, couldn't care less about flowering because it brings the sulfur pole all the way down into the salt pole. Its formative gesture is to have a seeding process down in the root.[9] Plants that specialize in this strategy are often the most pernicious weeds. Alternatively, a plant may pull the sulf all the way down into the root but, instead of becoming a weed, it also draws down the phenolic processes typically reserved for the formation of fruits, flowers, and fragrances. Flowering processes in a root often indicate a medicinal root. For example, you can actually think of a plant such as ginseng as *a flowering process in the root.* It brings a kind of volatility to what would be the root part in a human being, which is the nervous system. Ginseng is an adaptogen that harmonizes the nervous system.

These ideas illustrate the actions of medicines or food combinations. This way of thinking is very interesting in terms of a whole alchemy—putting food by, preparing food, dietary restrictions, homeopathic medicines, behavioral traits, and even autism. A lot of research into autism is targeting digestion. The salt/sulf/mercury imagination becomes a way of viewing autism as a nervous pathology based on a metabolic disturbance. We can see

9 *Cyperus rotundus* is a perennial plant that may reach a height of up to 140 cm (55 in.). The names *nut grass* and *nut sedge* (shared with the related species *Cyperus esculentus*) are derived from its tubers, which resemble nuts, although botanically they have nothing to do with nuts.

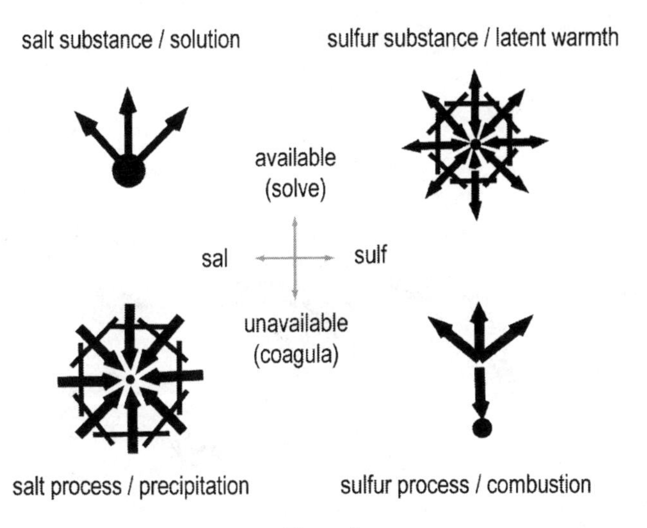

Figure 4

the value of the analogical method in these examples. This way of thinking can, based on the formative gestures, create bridges between things that science has categorically separated. To do this correctly requires greater flexibility in the way we think about divisions in the natural order.

Now, to get a sense of this, look at *figure 4*. It has the four arrow diagrams in it. On the left is salt, which basically has two modes. The substance mode consists of a salt as substance going into solution. The process mode is salt in solution going toward precipitation. These are the two polarities. Precipitation is a gravity state in the gravity pole of sal, gravity staying in gravity. A solution is a levity state in the gravity pole of sal. Solution is gravity going into levity. There is a way through the wilderness of thinking about these complex relationships in soil and plants. It took me a couple years of using these concepts as a meditation. I kept trying to picture it, and it would just fall apart in my mind. Then I started taking long walks while reciting to myself, sal is this and sulfur is that, until I found this little chart in *figure 4*. It took me about twenty years to find it. And when I shared it with people, they said, "That is ridiculous." So I knew I was on to something.

There is always polarity within polarity. To think in this alchemical way helps us to form pictures of *dynamics* rather than just concepts. So the

SALT POLE: FORMING AND DISSOLVING

Potassium (mobile)
- Creates potassium channels for ionic transfer. It is active in clay complexes in mineral transport.
- Unavailable in feldspars and micas.
- Available in clay colloid humus complex, composed of weathered clay interstices

Calcium (selectively mobile)
- Strongest fixing agent for phosphorus and potassium. It needs nitrogen to become active.
- Unavailable in feldspars and lime.
- Available in calc phosphate process. Phosphorus and nitrogen activate calcium. Calcium fixes phosphorus over time as the calcium phosphate process develops in the earth through weathering.

Figure 5

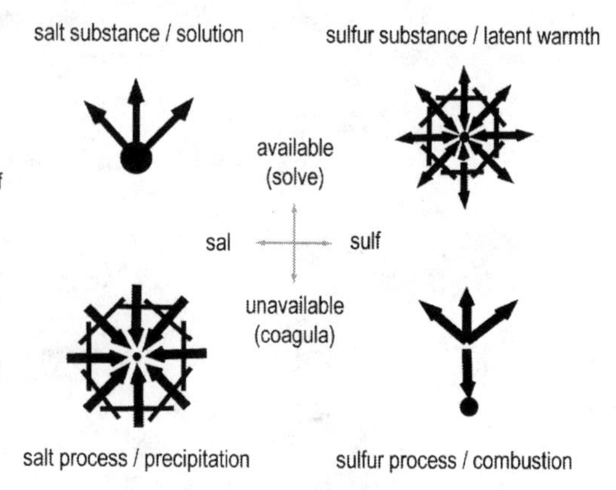

Figure 4

picture of sal shows the solution phase going from the center to the periphery and expanding. In *figure 4* the salt substance mode in the upper left has arrows going from the center to the periphery to illustrate how a salt cube that is dissolving as sal is going to sulf. Below that, in the salt process mode, the illustration shows something coming from a periphery and forming a substance in a center. In this salt process mode gravity reigns. Sal is a compound of levity and gravity. Sulf is a compound of levity and gravity. Gravity is one fundamental, levity is the other. Sal and sulf compound these two fundamentals in different ways in the formation of substances and processes.

So now we will look in the sal pole there in *figure 5* at potassium. Potassium comes from the word potash. If you look at homeopathic remedies, potassium is called kali, K-A-L-I. Kali comes from Mother Kali, the great destroyer who destroys everything and turns it into ash.[10] So with her divine fire nature she is the great transformer of karma, the great goddess of karma. Potassium has something to do with fire and ash. In the old days when people wanted to make soap they would take plants, burn them, and put the ashes in a pot; this was then potash. They then put the ash in water and boiled it. The ashes would release caustic lye into the water. The lye

10 *Kali mur,* (also *Kalium chloratum, Lali muriaticum,* and *potassium chloride*), is found naturally in the mineral sylvine.

solution was then boiled with pig fat and it would emulsify the fat to make soap. Potash salts are strongly alkaline and cause strong ionic exchange.

Potash has a kind of death-and-resurrection quality. One side of potassium is the mobile, soft, or active levity side. It creates potassium channels for ionic transfer of materials in solutions. It is active in clay complexes as a foundation for mineral transport. When acid solutions go into clay complexes or into feldspathic rocks, the source of clays, the acids in the solutions draw out potassium. You can actually see microphotographs of the potassium channels forming in the flow of the fluids within the clays and feldspars; channels are actually being built—crystals with little grooves, channeling, building, and flowing. Potassium supports flow in a plant. It is a salt that always interacts with solutions. It always goes toward the solve pole to keep things moving along. This is why potassium is so active at the beginning of the growing cycle in spring when there are new juices in the soil. It builds the green part of the plant to grow, pushing the stalk and leaf formation. Potassium is always moving the ions to the next growing point. That is potassium when it is mobile. *Figure 8* explains this a little bit.

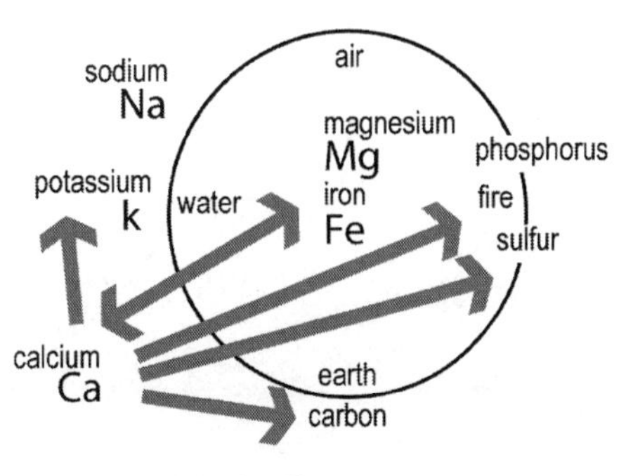

Figure 8

The potassium (at nine o'clock on the diagram) comes into contact with calcium and magnesium and, in the process, there is a kind of sal interaction in which the potassium forms compounds that settle out of the solution. The active minerals are in danger of being pulled down and fixed; calcium, the brother of potassium, wants to fix it all the time. In *figure 8*, calcium (Ca) is in the lower left.

In the *figure 7*, (next page) the stubby gray arrows show the relationships of fixedness. The little diagram in *figure 8* shows how calcium fixes all of the minerals (carbon, sulfur, phosphorus, magnesium, and potassium) that the

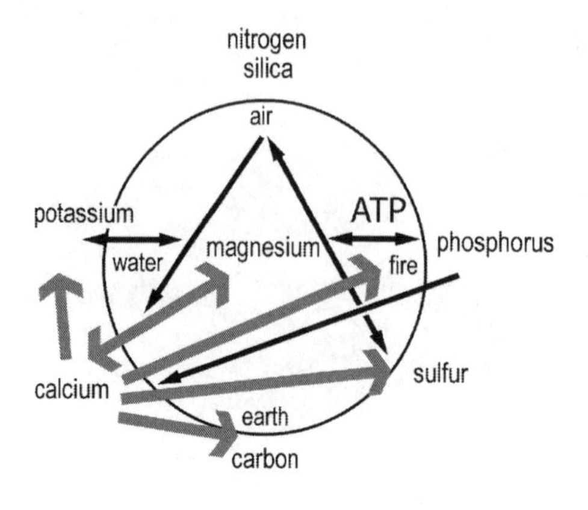

Figure 7

stubby arrows are pointing toward. In *figure 7,* we see the stubby arrows are interspersed with sharp double arrows that show either flow or reciprocal relationships. The pointy arrows show solve relationships. These diagrams represent relationships between two elements that are either coagulating or flowing. Where there is a stubby arrow, the substance issuing the arrow is producing a coagula reaction with the one receiving the arrow. With the pointy arrows, the substance issuing the arrow is activating the one receiving the arrow. This is usually an energizing influence. Sometimes the relationship is reciprocally stimulating; this is shown as a pointed double arrow. These reciprocal influences often fluctuate, depending on the pH and other factors.

Figure 9 shows the dynamic relationships usually with the non-metals, and *figure 8* shows the fixed relationships of the metals, while *figure 7* is a combination of both of the other two. These three diagrams (*7, 8, 9*) represent many bleary-eyed headache hours of me trying to present at least something reasonable to you.

These interactions sometimes neutralize and sometimes stimulate each other. This is why it is difficult to be certain that the solution to a particular difficulty is to add one substance or another. The kind of flowing and congealing shown in *figure 7* is the totality; *figure 8* shows the metals that tend to be congealers, some more than others, which requires a judgment call. In *figure 7,* we see calcium as the ultimate congealer, sending out the gray arrows to practically everything. This is how we read that chart. Calcium tends to fix things, with the result that, in physiology, calcium is the scaffolding of cell walls. Without cell walls, there is no organism.

In general, potassium tends to be within the cell protoplasm and creates a kind of pH balance through the cell wall. Calcium tends to be found in the

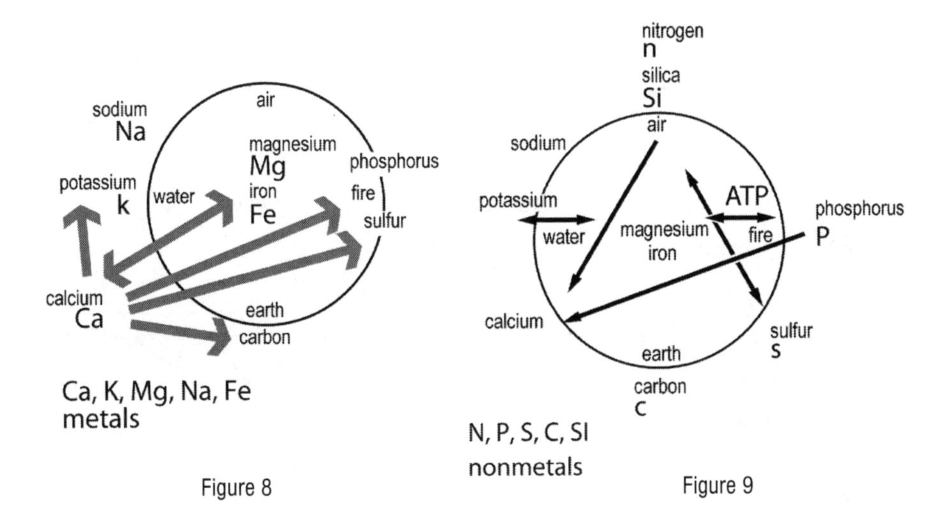

Figure 8

Figure 9

cell wall itself, maintaining integrity. Thus, calcium gives a plant the ability to ward off infections and invasions by creating cell integrity.

However, we see in *figure 8* reciprocal gray arrows between calcium and magnesium and iron. Fixedness works from calcium to magnesium, and magnesium works in a fixed way back on calcium. Together they create a fixed condition that makes potassium unavailable. If potassium is unavailable, it forms into feldspars and micas and loses its fluid interface. You can have all the feldspar in the world in your soil and still have a problem getting something from it if it is unavailable. When we want to make it available, we need to find what will make calcium available.

Now look at *figure 9;* it takes a nonmetal to activate calcium. Nonmetals tend to be more or less stimulating and solve. They are nitrogen, phosphorous, carbon, sulfur, and silica. They are activators. Nonmetals are also called metalloids, activators of things.[11] The metals are calcium (Ca), magnesium (Mg), sodium (Na), iron (Fe), and potassium (K). They tend to fix other minerals, some more (calcium), some less (potassium). Metals tend to be on the fixed side, while nonmetals are involved in organics—sulfur, nitrogen, and carbon. Looking at potassium, we can say it has an available

11 A metalloid is a chemical element with properties in-between or a mixture of properties of metals and nonmetals, which makes it difficult to classify it unambiguously as either a metal or a nonmetal.

side that is more volatile when it is not fixed to calcium. In *figure 9*, we see that potassium has a relationship with magnesium with phosphorous, as well as with ATP. Adenosine triphosphate (ATP) is an important compound in plant physiology.

Adenosine triphosphate is a phosphorous compound that opens the door to the energy cycle that enables the plants to utilize sugar to release energy to make new substances. ATP is a part of the Krebs cycle and it keeps the energy flowing and things transforming in the plant sap. Phosphorous is key in ATP, since the formula shows an amino acid, adenine, linked to a triple phosphate. The phosphorous in ATP is readily available, very fluid, in the solve pole. It is useful if I want to get the potassium and the magnesium operating. In *figure 9* we see stimulating arrows going between potassium in the water pole across to phosphorous and the fire pole, with ATP as the agent of transfer. This is good metabolism, good energy, good building.

Because it is so active, phosphorous in the soil needs a guardian to prevent it from overindulging connectivity. To see what that guardian is, look at *figure 8* (calcium to phosphorous). We see that calcium locks phosphorous and fixes it. In *figure 9*, we see phosphorous, which stimulates calcium. They have kind of a push/pull of reciprocity. This kind of thinking has to do then with the ability to manage soils that produce the other kinds of feldspar.

Looking again at *figure 5*, we have calcium in the salt pole. Calcium, though it is actually very fixing, has an available and an unavailable form. All of these minerals have an available and an unavailable form. That is the way soil science describes it, available means it is active, unavailable means it is already interacted and been fixed into a state of inactivity. In *figure 5*, we see that calcium is selectively mobile. Calcium, the great fixer has difficulty getting out of bed in the morning so it needs a little help. Calcium, the strongest fixing agent for phosphorous and potassium, needs nitrogen to become active. It needs a really good dose of the solve pole in nitrogen. Calcium needs to move up to the air to become more mobile, or solve. The sources of nitrogen are very much the same sources as phosphorous, those being decaying animal residues. Decaying animal residues, the remains of both micro and macro life, create acids that go back in and liberate the potash and the calcium from the feldspars. This is one reason why composting is so effective; it makes the dynamic stronger. Synthetic nitrogen in the form of urea is just too concentrated and has to go through a whole chain of interactions to break down. Thus, nitrogen, phosphorous, and sulfur are in the sulf pole, or solve pole. They are activators and keep things moving along.

SALT POLE: FORMING AND DISSOLVING

Potassium (mobile)

- Creates potassium channels for ionic transfer. It is active in clay complexes in mineral transport.
- Unavailable in feldspars and micas.
- Available in clay colloid humus complex, composed of weathered clay interstices

Calcium (selectively mobile)

- Strongest fixing agent for phosphorus and potassium. It needs nitrogen to become active.
- Unavailable in feldspars and lime.
- Available in calc phosphate process. Phosphorus and nitrogen activate calcium. Calcium fixes phosphorus over time as the calcium phosphate process develops in the earth through weathering.

Figure 5: Salt pole

So calcium is selectively mobile. The strongest fixing agent for phosphorous and potassium needs nitrogen to become active. In its unavailable form, it is a widespread constituent in country rocks, calcium feldspars, and lime. To make calcium available, the calcium phosphate process and nitrogen process must activate calcium. However, calcium fixes phosphorous at the same time. Over longer periods of time, the calcium phosphate process develops in the earth through weathering. Thus, phosphorous is very active in a process called *adsorption,* the ability for participles in the soil solution to attract other ions on the surface of the particle. This is critical to weathering. Phosphorous is active in making that happen, because it changes the charges in the soil solution. The shift in charges allows transfer to happen, and then phosphorus carries the exchange from the surface to the inside of the mineral particle, where that phosphorous nature gets up inside the mineral and exchange takes place. The phosphorous goes all the way into the soil particle, but once there it gets trapped. Phosphorus moves from the process of adsorbing on the surface of the soil particle to absorbing and becoming active within the soil particle. It then becomes trapped in the particle. However, if there are more acids created, then they go in and reanimate the process. When that happens the calcium releases and goes out into the soil solution. This is what happens in the soil when it rains; it is the push and pull of these things at a very intimate level. These same types of things happen in the fermentation vat, but in a finer way than it does in the soil. These reactions are much finer once all these things have been worked out in the juice of a plant.

Thus, calcium is selectively mobile and potassium is much more mobile. This is why the grape grower can use a little magic switch to shut down the potassium once growth has reached a certain stage. Potassium creates growth, and then we want to shut it down to keep the green-grass growth energies from going over into the more delicate stages of ripening. When all is working well, the acids within the plant regulate the shift between growth and ripening. These are represented largely by the tartaric and the malic acids. The tartaric regulates the potash uptake during growth and acts as a buffer to the malic acid that forms as the ripening process unfolds. The tartaric acid is just held in reserve once the setting of the berry is accomplished. It buffers the malic acid during the ripening process. Alchemically speaking, the tartaric would be more the coagula and the malic is more solve. Once the coagula is no longer useful, it is released to coagulate as the levity force of alcohol and phenolic oils dominate the fermenting action—that is the dropping of the acid out of solution into crystals. It is these balances of the pH and temperature that allow this dance to happen. It is happening in the plant and it is happening in the vat. Cooling the juice helps the acid to fall out; it is because cold is coagula.

Now we will consider the sulfur pole (*figure 6*). In the sulfur pole, we have the same type of polarities of available and non-available minerals. In the sulfur pole, we have the very available and unavailable in both phosphorous and sulfur.

We will begin with phosphorous. The sulfur pole is combustion and assimilation. The dynamic is assimilation, until substances becomes so assimilated that the compound combusts, after which you have a precipitate that falls out as an ash. Phosphorous is mobile and very available. It is ATP, adenosine triphosphate, the prime agent of active energy transfer for fruit and seed production. The phosphorous is the force that drives a plant to make a fruit. Potassium follows phosphorous and supports swelling, but once the fruit swells, this has to be followed by the cessation of growth and the production of sugars and proteins associated with seed production.

It is phosphorous that keeps growth moving toward the formation of seeds. This process involves essential oils and phenolics—anything that has to do with those kinds of alcohol–oil polarities in the flower–fruit pole of a plant. The phenolics and sugars are a kind of solve gesture. Phosphorous is very available to the plant and keeps it vigorous through to seed production. Phosphorous circulates freely in the whole plant during growth. With NPK we are always applying phosphorous, because it is always being used, especially when growing crops, where the crop is taken off. This includes crops for animal feed, because they need the phosphorous for their own ATP to make protein to form their bodies. Then they fix the phosphorous in bones, connective tissues, and so on. Bones are calcium phosphate and calcium carbonate compounds. You fix phosphorous in your body, as do animals. When you have rangeland or crop after crop of vegetables, you are always depleting the available phosphorus with the crop. You have to keep reapplying it. That is why people use NPK, or common nitrogen, phosphorus, and potassium mixtures. Nitrogen increases sensitivity to soil conditions; phosphorous provides energy; and potassium increases ion conductivity. Feldspars tend to be active in the soil, while phosphorous tends to be active above the soil, in the crop itself.

SULFUR POLE: COMBUSTION AND ASSIMILATION

Phosphorus (mobile, very available)
- Adenine triphosphate - active in energy transfer for fruit and seed production. Phosphorus circulates freely in entire plant during growth pushing it towards flowering.
- Unavailable in phosphate rocks; it combines readily with calcium becoming fixed.
- Available in animal and plant residues. The largest source of phosphorus is in calcium phosphate formation. It is an important pH buffer.

Sulfur (Selectively mobile in young tissues. Mobility is dependent upon nitrogen.)
- Calcium locks sulfur into mature organs (leaves). Nitrogen shunts sulfur in the plant from mature tissues to rapidly growing areas.
- Conditionally available: Availability is determined by environment. It is important in ribulose light antenna of photosynthesis (ribulose bi-phosphate) when carbon enters the biosphere.

Figure 6: Sulfur pole

A thousand parts of acid in the soil solution create precipitation as it has a tendency to be solidifying and congealing. On the opposite side, if alkalinity gets too intense it results in fixing tendencies right around pH seven where colloids are active. The colloidal state is how the particles get transferred in the soil and in the plant until they reach a cell wall. Colloids have difficulty crossing cell walls. This makes them good for transport. To be assimilated by the cell the colloid has to encounter a pH change near the cell. The shift in pH breaks the charge of the colloid and the minerals can then pass through the cell walls of the plants. This exchange is known as *cation exchange capacity* (CEC). A cell has a different pH inside than does the environment outside the cell. This change of pH breaks down the colloid and allows the minerals to pass in and out of the cell. CEC affects the ability of a cell wall to open and close in ways that are not completely understood. Scientists once thought that cell walls have holes through which substances pass. Today the cell wall is seen as a constantly shifting set of variables between the inner contents and the environment. A cell wall is actually a dynamic, not fixed as we usually think of walls. It is a selectively permeable dynamic connected with the relationship between inside and outside.

When we talk about soil solution, plants, root exudates, and mycorrhizae around plant roots, or the exchanges of gasses in the atmosphere with the plant, or exchanges of gasses in the soil, this language of sal/sulf and polarities within polarities helps us understand the concepts of available and unavailable minerals. Alchemical imaginations allow a clear picture-forming process that enables us to enter these complicated relationships, even for those who have little technical knowledge of these matters.

Sulfur is presented in *figure* 6. It is selectively mobile, much like calcium. Sulfur is selectively mobile in young tissues, and its mobility depends on nitrogen and calcium, which tend to lock sulfur. This is what we see in *figure* 6. Sulfur is selectively mobile in young tissues; its mobility depends on nitrogen. Calcium locks sulfur into mature organisms. This is the imagination I presented earlier by looking at whether the deficiency is in the vein of the leaf or the growing tip. In the plant, nitrogen shunts sulfur from mature tissues to rapidly growing areas. In the presence of nitrogen, sulfur "wakes up." Sulfur goes to where the most growth is happening and then starts to coagulate at that site. Nitrogen the sulfur can keep migrating if there is a little extra—that is, if there is nitrogen available. The sulfur keeps moving to help the plant toward flowering. Sulfur is always present in the formation of oils, just as phosphorous is. It has a kind of greasy quality. Sulfur is also active in processes such as essential oil formation. When it is not balanced, sulfur contributes oils that smell like turpentine and acetone. Sulfur is available conditionally. The environment determines the availability of sulfur.

Sulfur is important in the ribulose light antennae structure of photosynthesis pigments.[12] In the formation of chlorophyll in a plant, there is a structure known as a "light antennae," composed of red, orange, yellow, and green pigments in the chlorophyll of the leaf. The red and orange pigments pull in light and send it to where the real green pigments are. If the green pigments are strong enough, they mask all these reds and yellows in the leaf and pull light in toward a magnesium molecule in the chlorophyll. This light induction is a light antennae and the basis of photosynthesis. Once the green has been assimilated and disappears in the fall, we see the reds and yellows that serve to induce the light. They were always there but the green masks them during much of the year, because it is stronger; the magnesium receives the light and sets the stage for energy transfer in the plant. This process is helped by sulfur, which helps to make a substance called ribulose, which allows carbon, an inorganic mineral substance, to enter life.

It is the form of the light antennae with magnesium as the center that allows the plant to develop life energy. This is the basis of photosynthesis. In

SULFUR POLE: COMBUSTION AND ASSIMILATION

Phosphorus (mobile, very available)

- Adenine triphosphate - active in energy transfer for fruit and seed production. Phosphorus circulates freely in entire plant during growth pushing it towards flowering.
- Unavailable in phosphate rocks; it combines readily with calcium becoming fixed.
- Available in animal and plant residues. The largest source of phosphorus is in calcium phosphate formation. It is an important pH buffer.

Sulfur (Selectively mobile in young tissues. Mobility is dependent upon nitrogen.)

- Calcium locks sulfur into mature organs (leaves). Nitrogen shunts sulfur in the plant from mature tissues to rapidly growing areas.
- Conditionally available: Availability is determined by environment. It is important in ribulose light antenna of photosynthesis (ribulose bi-phosphate) when carbon enters the biosphere.

Figure 6: Sulfur pole

12 Ribulose ($C_5H_{10}O_5$) sugars are important in the formation of many bioactive substances. D-ribulose combines with carbon dioxide at the start of the photosynthesis process in green plants.

a plant's sap, magnesium is the energy producer for photosynthesis, or the energy transceiver. Plant sap has exactly the same constituents as human blood, except for the presence of magnesium. In human blood, iron performs the function of magnesium as a focalizer. Human beings and plants have a beautiful relationship between magnesium and iron, which we saw earlier in the mafic rocks. The relationship between magnesium and iron shows the relationship between human blood and plant nature. This picture has to do with the way that oxygen and carbon are exchanged in the respiration processes. That is how plants interact with the human lung.

This relationship between magnesium and iron is one of the great mysteries in the natural world. The activity of the real energy transfer in a plant is based on magnesium, and the activity of the real energy transfer in the human being is based on iron. In the sap, both can be found and are opposites in their relationship in terms of how carbon goes from a dead mineral into life. If you work with these diagrams, it is possible to see that the metals and the nonmetals are a valuable imagination for contemplation. In the relationship between the metallic nature of fixing the light and the nonmetallic nature of moving the processes, we get the two poles of coagula and solve.

In alchemical terms, how would fruit growers, for example, deal with a cold wet spring that promotes a lot of leaf growth but very little flowering? Where would the cold and wet be in the mandala of the elements? We can see right away that cold and wet are on the gravity side of the ledger. If we know where we are, it is easier to see where we need to go—in this case toward levity, toward fire. Therefore, we will probably need to use sulfur or stimulate phosphorous down the road to balance the cold and wet. We could certainly apply sulfur as a substance, but let's consider pruning. What pruning strategy in the early season would bring out a levity form gesture in the later growth? We can do levity-supporting operations even, during cultivation, by paying attention to the Moon's position. We also have to know our soil type in a cold and wet situation like this. During the growing season, we would have a problem early in the season of too much vigor, too much growth in the earth and water pole.

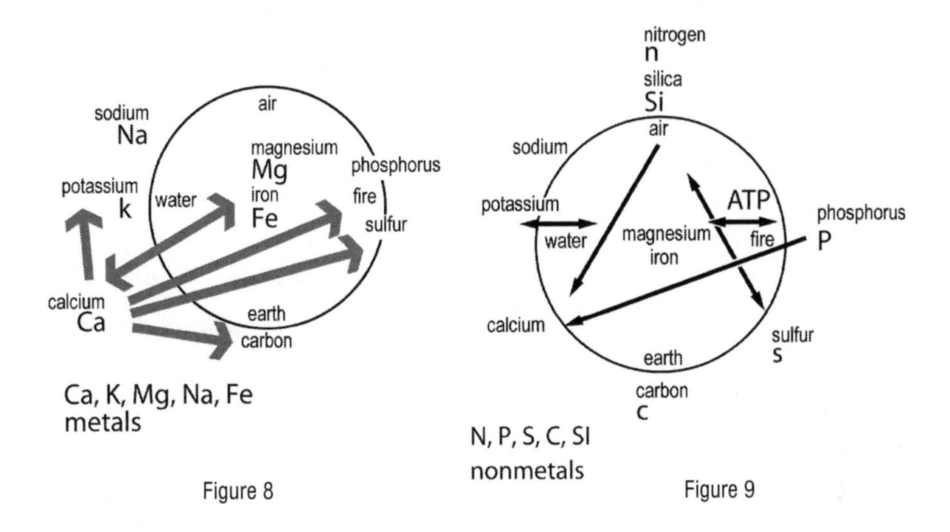

Figure 8

Figure 9

This means that, somewhere along the line, we would need to bring influences from the activity, or levity, pole. The picture here is that the cold, wet weather conditions support the dampening calcium forces that do not allow the plant to move past growth to flowering. Those are depicted in *figure 8*. As a metal, we have linked potassium to the water process, so we would not want to stimulate that. Calcium locks the potassium, but these combined forces lead to the formation of leaf and stem. Therefore, potassium, even though it is active, does not give enough of the fire and air elements from the sulfur pole.

Now look at *figure 9* for the nonmetals. In the air pole nine we find nitrogen and silica. Nitrogen activates calcium, but it also enhances vigor, so nitrogen is not a good choice. Silica is used in biodynamics to bring in the light pole. That would be a good start for a substance that would be serviceable for a cold wet climatic regime. In the fire pole of the levity side there is sulfur, which is a prime remedy for cold and wet in the grape industry. So far, we have sulfur and silica as agents to move our cold and wet scenario toward flowering and fruiting.

When it is wet and cold in spring, we get long, leggy random growths. When we see this, what is the plant actually doing? If spring conditions are overcast and cold, the plant will not get enough light to create a strong vegetative growth. Leggy, spindly growth during lengthy overcast is not the result of too much vigor but a symptom of too little light for photosynthesis. The plant is moving to the periphery and the levity side, it is skipping vegetative expansion as a foundation for growth. The picture is different, however, if we get a spring with seasonable temperatures and precipitation and the vine runs out and gets long canes. I would then ask about fertilization of the vine in the previous growing season. I would ask how and when the compost or green manure was taken up by the plant. We could even talk about the source of the nitrogen used for the compost. Do we want more or less nitrogen for a cold wet spring? Now we get down to the farmer who put the nitrogen down the year before we got all this rain. Now add to this variable the quality of the nitrogen in the compost used. It makes a huge difference whether I use cow or chicken manure for compost. Which has more nitrogen? Alchemically it is their food—grain and seed—which is sal, or sulf. A fruit and flower process produces the grains that they eat, and the mineral in our mandala that supports this process is phosphorus.

Chickens are collectors of phosphorous. If I apply mostly chicken manure, then next year a strong phosphorus potential is held in the cambium around each bud. If it is cold and wet in the early spring with little available light, the previous composting residues will drive the plants to shoot up. This is also true of compost teas applied during the growing season to the leaves. This year's growth can be enhanced through foliar sprays of compost. However, in the fall whatever has not been used returns to the cambium and provides the impetus for early growth the next year. A cold and overcast spring following a summer of foliar feeding can be a recipe for mildew as soon as the weather turns warm. The plant suddenly grows rapidly, and mildew provides a premature fire solution to the rampant growth. The problem is the application the year before of abundant phosphorus and nitrogen in the chicken compost.

Now suppose I mix chicken manure with cow manure. A good cow eats grass, whereas a factory cow may eat grain or fish meal, so when I think about the manure I also use this mandala of elemental relationships. I have to consider the source of the manure and learn whether it is a solve or coagula manure. This mandala is how I can imagine the levity–gravity polarity of weather, soil, manure, and actions of the previous year. Not every grower is hit in the same way by a cold and wet spring. These parameters of levity and gravity are the formative gestures that create the conditions for a plant when the rain comes. Grape growers also need to consider the kind of stock, scion, and sunlight exposure. *Terroir* is the term for this pictorial kind of thinking about all of the parameters acting on the plant. This kind of thinking can be applied through the whole process to solve problems. Growing problems are seldom simply reactions in that moment, because nature works in large time rhythms. Unfortunately, most people do not think that way and therefore do things that contradict nature.

Everything depends on balance, so an abundance of nitrogenous manure with a lot of phosphorous in it forces plants to gorge on phosphorous, which signals a push of nutrients toward the meristem. To sustain itself, a plant needs to be able to pull other minerals along to balance a push by phosphorus. Thus the plant pulls the calcium, potassium, and magnesium in the phosphorus flow, leading to very good vegetative growth. However, rich manure causes so much vegetative growth that the archetype of the grape plant gets indigestion. Ideally it needs to stop growing and start to flower and fruit by May, but if the pipeline is full of phosphorus, calcium, potassium, and magnesium, a signal is given to continue vegetating. Therefore, the call goes out and mildew shows up with a solution that is, speaking alchemically, a fire gesture. Mildew can take care of the over-production of growth, because what is in those leaves has gone beyond where it should be for the plant to use phosphorous to take it into the flowering and fruit stage. Phosphorous just tries to use the nitrogen and accelerate the action of the potassium.

When there is too much nitrogen in the pipeline, the plant cannot move it, so the call goes out for a solution to the excess nitrogen. This leads to mildew, stem rot, or caterpillars, depending on the type of crop. Every pest is really a fire principle that works on energies at a level where the plant cannot deal with them on its own. Basically, there is too much below and the earth and water come in and move it into solution. The plant archetype cannot assimilate the excess, so the pest is called in to bring an ash process that is premature. It returns the situation to balance, but we will not be happy because we were hoping for a crop. Thus, when the growth process gets out of balance, it involves more than the present year. We have to look at what was done a year ago, which is now being aggravated by a wet spring. Alternatively, perhaps you put out sulfur, and then there are three days of hundred-degree weather. Whose fault is this? The torched and untorched areas are not crop's fault. Maybe grandpa always put sulfur down on April 12, so you also do it on April 12 because it is cold. Then April 15 comes with hundred-degree weather.

The qualities of fertilizers depend on the forage that produced the manure. The gesture of what an animal craves as food is thus carried into the crop. Pig manure brings in a lot of potassium, because they are rooting animals. They love stems and stalks, which is where the potassium gathers and is most potent in the plant. A pig's desire to eat roots and stems makes it a potassium gatherer. By contrast, a fowl is a phosphorous gatherer. Chickens love to eat grains, bugs, and other high-protein phosphorous substances. Their desire drives them to concentrate phosphorus in their diet, which is passed on to the manure. Thinking alchemically about the desire life of the animals can help us determine what to include in a compost heap. I can go back a year and look at my notes about what went where to solve problems in the present crop.

It is truly important to keep records. Where did the manure come from? How did the compost go through its cycle? Was it a quick ferment or did it stall in the middle? Compost is similar to ferment in a wine vat.

If you have a high abundance of potassium and phosphorous and those driving forces of growth, you could use calcium to slow the plant, since calcium will bind with potassium and phosphorous. Steiner did indicate a specific application to slow the growth processes, because he worked in Europe where the summers are usually cool and damp. Therefore, preparation 501 was designed to help the light. In California, where it is generally dry and hot, I know people who have torched their whole vineyard by using 501 during the summer. They were looking for a safe alternative to sulfur. They thought that spraying 501 against mildew would not burn the foliage, but the results did not support that line of thinking. The foliage was damaged by the silica spray. This is the reason I worked to develop the amethyst spray. It brings a light gesture, but it brings a moist light. I am currently experimenting with tourmaline, and I think that for phosphorus there is an interesting mineral apatite. *Apatite* is a calcium phosphate gem. In apatite, it seems there is a potential to make a homeopathic gem preparation that would help to retard rampant growth a bit.

Now, what if the problem has to do with the soil itself? Suppose, for example, that there is too much potassium. When it gets wet, a lot of potassium comes into the soil solution and into the plants. To deal with this we are working on a spray using potassium bitartrate as a solution to spray on grapevines. In the vine's early growth, the tartrate holds back potassium. So we are experimenting with certain animal organs to see if it will actually hold back too much potassium uptake in the spring. The initial tests are positive using a tartrate solution using alchemical methods.

Next, I turned my attention toward the rhythms of these processes and applications, because we can apply all of the substances in the world to plants, but in the end it is the rhythm of the application that makes it possible to turn something in another direction. We take the tarter, and then there is a rhythmic process of burning to make an oil from the tartar. Using one drop per gallon of that oil to spray on tomatoes and cucumbers, it almost shut them down for good. The sprayed branches turned yellow in a day. I had to cut them off and spray the rest of the plant with apple cider vinegar, which is malic acid. This led me to do more research, and I found that malic acid is the first part of the Krebs cycle that goes through many changes to allow sugars to be assimilated. At the very end it returns to malic acid, but it is right in that spot where the tartaric acid becomes active in the cycle and adds a balancing note to the rush toward ripening. In a grapevine, there is not a lot of malic acid production in the beginning, which is when the tartaric acid is developed. Therefore, when the plant reaches flowering and fruiting, the malic will come in to support the fruiting process. As the alcohol develops in the fermenting wine, the tartaric acid that has buffered the influence of the malic acid is no longer needed, so it drops out of the wine as crystals of potassium bitartrate, or tartar. My research tries to use that natural cycle as a key to supporting grapevines when adverse weather conditions threaten the natural ripening cycles late in the year.

A plant such as a tomato does not do well with a blocked Krebs citric acid cycle. That plant is born to grow, fruit, and check out. It is the same with a cucumber. However, when we spray the grapes with the tartar solution, they benefit from the extra tartar during ripening. If there are any discontinuities, such as a heat spike in the middle of ripening, we find that tartar sprayed early in the growth cycle controls uneven ripening. The idea is that the tartar holds back the potassium so that the vine gets a signal that growth is finished and now it is time to focus on the fruit. This is the imagination behind the experiments because that is its function in nature. Such experiments use these models to say what is going on in a plant that is solve and coagula. I try to determine a substance from a plant, find an analog to it as an animal organ, and then apply it back to the plant to help it during adverse conditions.

I was recently told by a grower in California that growers there are beginning to panic because climate change will make their particular climate much warmer, which could push the premier wine-growing areas out of California into Oregon. They are in a panic because the grapes they have acclimatized to California would no longer be able to grow there. In Davis, California, people are doing an experiment, planting grapes that grow very rapidly and have very early fruiting and bud break. Then, when the fruit is first set, they go out and cut all the fruit off. The plant then produces a second flowering with more fruiting about forty-five days after the first set. This pushes the ripening period past the hottest part of summer into later cooler months. With this approach, growers will harvest in November instead of October. This is an interesting picture. The much later ripening is an effort to avoid the possibility of having everything fry in early October. Our hope is that the tartar spray will do this without having to cut off all of the fruit.

Issues in chemistry

The four-elements theory that is fundamental to alchemy is often seen by chemists as a throwback to an inferior way of thinking about physical chemistry. From the perspective of the periodic table and its complexity and practical applications, the idea that the world is composed of four elements is understandably seen as retrograde. The four elements theory is not useful for analyzing atomic structure, but it is useful in allowing researchers to integrate human consciousness into the experimental method. Contemporary analytical approaches in science based on abstract reasoning focus on excluding human consciousness as a parameter in experiments, which is meant to minimize bias. Contrary to this attitude, alchemists understand that substances are linked to human states of consciousness and they seek to harmonize their inner states with what could be called the lines of emergent substances. In this worldview, substances are the corpses of spiritual beings' creative activities. In alchemical thinking, the goal of the worker is first to determine the state of consciousness of the spiritual being behind a substance and then harmonize one's inner state with the consciousness of that being. This is accomplished by using feelings as a part of the experimental method. The task is to transform intensely personal feelings into what could be called the feelings of the world soul. The following is a short example of how the feelings of the world soul might be imagined.

In contemporary plant chemistry, the metalloids silica, carbon, magnesium, sulfur, and phosphorus are considered to be activators, whereas alkali metals such as potassium, sodium, and the alkali earth calcium are considered to be fixing agents. These polar differences are based on the valences of the atoms of these substances. Physically—that is, in atomic theory—the valences of these substances are thought to indicate their ability to interact with other substances. In alchemical thinking, these valence tendencies are understood to be interactions in polar relationships, resembling subtle feeling states of being either connected or estranged. In this symbolic language, feeling connected is sympathy, and feeling estranged is antipathy. Opposites exhibit sympathy for each other; similarities exhibit antipathy. Antipathy tends to move potential states into manifestation; sympathy tends to move manifest substances into rarified states of potential. The idea of charges and potential, which is the fundament of the periodic table, can be seen from these feeling perspectives. A look at the active relationships between metalloids and the metals can reveal these feeling polarities.

This imagination is based on the alchemical mandala of the four elements as a basis. The face of a clock is an analog of the mandala. To simplify the descriptions, we will use noon as north, or air; three o'clock is east, or fire; six o'clock is south, or earth; and nine is west, or water. No assignment of elements and directions is intended.

Phosphorus is a highly reactive element. It is so reactive that it is not found in a pure state in nature, since it will link immediately to other elements to form compounds. This volatile quality may be seen alchemically as the fire principle, a state of great sympathy, since phosphorus as a substance exists in a highly rarified state of potential. In plant saps, this quality of phosphorus makes it readily available to plant organs in the growth process. However, the phosphorus fire quality is so reactive that it is in danger of damping down as soon as the phosphorus becomes available for a life form. Its sympathy is so great that it is constantly moving into a state of antipathy or manifestation, or alchemical earth. Alchemically, this strong damping down, or movement from potential to manifestation, is an earth signature. That would place phosphorus as an element in the east/fire, but phosphorus locked in a compound is in the south/earth. This highly reactive fire property, combined with an almost immediate damping down into an earth compound, gives phosphorus a particular feeling nature that sets it off from the other metalloids. To understand this idea it is useful to look at sulfur as an element akin to phosphorus.

Alchemically, sulfur is also found in the fire/east pole. However, in contrast to phosphorus, which is not present in a fixed form as an element, sulfur is found abundantly in nature as an element in mineral form. Alchemically, sulfur represents the element of fire, because as a mineral it can be ignited by a flame. Through ignition, it easily moves from manifestation into an unmanifest state. This means that it has a fire principle active in it, yet the periodic table has it in the form of a mineral substance. To alchemical reasoning, this signifies that the fire in sulfur is naturally earthed. In a way, sulfur is a compound moving in the direction opposite to phosphorus in relation to their mutual activities. Sulfur's sympathy is less profound than the sympathy of phosphorus. It has satisfied its sympathy by uniting with the earth in its native state. What does this mean?

From the view of analytical chemistry, in a plant's life sap, sulfur is said to be selectively mobile. This means that, compared to phosphorus, sulfur is a bit more sluggish in its reactive sympathy. Sulfur requires the presence of nitrogen before its reactive sympathy can be unfolded. The fire of sulfur and the fire of phosphorus reveal subtle differences in their activity as metalloids. Alchemists would say that these differences resemble properties akin to feelings of the world soul, which are seen as potential deeper links between an experiment and the consciousness of the experimenter. Alchemically, these kinds of feelings can be used meditatively as doorways to deeper insights.

To take these ideas a step further, silica and carbon, too, are seen as metalloids in the scheme of the periodic table. By definition, this makes them activators compared to the metals. However, carbon is active in its intractable bonding gesture with itself. The covalent bonding of a carbon atom to itself is the structure for innumerable compounds in organic chemistry. However, this great activity arises because the carbon is, in a sense, not so giving of itself but clings to itself with such tenacity that other elements are dragged into its chambers once other linkages are created. This unusual activity gives carbon a distinct personality that runs the gamut from the ubiquitous charcoal through the less active graphite to the non-interactive diamond. In the periodic table, all of these substances are carbon. According to chemists, such variations in morphologies do not change that analysis, whereas alchemists see very different properties to explore in a range from sympathy to antipathy.

In the periodic table, the octave of carbon is silica. Silica is also a metalloid and therefore considered to be an activator. It, too, however, runs the gamut of the quartz crystal, from being highly resistant to chemical combinations to amorphous forms of silica gels that create a rich array of dynamics and potential compounds in industrial applications. All forms of this element have the same chemical analysis, but its numerous morphologies speak of very different dynamics and potentialities. Meditating on these differences, an alchemist hopes to gain insight into healing modalities and applications in the life realms.

These simple examples show us that, even though these elements of the periodic table are in the same group, their fire nature is modified by an earth or air "feeling" quality, which gives the element a palette or spectrum of activities, even though its actual chemical formula is the same through all of the morphologies. Similar perspectives could be developed for calcium, which could be placed alchemically, as an alkaline earth, in the earth pole; it is certainly considered to be an earth by its nomenclature in the periodic table. However, calcium in context of the living forces present in plant sap exhibits a strong relationship to water and the ability to form solutions. We could say that calcium, as a calcite crystal, is an earth, but calcium in the mucus of an oyster, as a calcium-rich colloid, is active in the water pole. Alchemically, calcium has an inherent sympathy for the water/west pole of the mandala. Is calcium in the solution of a plant the same calcium in a crystal? The answer must be yes to chemical analysis. From an alchemical point of view, the crystalline form of calcium is in the south/earth pole, but it would be found in the southwest because of its relationship to water. A true south/earth position would arise when calcium links to carbon to form calcium carbonate. In this substance, the calcium is linked to the fixed nature of carbon. Alchemically, this form could be seen as characterizing a strong south/earth pole on the alchemical mandala, whereas the calcium mobilized in plant sap is found in the southwest.

In the alchemical language, the dynamic form of calcium that is present and moving toward the water pole would be a calcium process in contrast to a calcium substance in the fixed crystal of calcite or limestone. This type of reasoning can go much further. For instance, it could be used to contemplate the deep mystery found in the horsetail plant. Analysis of horsetail shows it to be remarkably high in silica as an ash. However, doctors prescribe it for patients who need stronger mobilizing forces for calcium, such as in cases of osteoporosis. Similarly, chemical analysis of oak tree bark growing in nothing but sand reveals an abundance of calcium. Where does the calcium come from in this process?

When seen from the perspective of the periodic table, the minerals and metals of analytical chemistry have clear divisions based on the atomic shell theory. These divisions are quite useful in a chemist's laboratory, but in life other properties emerge that seem to go beyond abstract analysis. Understanding these life properties is the goal that alchemists seek. In the living forces of plants, animals, and human beings, other levels of healing can be discovered by harmonizing our souls with the subtle feeling states of chemical elements.

IV. Planet Loops • Etheric Forces • Using Planetary Rhythms • Advanced Preparation Work • Dandelion • Chamomile • Polyrhythms

There is an old saying in Polish that goes something like this: The stupidest farmer grows the best potatoes. It means that a farmer can do everything wrong and you still get a good crop. There is another old saying in alchemy: What you do may be important, but *when* you do it is more important. So, as far as agriculture goes, the failsafe is that the good Lord loves us all and provides sustenance even when we are not doing the right things agriculturally. You can do things and have absolutely no clue about the finer aspects of working with nature. When agriculture involves slash and burn, you slash, you burn, and you get free fertilizer. You can then just put the seed in, and the plants grow like crazy on the newly fertilized land. Then, when that plot is depleted in three years, you simply move on. Most farmers no longer use that method, which means there has to be a different way to keep our land sustainable.

The great gift to human beings is the ability to create conditions that nature does not. This includes selecting seeds to save, which is not natural. Nor is pruning natural. Fertilizing by concentrating manure in one spot is not natural. When we do things that nature cannot do, the real issue becomes timing. Alchemists said that it is not just a matter of the process you put a substance through but the rhythmic nature of that process. This comes through imagination of the best points in time for transformations. Alchemists understood that, if you work consciously in time, your activity becomes a sort of heartbeat that nature will eventually gravitate toward. All of nature is already orchestrated in time, which has to do with movements of the Sun and planets. If wise workers look to those planetary rhythms, doors in time could be opened for operations. Moving one degree each day like clockwork, the Sun is the great chronometer. That rhythm, the figured bass line, so to speak, of the great symphony. This rhythm is fundamental to the harmony of everything.

Within the solar year are hidden forces of timing that spread through the year as ways of working onto time and then focusing it in a particular way. The Sun has much to do with how surfaces form and react to one another as sensitive membranes. The formation of energetic membranes is an action of the etheric sphere, which is the specialty of the Sun.

All energy transformation in nature is related to energetic membrane formation, which precedes the formation of an actual physical membrane. A physical membrane retains the qualities of the energetic membrane in the ability to selectively transmit across a physical membrane, or what we call semi-permeable. The "semi" nature of the permeability has to do with temperature, pH, and so on; we could say that selectivity is a kind of consciousness. Hidden in this selectivity is the great mystery behind the fact that light comes to us as energetic planes or membranes, which is a concept that comes from physics. It is a difficult concept to understand, but I will try to explain it because, we are talking about Sun. Every star is a sun, every sun generates light, and the light from every sun travels through space in energetic planes.

The great mystery of this is that the Earth is constantly bathed by the energizing light coming from countless suns streaming for countless years from all directions in space. Science tells us that our Sun's light takes about nine minutes to get to Earth, but there is an infinity of suns out in the universe, and we are also receiving their light on Earth.

Scientists did an experiment. The darkest spot in the universe is a spot in Pisces where nothing had ever been seen. Scientists focused the Hubble telescope and took 270 pictures of that spot the size of a pinhead where no light had ever been seen. When they finally processed and assembled the images, they found that it is filled with galaxies—not just stars, but galaxies that look like organisms in a microscope slide of pond water. They saw galaxy upon galaxy in the darkest spot they could find. From the stars in countless galaxies in all directions, starlight is streaming through space. Rudolf Steiner calls this the "weaving of light." That weaving of light has been understood by alchemists as the "action of the crystal heavens."

Our Earth is moving through a sea of light. That light, according to physicists, is measured in "wave fronts." Light travels in waves that cannot be measured sideways or diagonally. We are told that the waves of light propagate in two directions simultaneously. They move up and down and side to side in a form called electromagnetic propagation. The two waves propagate together by interweaving through space around a common axis, which forms the general direction of travel. They actually rotate around that axis, forming a common plane perpendicular to the direction of travel. The common plane they share in propagation is considered to be the front of the wave, or the wave front. The wave front of a light ray travels through space from the light source as a plane rotating perpendicularly to the direction of travel of the ray.

Alchemists understand that the archetypes of all life are contained in that light. It is a kind of sea of life and potential—planes of light creating potential for points. A point is created where three energetic planes of light meet. In projective geometry potential points are called *star points*, where potential life is focused into a point. The star point is the place on the ovum that is receptive to a sperm. Our universe is permeated with star points. For alchemists they represent the web of potential that is the *pleroma,* the great web of the fullness of life.

Figure 10

Look at *figure 10.* On the left side is a circle with an angle in it. The horizontal line moves from the center to the periphery, where it touches the circumference. A line like that from the center to the circumference is the radius. In the second circle is a perpendicular line, or *tangent plane,* whose radius touches the circumference. A tangent, or touching plane, can be drawn where the radius of a circle meets the circumference of the circle. It is at right angles to the radius. Euclid said that any line drawn from the center of a circle to the circumference can be extended to infinity. A tangent plane that touches a circle where the radius meets the circumference is a finite representative of what he called the "infinitely distant plane." The ancients saw the infinitely distant plane as the source of all life. Buddhism calls it "beyond the beyond."

However, the infinitely distant plane is at a particular angle to my line of sight as a human being on this planet, which geometrically is a sphere with a center and a circumference. The surprising thing about this line of reasoning is that, if I look out into space and try to imagine that infinitely distant plane, the infinite is a totally flat plane of infinite extent. Physicists say that the universe, according to very demanding measurements and mathematics, is a totally flat plane of infinite extent. Right, I know, it is a little strange.

Imagine in *figure 10* that I am here in the center and I draw a line out and I make a little circle. The circle represents the infinitely distant plane as I know it from my perspective as a perceived center. Now, recalling the planes of light coming toward my center from everywhere in space, when I take the infinity of all these planes of starlight and make a center within it, I constellate it into a star point with great potential to become something. I constellate the cosmos. Again, on the left of *figure 10*, I have a circle with an angle. If I draw a horizontal line out to the circumference and then draw another line at, say, a forty-five–degree angle to the first, I form an angle within the circle, and I can extend both lines out to the infinite. When I extend both lines of the angle to the infinite, does the angle become larger? No, but the "space" does become larger. This means that any lines I draw to the periphery or any lines making an angle in a circle reference a circle at the infinitely distant. A ninety-degree angle will be the same. Any angle will stay the same when the lines are extended.

Though strange, we can take this idea much further. Suppose I draw a hundred-eighty–degree angle and take it in one direction out to infinity. Suppose I say that the direction of that infinitely distant point is east. Now, what happens when I go in the opposite direction, which we call west, to infinity? Is the infinitely distant in the east any different from the infinitely distant in the west? No, it is the infinitely distant. Can I go up, down, or anywhere else to the infinitely distant? No. This is a little thought experiment, and the conclusion is a bit unsettling. When we go in any direction from the center of a circle we always end up at the infinitely distant, which is always the same "place."

When I establish a relative center, I experience the infinitely distant everywhere as the polarity. But we can go further. We have seen that a tangent plane is formed where a radius crosses the circumference of the circle (*figure 10,* third circle from left). This means that, at the infinitely distant circle, any line drawn from a relative center encounters a tangent plane at the infinitely distant circle. Since I can draw an infinite number of lines from the center of any circle, there must be an infinite number of tangent planes at the infinitely distant circle. Now expand the circle into a sphere and the same holds true; our infinitely distant sphere is composed of an infinite number of tangent planes. Science tells us that light travels in planes, so we can imagine an infinite number of planes of light streaming in toward the relative center of the Earth. In the infinite crossing of those planes of light, countless star points are generated where life seeks an entry to manifestation.

Those planes and points of light are breathing in and out toward relative centers all of the time creating a web of potentialities for life. The web of planes and points is what was called the entelechy or the pleroma, the source of all archetypes of the web of life. Here I call it a grape, there a cabbage. I call it a finger, an elk, or a crow. The web is composed of infinite planes and points, breathing in and out in space as a kind of backdrop, a drone in the cosmic music, or the etheric, as Rudolf Steiner called it. This is all a pretty good description of what he calls the etheric realm. The etheric is limitless potential organized into particular patterns. We can imagine planes of light moving toward us from all directions, and then we can go out in the morning and look out at light glistening off drops of dew on the grass. This kind of imagination can give you a sense of the oneness of everything.

Within the context of the etheric potential is the movement of the planets. From my perspective on earth, they appear sometimes to go in one direction against the starry background and sometimes in the opposite direction, creating a kind of looping process in the etheric that Steiner calls the astral realm. This apparent looping of planets creates "wrinkles" in time, changing etheric forces into what he calls "etheric formative forces." These give rise to the forms of organs in human beings, animals, and plants. They give form to life because, from the perspective of Earth, some planets appear to go backward when in fact they do not. However, human beings are constructed so that their consciousness changes the cosmos through acts of perception. This change of the cosmic etheric realm creates the astral realm. We change the ethers into astral forms so that we can have bodies that allow us to experience ourselves as entities separate from the cosmos. Because I can reach out to you and touch you as a separate person, it is clear that we are separate, even though we are not in soul and our spirit. This is the great human conundrum, especially in the realm of perception. Our perception changes the structure of how the spiritual realms appear.

Sometimes there are more *ether forces* available to physical entities and sometimes fewer. To alchemists, the rhythms of availability or non-availability of the cosmic forces is the music of the planetary spheres. They understood that the movements of the planetary spheres sometimes give a special impetus to a level of forces that creates a solve condition. At other times a block is established in the way those forces come in from the periphery. These blocking periods create a kind of coagula. In these great life rhythms of forces, alchemists knew that there are doors in time whereby we can act to bring things together, and there are places where things are best left alone to simmer. Sometimes it was appropriate or advantageous to be active in the work, and sometimes it was advantageous to let the work alone. They recognized that there is a breathing process in the rhythmic oscillations of the coming together and the leaving alone, and they recognized this breathing process as a significant part of the success of the work.

The rhythms of this breathing process were significant, because the rhythms followed the planets motions. As the planets come together and spread out from one another in space, alchemists recognized that similar rhythms are generated in time that can be discovered by observing nature. These rhythms constitute what Steiner calls the astral realm, which brings the breathing process of coagula and solve into the etheric realm. The rhythms are characteristic of natural phenomena, the Sun, because it has a breathing process of coagula and solve.

In a solar year, the Sun moves through Gemini, Cancer, and Leo to Virgo. Gemini is an air sign, Cancer a water sign, Leo a fire sign, and Virgo an earth sign. The Sun moves one degree each day and stays in one sign for about thirty days, more or less our average month. For thirty days around the summer solstice, with the Sun in Gemini, the elemental quality of air dominates natural phenomena. If you are making compost at that time, it will not develop well. There is too much light and air. Compost, which needs water and earth to support its action, just sits there and waits. The Sun then moves to Cancer, and if you are observant, you will notice that the morning dew is a little more abundant, but the Sun is still caught in the time of little change around the solstice. Then the water quality shifts to the fire sign of Leo, but the Leo's heat is still not good for compost, though the Sun is moving away from the solstice doldrums and toward the fall equinox, which is in the earth sign of Virgo. Now if you start making compost in late Leo, late

September and early October, you get composting activity. Microbial action comes to life aournd the equinox, and the Sun supports the earth element.

The cycles continue with the elements rotating in an earth, air, water, fire sequence, until we get to the winter solstice, when transformation and composting is available again. *Solstice* means the *Sun* (*sol*) and *stays* (*stice*). The Sun stays near the same degree of latitude for two months. During this time, the Sun is giving nature a signal to settle down or coagulate. Just do the status quo. Do not change anything. Let's go to sleep, but during equinoxes the rapid motion in latitude every day promotes change, transformation, and solve. This is the typical solar rhythm. Things do not transform for a month and a half around the solstices, but for a month and a half around the equinoxes things transform very rapidly. At the spring equinox, during the water sign of Pisces, the microbial world is very happy. Sponges on the counters start smelling, and produce from the winter starts rotting, because the microbes are activated by the Sun rapidly moving in longitude each day, energizing the etheric spheres.

It takes a year for these Sun cycles to unfold, but the Moon follows the solar rhythm every month. Every thirty days, the Moon passes through the position where the Sun would be during the summer solstice. It remains in that sign for two and a half days, because it moves thirteen degrees of longitude every day. Two weeks later, the Moon is at the position where the Sun would be during the winter solstice. These biweekly rhythms give a resting cycle to the lunar month every fourteen days. Following the solar cycle, the Moon moves through the position of the equinoxes while between the solstice positions. While the Moon passes through the equinox positions, there is an acceleration period for about four days. Then, two weeks later, the Moon is at the position of the other equinox.

Calculating the thirteen-degree motion each day, the Moon passes through one sign every two and a half days. This means that the Moon will return in about ten days to a sign that has a similar elemental signature. So, if the Moon starts in air, two and a half days later it is in earth, then two and a half days later it is in water, then two and a half days later it is in fire, and then two and a half days later it is at another air sign. Therefore, every ten days the moon returns to air, or has what we call ingress into air, water, or whatever sign it starts with. Ingress of the Moon into a sign is important for agricultural activities.

Regarding lunar influences, alchemists would say that if I do some operation in which I want the element of air to be active I begin that operation on the day when the Moon has ingress to that quality sign. Then if I want to enhance the air influence, I wait ten days again until the Moon is having ingress to air and work with the operation again. In this way, I can radically enhance the elemental quality of my choice by working with what is known as the Moon of the trines. I try to make sure that every time I work with that operation I work with the Moon having ingress to the element of air. My consciousness, or *akasha,* allows me to look through the calendar and plan when to instigate an operation. I pick the most potent or supportive places in time when I want to do something.

There are also other planetary rhythmic techniques to enhance operations. When making certain preparations, it is useful to work with a forty-day "alchemical month." Now, you can search everywhere for planetary references to forty-day rhythms. I have asked friends who are masters at esoteric astronomy what planet has a forty-day cycle? Nobody knows. Online, I finally found a researcher who did the math. His calculations showed that an interesting rhythm emerges if you count the days in an alchemical month at forty days, then skip a day, and then begin counting again. If you do this for nine years to the day, the Sun will be back exactly at the starting point. So, the forty-day rhythm—forty days in the desert, forty days for Noah on the ark, and so on—are actually an extended nine-year period of a solar or etheric cycle. Something in the etheric solar cycle allows life forces to mature in a different way. This is the alchemical month.

Therefore, when making preps, it is useful to start forty days before the day chosen to put the prep down into a heap. At that time, we instigate the first phase of an operation. Look for the solar sign when choosing an element that will support the operation, and then try to pick times during the next forty days to undertake the other phases of the operation when the Moon is having ingress to the elemental quality you wish to enhance. In an alchemical month of forty days, it is possible to enhance the elemental qualities four times. Then, after forty days, the final burial is possible, and it is often a good practice to leave the buried substance in its resting stage for three alchemical months.

When the substance is taken out after three alchemical months, any spray program can also follow the lunar cycles, taking the lead for spraying from the trine moon. To do this, spray your substance when the Moon is getting ingress into the sign used for burning, grinding, chopping, gathering, stuffing, or whatever. Thus, the day when you stuff the horn is not merely random—of course, people are free to do as they wish. Nevertheless, what I am sharing here is a powerful key to the possibilities of working creatively in time. By paying attention to such rhythmic patterns, it is possible to take advantage of how timing things brings hidden power to the preparation of a substance, the application of a substance, and even in the harvest of the materials that will undergo the transformation you are trying to develop.

A practical example of how these ideas can work can be found in some experiments with spraying, harvesting, and fermenting cycles in a California winery. A key rhythm for this experiment is to take advantage of the elemental transformation that can unfold when the Moon moves from air to water—that is, on the day that the Moon moves into a water sign or has water ingress.

The problem at the center of the experiment had to do with the extraction of phenolic substances from the wine grape skin in the ferment stage of the winemaking process. In California, wine grapes are typically harvested when the sugar content is measured at twenty-five or twenty-six brix.[13] Owing to California's warm climate, the sugar content of grapes can be in this brix range, which is rather high. The problem is that, with this much sugar in the juice, once the ferment starts the yeast rapidly consumes the sugar. At a sugar content of eighteen brix, the wine yeast is at maximum consumption and reproduction, which normally happens about two or three days after the juice begins to ferment. As the ferment develops, the

13 *Brix* is a measurement in the wine industry of how much sugar is in the grape juice, which supports the yeast in the fermenting process.

warmth of yeast reproduction picks up speed. At eighteen brix, the feverish consumption and reproduction of the yeast draws valuable phenolic pigments from the grape skins that contribute the qualities prized by wine lovers. Unfortunately, maximum extraction comes exactly when the process is slowing down as the wine yeasts die with the increasing alcohol content. Ideally, there would be a slight hesitation in the frantic reproduction of yeast so that maximum extraction of phenolics could be extended.

To produce uniform conditions in the winery, winemakers hold the juice, or must, in a vat surrounded by a sleeve of cold water in a process called "cold soak." It is possible to hold the must in cold soak for a few days, until the disturbance of the pressing is settled. Warm water is then introduced into the sleeve and the yeast wakes up. It normally takes about a couple of days for the must to go from twenty-six to eighteen brix.

The idea for the experiment was the idea of biodynamic wine grower Matt Taylor of Healdsburg, California. He thought that, if the grapes are taken out of cold soak when the Moon is having ingress into water—that is, when the must is at eighteen brix—the water moon would help the extraction of the phenolics by making the must more watery and less prone to the vigorous sugar consumption, which is characteristic of the yeast at this stage of the ferment. He reasoned that in water the maximum extraction would occur because water is the universal solvent. Therefore, the experiment was done with a couple thousand gallons of wine.

In the end, the must behaved as predicted but, during the extraction period in the water moon, an influential taster from a popular wine magazine vised the winery to taste the must. The expert tasted the water moon must and declared that the winemaker had created an unmitigated disaster, suggesting to the owner that the winemaker should be replaced. However, at the end of the ferment period, the wine turned into a most beautiful bloom of phenolics. When the usual blind taste test was done before bottling, this particular wine was declared the best of the harvest.

This is an example of the kind of thinking we can do. We did the spraying as the moon moved from air to water to let the grapes know this is what will happen during the crush. Biodynamic growers typically avoid water days, which is what I recommended when he asked me for advice. He called me up and asked me about what he wanted to do. I told him I do not think that is a good idea. He agreed, but told me that it takes about two days between turning on the heat and the beginning of ferment, so I suggested he go ahead and try it. It seemed counterintuitive, but I think the exciting part about this is that it shows a deeper aspect of the traditional way of working. Traditionally, we plan the work so that the first day should mirror the elemental quality that we want to enhance. Matt's reasoning was to let the wine yeast determine the elemental quality that he wished to enhance. This was a ferment honoring the state of the yeast. It was an elegant and deep kind of alchemical reasoning that seemed to pay off in the end. I cannot say that this is the only way to go, but it seems worth a couple hundred gallons of wine to do such an experiment. It is the kind of research that could produce a whole body of knowledge built around planetary rhythms. I think this is the kind of thing Steiner had in mind when he developed the preparations.

Figure 11 is a little diagram of how Steiner's agriculture course describes the mineral polarities that are activated by dandelion. In the diagram, we have K (potassium), Ca (calcium), Si (silica), and P (phosphorus). Silica and phosphorus are the two activators on the sulf side, and potassium is an activator on the sal side. Calcium is the anchor.

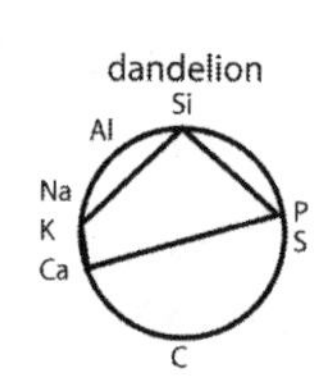

Figure 11

Consider a dandelion. As a taproot, it is very alkaline, which is your calcium influence, but has a basal rosette that it forms. Rosettes, especially in biennials, are a signature of potassium influences. Moreover, it has silica and phosphorus, two very active substances that move it to the flowering pole. Think of how it flowers. There is a slow buildup for a whole year, and then in the second year there is a remarkable explosive shift right into flowering in an almost exhaustive flowering phase. The flowering gesture is the dynamic of the substances from the view of solve and coagula. There are three dynamic activators in the substance profile along with one calcium coagulator.

In *figure 12* is a picture of that flowering process. It is a diagrammatic picture of the way seeds are distributed in the flower head. The little black part under the seeds is the receptacle. In the composites, the receptacle is the organ that determines how the composite family holds the seeds. The receptacle is the part at the top of the peduncle, the terminal end of the flower stem. It is the part in the flower head that receives the seed-forming process. What we can see in *figure 12* is that the little seeds have their little parachutes at the top of the seed, and they use those parachutes to fly away to the periphery. This is a picture of the calcium, potassium flying away to the phosphorus sulfur. The seeds are held in and then, suddenly, they are released and they go far out to the periphery. This is a picture of a very intimate center and then a search for the periphery. Silica seeks the periphery. Phosphorus seeks the periphery. Now, why does Rudolf Steiner connect the dandelion with the mesentery of a cow?

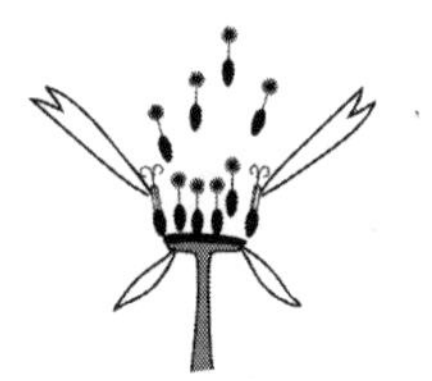

Figure 12

Figure 17 shows the mesentery, a membrane that stretches from the wall of a cow's digestive cavity. It is the muscular cavity that holds most of the digestive organs and is lined with the mesentery. The intestines hang from the wall of the cavity like suspended sausages. The mesentery is a very thin, transparent membrane and often exhibits rainbow colors as though made of siliceous materials. It is inside a bag inside a bag that is inside a bag. In alchemy, any process, substance, or form that becomes enhanced to such an extreme always suggests that the process, substance, or form is about to transform into its opposite. The mesentery is a membrane inside a membrane that is inside another membrane—completely inner space. The more a substance is completely in, the more the process is completely out. This is substance, process, and polarity.

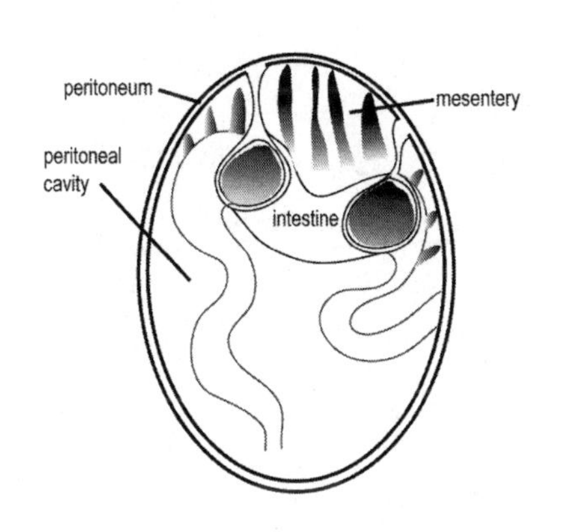

Figure 17

Steiner chose the membrane deepest inside the digestive organism so that it would attract the forces

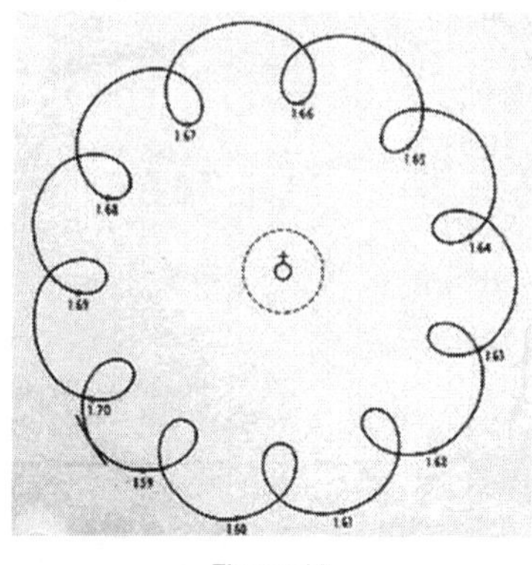

Figure 13

and processes from the furthest reaches of the cosmos. This is a great law of form and function in an organism. Alchemically, it is kind of a rationale of why he would use a mesentery as a membranous sheath for the dandelion. In esoteric lore, the dandelion is connected to the forces of Jupiter, which is considered the doorkeeper to the stars. Saturn is the actual doorway; Jupiter is the doorkeeper. Jupiter takes what is cosmic and introduces it to the earthly, thus bringing the cosmic and the earthly into contact once the cosmic comes through the doorway of Saturn. In a sense, Jupiter traps what Saturn allows in. Esoterically, such a trapping process is done through the looping processes of retrograde planets.

In *figure 13*, we can see a picture of Jupiter's looping movement. In the center is the symbol for Earth. The dotted circle around it is the apparent orbit of the Sun as seen from Earth. In the diagram, we can see that the loops of Jupiter happen way out on the periphery, far from the Sun. The action of Jupiter moving in front of the stars has an astral-trapping action or signature way out at the edge. That is where Jupiter forms images of loops in space and time. In the diagram, the loops are actually a little deceptive, because if you actually could see them as they exist in time and space, they are almost flat like a skin or membrane out in space. If we could take these movements of Jupiter and render them visible, we would be surrounded by a kind of pocketed skin membrane of Jupiter loops out on the periphery of our solar system. If we could see it in space, there would be temporal–spatial loops out on the periphery. Esoterically, when a planet moves out on the periphery in this way, its action is considered to have a resonant effect on things here that are images of that movement. The image of Jupiter's temporal–spatial looping forms on the periphery is reflected in the cow's mesentery. The action on the periphery is an analog of the seeds' motion as they are blown from the dandelion seed head. Looking for analogs, these actions of center and periphery can be seen in the picture of the relationship of potassium to calcium, silicon, phosphorus in *figure 11*. Steiner envisioned what the dandelion does as an image of Jupiter is analogous to what the mesentery does for the cow. That is the imagination behind preparation 506.[14] Even though these ideas are remarkably elegant in their own right, I believe that such planetary analogs are real seeds for further research. They present portals for inspiring people to look for ideas for other links among substances, processes, and planetary influences.

Without acknowledging planetary rhythmic signatures, there is very little to support the delicate results that most biodynamic research represents. I do not think the results are ever going to be as robust as chemical or technical proofs, but they have the advantage of being non-intrusive. Within this delicate realm, forces are so fine that it is often impossible to determine whether what you have done is actually working. There are few robust parameters to grab on to; when we work with life, we can rarely say that one thing causes something else. We need to record our actions on a calendar so that we can go and check where, for example, the moon was when something was done to see if there is a change connected to it. This kind of recordkeeping really helps. Then, as you get an idea about the nature of Jupiter, you ask yourself where Jupiter was and what it was doing. If you ask such questions, the being behind Jupiter may indicate in some way where to find the answers you need. When someone comes to you with an inspiring story or source of information, this is a good signal that you are being tapped to do something in connection to Jupiter. When an impulse comes from someone else, it is a failsafe that one is not simply inventing answers, but if it comes through me personally, I need to check my belief system.

We find the dandelion at the one pole of the biodynamic preparations. The mesentery is a membrane outside the intestines, but it is deep inside the animal. The intestines are inside the mesentery, which we would think puts them even more on the inside, but the intestines are actually a part of the

14 Preparation 506 uses cleaned fresh cow's mesentery filled with dandelion. It is buried during autumn in an unglazed earthen jar and lifted and prepared for use in early spring (http://cityfoodgrowers.com.au/bd506.php; accessed Feb. 2013).

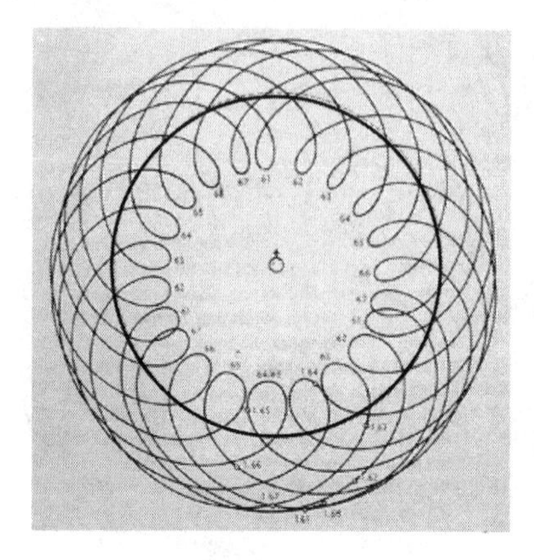

Figure 14

digestive system, which is connected to what is outside the organism. What you put in your mouth comes from the world, goes through the intestines, and comes back out into the world. Even though they are deep inside, our intestines have evolved and are linked to what is outside the organism. Here we have the great alchemical reversal; a space that is completely inside is functionally linked to the outside.

This is the great inside–outside polarity of the digestive system, as well as many other systems in the human being. The inside "space" of the intestines is very different from the inside "space" of the mesentery. The inside of the intestines is in direct contact with the outside world. By contrast, the mesentery is so hermetically sealed within, that the digestive cavity itself is tuned to the infinitely distant periphery, and not just to the immediate outside environment. This same spatial motif can be found in many other places, including the *Agriculture Course.* Steiner talks about both the inner and the outer planets in relationship to soil. Here the soil is seen as a membrane, or diaphragm, that separates the realms of the outer and inner planets. The influences of the inner planets enter the soil indirectly through the atmosphere and the absorptive qualities of lime, whereas the influences of the outer planets enter the soil directly through the action of silicates.

In *figure 14,* we see the Earth in the center. In *figure 13*, the Earth is also in the center, but the dotted line (the apparent orbit of the Sun) is much closer to the Earth. *Figure 14* shows the dotted line as the Sun's apparent orbit based on Mercury rather than on Jupiter. Observe Jupiter's loops, which are way outside the Sun's apparent orbit (*figure 14*). Look at the loops of Mercury; some come in all the way; some loops come completely inside the Sun's apparent orbit; and some of them extend both outside and inside the Sun's apparent orbit. This is because Mercury has conjunctions that are inferior between us and the Sun and outside the sun, superior and inferior.

Mercury oscillates inside to outside, but mostly it does not loop inside. Look at the alchemical substance diagram of chamomile (*figure 15*). Potassium is active and moves, and both calcium and sulfur, as we have seen, are coagula, or holding. There is a kind of internalizing of this Mercury principal, creating a sort of inwardness rather than complete outwardness.

Figure 15 Figure 16

Now look at *figure 16,* which shows a cross section of the chamomile receptacle with an empty space in the middle. Chamomile creates an inner space here, while the dandelion goes all the way out to the periphery. Chamomile substance is put in the intestine, and the dandelion is put in the mesentery. These two organs are inner and outer, the two plants are inner and outer, the substances they activate are inner and outer, and the planets associated with them are inner and outer.

Steiner's alchemy in this area is very consistent, very rigorous, yet highly imaginative. When we put them in a heap, we are making a mesentery and intestines so that the plants get a full spectrum probiotic digestive. They fully complement the inner and outer relationships in the digestion of humus and so on in the soil. It is based on cosmic pictures of the planetary rhythms. Therefore, working with planetary rhythms is really the future, because this is where we can share with each other.

However, because of something called polyrhythmy, when it comes to rhythms we discover that we never know if what we are doing is correct for the long term. A good way to explain polyrhythmy is to imagine that you are in a car, it is raining, and the windshield wipers are on. You stop at a red light, a song is playing on the radio, and the windshield wipers come into sync with the song. Then you notice that the turn signal of the car in front of you is also in sync. That syncopation is an example of polyrhythmy; individual rhythms come together momentarily, but then one of the rhythms drops out and then the other. They drop out because the separate rhythms are slightly out of phase. They are not a beat but rhythms. If they had the same beat they would stay in sync. A beat does not breathe like this, but rhythm does. When learning to work with rhythms, we learn to look for the breathing cycles in which the rhythms connected to particular phenomena breathe in and out of sync. Polyrhythmic breathing is the reason that I said we may think we know what is going on, but then we realize there was another rhythmic layer that we did not observe, and now we do not know again.

I have been doing this kind of rhythmic research in climate studies for thirty years, and there is always one rhythm behind the one I think I am studying that is tempering what I am studying in ways that are not yet apparent. When we start studying these rhythmic cycles, we are usually quick to identify some relationship. Then we watch that relationship dissolve into the rhythms of time. For example, the young man who did the experiment with the water moon and the fermentation began with the good insight about the lunar ingress into water as the wine must reached eighteen brix. That was a good insight. Based on his initial success, the following year he repeated the same protocol. He waited to take the juice out of cold soak so that the yeast would reach eighteen brix in two days. However, even though the wine yeast was warmed, it did not take off. The brix number did not move for two days. He thought he would have to kick the must to get it going, but then it suddenly started moving, though very slowly. He called me in a panic and said, "Dennis, I don't know what to do. I did exactly the same thing I did last year, but it is stalled." I told him to look at his calendar from the previous year. When he started, there was nothing else in the planetary spheres going on around that date. So, I said, "When did you take it out of the cold soak this year?" He told me the date and I saw that there were eight conjunctions when he took it out of the cold soak and on the following day. I said to him, "Do you see that huge cluster of conjunctions on the calendar there where you started? That has a meaning."

He had repeated his action from the previous year, but it did not work because a conjunction gives a halting force to any interaction. It is a forceful coagulation event. To have eight forceful coagulation events in the two days when you start the ferment tells the wine must to wait a while before commencing the ferment. When we went through each conjunction, we saw that there were some potent events and he realized the need to study polyrhythmy. Because of his initial success, he thought he had the answer. This is the nature of studies linked to planetary motion.[15] The study of planetary rhythms allows us to see impulses move through the natural world, accompanying the planets through time. Being able to see these rhythms unfold in the natural world becomes very interesting when doing things such as planting vineyards. With insights into rhythm research, we can begin to see into the long-term patterns that create difficulties in working with nature. However, we need to be committed to recording all our observations on an ephemeris to see these rhythms reflected in nature.

Working with natural phenomena leads to insights and unique observations of unusual events. It is important to record these events as a regular practice, but beyond just recording, it is also very useful to keep records in the context of planetary motion. Discipline yourself to record observations every day on an agricultural star calendar. When you do this, your mind begins to contact the beings behind the phenomena, and they will speak to you through the phenomena by helping you see more deeply into the rhythmic structures of the natural world. Then the phenomena will speak to you in a deeper way. Record keeping with an eye to planetary rhythms helps us change soil science and agricultural practices into something we could call an initiation process. This is why I think the grape growers have a destiny to pick up the challenge of biodynamic research, since they already have laboratories.

15 See my book *Climate: Soul of the Earth*. It contains some 300 charts describing rhythms. Most have to do with climate, but the idea can be applied to our subject.

One young grower I spoke of has added a column for the position of the moon to his spreadsheets used to monitor growth and fermenting information. When we get the readouts from the vats, we know where the moon was on a given day. He just puts an extra column in his spreadsheet for the lunar movements. The lunar information is right at our fingertips when we sit and have a meeting. We get our calendars and begin to see the planetary dynamics in the data. Then, the secrets that Steiner brought begin to be more like a language and less like recipes. His insights become challenges and inspiration for new research.

CHAPTER 5

CLAY

Body, Soul, and Spirit in Nature and in Human Beings • Acid and Base Polarity • Soil Formation and Granite • Clay Singing Ritual • the Solar Calendar

The relationship of the human being to the natural world is an important aspect of the work in biodynamics. Rudolf Steiner says many times in the *Agriculture Course* that we have to make this work with nature *personal.* I would like to compare the idea of the human role in the natural world to the role of clay in the mineral world. It is often said that the Bible records Adam, the first human, as having been made from clay. In fact, the Bible says that Adam was made from dust (Gen. 2:7). However, in the Quran, there is a reference to making a little human being from clay, but one of the angels gets jealous of God's little mannequin, which makes for an interesting picture.[1] The relationship of human beings to the natural world is as a kind of middle ground between the spiritual animating force behind the natural world and the manifest world of matter. The human being is a kind of hybrid. In his book *Cat's Cradle,* Kurt Vonnegut quotes God: "Let Us make living creatures out of mud, so the mud can see what We have done." Here is a drama that has to do with why the natural world had to fall into matter, and why it has to be lifted out again. The human being is the reason for both the fall and the lifting. The human being is involved in both. We could even say that

1 "...your Lord said to the angels, 'Indeed, I am going to create a human being from clay. So when I have proportioned him and breathed into him of My [created] soul, then fall down to him in prostration.' So the angels prostrated—all of them entirely. Except Iblees; he was arrogant and became among the disbelievers. [Allah] said, 'O Iblees, what prevented you from prostrating to that which I created with My hands? Were you arrogant [then], or were you [already] among the haughty?'" (Quran 38:71–75).

the essence of being human is to be the reason nature falls and the reason nature will be redeemed. It is our sacred task to redeem nature.

The sacredness of our work on the land has to do with the fact that nature has fallen so that we can rise. However, in our rising we must transform nature back again into a spiritual condition. As I understand it, this is the very essence of the work that Steiner brought late in his life as biodynamics. What work on the land, the therapeutic work, and the arts have in common is "the healing of perception." It is the perception that, for human beings, work is the problem and the cure. The problem and the cure are the same issue—transformation of the senses. Therefore, we will begin by looking at this task from an esoteric point of view that sees the human as having to maintain a living relationship among body, soul, and spirit. Our body is the seat where matter has fallen; spirit is the transcendent consciousness that inhabits our body; and soul is the flow between these two. I can have an awareness of my physical body, which is awareness of the way matter behaves. A sense of my body is based on the forces of attraction and repulsion that govern such phenomena as bodily processes. In fact, such perception is unconscious, because matter in my body is a corpse fallen from the spirit world.

If we try to go into the consciousness of the body, a common experience is a deep fear of dying. We could say that this arises as an experience of the matter trapped in the body. I must give the matter back to the earth when I leave. My vehicle is composed of matter that will return to the earth. My vehicle is not me, but it does deeply impact my sense of self. To develop in life, my consciousness has to engage my vehicle. My dilemma is to separate who I really am as a spiritual entity from my experience of the body, which contains a kind of fear. However, to do that I have to engage the fall into matter, since that is my body's experience that helps to form my soul. My physical body contains a corpse of what used to be a spiritual condition. We call it "matter." I can be aware of the difference between the matter in me and *me as a soul,* because my soul is the source of my awareness. My soul is sensitive and aware, and it can look at the matter in my body as separate from my soul.

I can look inwardly at my soul activity and out at the world. I can engage the world and try to transform it. That is an artistic impulse hidden in the work on the land. You take matter and try to bring it to life, which involves being sensitive to life and aware of life—being aware of the difference between a living plant and a dying plant. A gardener's art is being aware of a plant that is going to give good seeds and one that will not produce good seeds. To those who are not aware and sensitive in this way, this seems like magic. How can you know that plant will give good seed? A "green thumb" is sensitivity and awareness carried in what Steiner calls one's "soul life," or "soul body." The soul is a liaison between my sense of who I will be when I become part of the earth as a corpse and who I am as an eternally living spiritual being. I also have a component that affects my soul from a transcendental level beyond the soul. This is my spirit that is aware of the transcendent level of consciousness that everything is made of consciousness rather than matter. I have an ability to perceive myself as consciousness or as a vehicle packed with matter. I also have the ability to see myself as a liaison composed of various states of sensitivity and awareness flowing between myself as a spiritual being and myself as an incarnated body. This is called "living."

Sometimes, when I want reasons for why I am not well, I go to see someone who can tell me about my body. Sometimes I want reasons for why I am not well, so I see someone who tells me about my soul. Sometimes, when I want reasons for why I am not well, I go and talk to someone like Rudolf Steiner, who tells me about the spirit. I am free and can choose where I place my consciousness. That is the great gift of being human. The unconsciousness of the physical, the aware consciousness of the soul, and the transcendent consciousness of the spirit are the stage on which the human drama is played out.

In my physical body is matter, in matter science finds "force," and the force is a particular kind of movement. We call those movements "natural laws." That is the way science would look at it. Matter is moved by forces that manifest the laws of nature. Those laws are transcendent to the forces. So in an abstract consciousness—the consciousness of modern science in which forces kind of rule things—this would be the trilogy, or the trinity.

I would have matter at one end, the laws of nature at the other end, and forces moving between and liking them. This is a contemporary scientific view. However, if I want to look at an esoteric scientific view of this, I would say that we have something called "elements" in the realm of matter. Science would also say matter has elements, but the esoteric perception of elements is different from the scientific concept of elements. Esoterically, the concept of elements is more of a sense of processes than substances.

In conventional scientific consciousness, I find processes in the force realm where substances are bases, whereas esoteric consciousness sees processes as the bases. Alchemical elements are processes, and matter is considered to be a corpse. In alchemy, the most fundamental state of being is an elemental process of earth, water, air, and fire. Moreover, the elements are ensouled for esoteric consciousness. In esoteric science, the elements are not merely abstract processes and forces but beings. Each element has a particular consciousness. Alchemical elements are ensouled by *beings*. They are no longer just abstract forces but elemental beings.

The soul of the natural world, or the equivalent of the soul of a human being, is consciousness of sensitivity and awareness. In esoteric science that deals with the natural world, the place of the world soul is the elemental world, the groups of elemental beings. They carry sensitivities and awareness of the forces of nature. They are, as Steiner called them, "the workers." They get the job done. They are the forces of nature and move in particular ways. They have patterns that reflect particular states of consciousness.

In the ancient world, it was recognized that natural forces have great intelligence, and if they have great intelligence they must be the function of a being—an elemental being, the soul of nature, the forces of nature. Then, however, it was understood that deep guiding principles are behind the sheer forces and their interactions and that they are much higher and manifest a more sublime consciousness. This consciousness was transcendent to the play of forces. There is a whole hierarchy of elemental beings that Steiner calls "elementary beings." He said that those elementary beings are the laws behind the forces. He calls them the "princes and the princesses" or the "kings and queens" of the elementals, and they carry the lawfulness of the forces that eventually become the elements of earth, water, air, and fire. Modern science calls those beings the laws of nature.

Here we start to see in the esoteric approach to the natural world, which sees the consciousness of the operator determining some aspects of the outcome. This is considered dangerous territory for conventional science. If I say that my state of consciousness affects the outcome of the experiment, such an effect would be considered caused by "bias" or "subjectivity," and science wants to eliminate this. Thus, natural science and natural scientific proof does not *want* consciousness to be part of the proof, whereas esoteric natural science understands that the operator's consciousness has some type of interaction with the phenomenon. This is a critical point, because an experiment for the esoteric scientist is much more a question of consciousness than empiricism.

To complicate things in this realm, Steiner uses the term *spiritual substance.* We have scratched our heads about this for years. What is spiritual substance? We could say it is akasha, or consciousness, including all potential states of consciousness that lead to manifestations. In Latin, *substantia* refers to the presence or *essence of being.* The words a*kasha* and *substantia* are similar in that they refer to a state of total potential being that has not yet manifest. Akasha is a substance, but is totally spiritual. The potential consciousness of akasha needs a *particular* consciousness under its complete universality before it can manifest a particular. Akasha as *substantia* needs a being to render it into manifestation. This may seem like splitting hairs, but these ideas form an outline of threefold nature, or the trinity, of physical body, soul, and spirit.

Now, I would like to use that trinity—the threefold gesture that Steiner associated with many things—as an idea to illustrate how human consciousness interacts with the soul and spirit of the Earth. What does our consciousness have to do with the substances we use in the biodynamic preparations or, more specifically, in the realm of clay?

To understand clay better, I will give a couple of examples from the mineral realm, especially the minerals as they appear as a dynamic. Making a pottery glaze is one instance that involves a useful understanding of mineral

Glaze constituents

flux + amphoteric + glass former

Amphoteric means that something acts as a base or acid. Many amphoteric substances are metals that can link between silica (the glass former) and the fluxing agent. The amphoteric substance balances the flow, inducing the force of the flux and the resistant nature of the silica to the flame. The amphoteric substance causes the silica to melt at a lower temperature and helps the flux remain stable at a higher temperature.

interaction as a threefold event. There is a component known as the glass former in a pottery glaze formula. The glass former is a necessary stabilizing agent in the glaze owing to the high melting point of silica. The second element in a glaze is called a flux. It is the component that melts at a low temperature. Fluxes are often metals with a low melting point. The most common fluxes are feldspars that contain potassium or calcium in the context of alumina. The third element of a glaze is an amphoteric. The flux induces flow, the glass former promotes congealing, and the amphoteric makes things flow correctly and then congeal, so that the glaze does not run off the pot when it melts. For balance, a glaze needs flux with a low melting point, a glass former with a high melting point, and an amphoteric that brings the high melting substance into interaction with the low melting substance. The amphoteric is the control component in the glaze and regulates the pH of the glaze. The whole science of glazing has to do with this kind of balance. The dynamic of glaze formation is a wonderful analog for the development of the soul.

How much heat are we willing to apply to our own consciousness to get it flowing around the issue preventing us from growing? How much heat can we stand in the name of changing ourselves? Do we need a Buddhist immolation technique, or can we just put our feet near the fire? How much heat is it going to take to flux? Moreover, will we flux so much that all of our good stuff ends up on the floor of the kiln? How many people do we need to put us into the cauldron before we understand that we do this inner work of transformation ourselves.

The human soul has amphoteric properties that try to balance between flowing and staying, between spirit and body. In terms of fertilizing your garden, this is called putting a lot of nitrogen on your lettuce and then watching it turn into confetti. We need to bring it back into balance, so nature sends in agents such as rot and insect pests to serve as an amphoteric and bring harmony.

Balancing substances in the outer world can be seen as an analog of inner soul processes. This seems to be what Steiner means when he says to get up close and personal with nature. Nature is *us out there*. It is not just nature out there and us in here, and the two are totally separate. It is us out there and nature in here as complements. What you think of as your digestion is actually a teeming jungle of various organisms that have their own world going on. You do not really control it any more than workers in a beehive control production. They bring whatever is out in the fields and give it to the queen, which determines how many eggs the queen will lay. This is nature. The queen does not decide. The queen serves the hive and the whole of the natural world. How does she learn to serve it? She responds to what she has been fed by the workers. That changes her whole physiology toward laying the eggs.

This is the flow between too much and not enough in nature. This is also the governing principle in the human soul. These types of things in the natural world are in us, too, as our soul and that of other ensouled beings. They are analogs of the way the Creator designed the soul for human beings to learn. That is why for millennia we have been dropped into the very lap of nature, to depend entirely on nature for our sustenance; nature is the great primer for learning who we are as ensouled beings living in physical bodies. However, nature cannot teach us about how our spirit lives in relation to our soul, because we have freedom from many of the constraints of nature, or what we know as biological necessity. As evidence for such freedom from biology, we have a modicum of control over some of the forces of nature. Therefore, we tend to think that we do not have to listen to the laws of nature, leading to climate change, pollution, and the spread of genetically modified organisms. Human beings are in conflict with nature, though our spiritual destiny is to be stewards of the Earth. Our soul life is a kind of amphoteric between our spiritual destiny as human beings and the Earth soul.

Now, returning to the subject of clay, in terms of mineral transport in the earth, metals are unique in nature's household. Alchemists consider metals to be essentially fluid but in a suspended solid state. That is why you can pound lead with a hammer and it reacts as though a stiff jelly. We can do the same with copper, because they are fluid in their essence. They are oxides and they are amphoteric, since they can bring silica and other metallic fluxes into relationship. They can be either solid or liquid. Just apply a little heat to lead. Following an ancient custom, at the New Year we would heat a small piece of lead, drop it in water, and look at the congealed lead for indications of what might happen in the coming year. This old custom is evidence that lead is connected with a soothsaying capacity. Alchemically, lead is Saturn, and in this practice we were asking Saturn for a bit of foresight.

In the natural world, especially in the mineral realm, metal oxides are usually amphoteric, because they represent the ability to be both solid and fluid, to be both acidic and basic, with fine shades of difference. When they get into parts per million and finer parts, they are an amphoteric even in very delicate relationships with enzymes, proteins, and so on in organisms. The iron in food and the digestive process, as well as the calcium, magnesium, and phosphorous in plant nutrition, have the same type of congealing or flowing amphoteric pattern influence. The alchemical secret of metal is that it is actually a code for sensation. When alchemists talk about metal, they may mean the metal, the thing I have in my hand, the corpse of the metal. They would say that the metal in your hand is the corpse of the true metal.

Alchemists call metal, or the activity that gives rise to the metal, an aeroform. According to alchemy, metals were actually disbursed throughout nature in a kind of aeroform, or gaseous, state. The source of light is aeroform metals. Alchemically, light decays into air, which is the corpse of light. In Chinese alchemy, metal is the element linked to the lungs. In that system, the air or metal element represents the hidden essence in the senses. It is associated with nature passing away during autumn, revealing the hidden essences that attracted and supported the life forces during spring and summer. Metal is seen as a hidden essence and a corpse of that essence in one entity.

In Western science, there are particular qualities of light that we would call metals. When we try to judge the substances that make up a distant star, we run the star's light through a spectrograph, a gas chromatograph, which displays various lines that indicate which metals comprise that light. This tells us what the gases are on that star. This is possible because the metal is a gas that is emitting light. Thus, aeroform metallic light is the alchemical metal that is present but nevertheless hidden from direct human consciousness through sensory experience. Lead is one kind of sensation, iron is another kind of sensation, and tin is yet another. Metals are a quality of soul perception.

Lead is the sad-sack consciousness of Eeyore in Winnie-the-Pooh stories. Mercury is the consciousness of Loki, our crazy little dog that runs around barking at everything that moves on the farm here. He walks into the room, and everyone knows Loki is here. In alchemical code language, sensory experience is a metal. If I have a particular sensory experience, I am participating alchemically in what we could call the "metalizing" of the world, but I am not participating in the physical corpse of the metal. I am participating in metal as a form of consciousness, an atmosphere, a particular quality of akasha.

Granite

feldspar + mica + silica

Clay is a phyllosilicate, meaning that its form resembles plant leaves. Mica is the central ingredient in granite and is a phyllosilicate. This is the relationship between mica and clay.

Now it gets very interesting. Granite—what Goethe called the protean rock (Proteus was the god of change)—the basic rock, has three constituents, quartz, mica, and feldspar. I can look at quartz and ask what it would be as a glaze. The equivalent of silica in a glaze would be quartz, the glass former. Feldspar is traditionally the most fundamental flux for a potter, because it has a calcium or potassium quality in it. Then, in between in the granite, mica is a unique silicate that has a structure like the leaf of a plant. If you get a little "book" of mica, you can unfold its leaves. You can actually see *through* a leaf very easily. It is similar to a piece of cellophane. As a phyllosilicate, the leaves of mica provide surface after surface inside the mineral.

Recall the law of minimal surface discussed earlier and that all action happens at the surfaces of things. All chemistry happens at some sort of surface or membrane. All pathology happens at the surface of things. All healing happens at the surface of things. Surface in nature is the place where congealed stuff meets flow. Between congealing and flow, some type of surface gradient, boundary layer, or sensitive membrane shows up and allows an exchange from one side to the other. Surfaces are the business end of the natural world.

In mica, therefore, we have a silicate mineral, but instead of having inner surfaces as in a quartz crystal, in which all of the boundary layers are locked into one another to make a solid substance, in the mica, I can peel that whole mica book into numerous little transparent leaves. It has a silica quality of infinite inner surfaces, but each surface is separate from the others. Water can penetrate the mica found in the granite, because the surfaces are so close together. Being so close, capillary action is really active, pulling the water into the spaces between the mica plates. The water, which has a weak charge, or weak acid, arising from protein degradation, starts to pull potassium and calcium out of the mica. Potassium and calcium are the salts that need to be mobilized with phosphorous to make a soil solution for plants to live. This process is called weathering.

When water gets pulled into mica, the spaces between its layers are very active on a molecular level, at which the inner surface is rich with even smaller surfaces. It is a radically amplified reactive surface that produces an ion exchange with the water. The water pulls out potassium that then builds little ion channels. On the web you can find videos of these little potassium channels growing from the surface of mica crystals, carrying ions that support the growth of plants. The most fundamental chemistry happens in mica.

Silica creates the mica. It is a phyllosilicate. Feldspars are also silicates; when all the potassium and calcium comes out of the mica, feldspars form and some evolvements of that—calcium feldspar, sodium feldspar, potassium feldspar, or clays. Mica is a kind of intermediary between the calcium pole and the silica pole, because it is active in the presence of water. It is sensitive. We could even say it is dimly *aware,* or sensitive to water. It is sensitive to pH of the water. If water does not have a weak charge on it, there is not much action. It is not that the physical rock in the ground is aware, but that somewhere in the cosmos a spiritual being, the archetype of mica, is aware of the pH of water. The activity of that archetypal mica being is present in the rock corpse as a pattern of forces we call weathering. The patterning creates weathering that gives rise to soil. Once soil is created, the clay content in the soil mediates between what comes in from outside and what comes up from below.

Clay arises from the mica content and continues the transport of energies in the soil solution. Clay is an amphoteric between the insoluble silicates and the soluble lime elements, between spirit and matter. It is, we could say, "the soul of the soul of the soul." It is the center of the action. Technically, clay is the smallest particle size in the soil. In adherence to the law of minimal surface, clay has the greatest surface area of any mineral in the soil solution.

This makes it very attractive to organic acids that form proteins and humic acids used by plants and soil animals in the development of stable, humus-rich soils. In the clay humus complex, the dead minerals and life forces of the colloidal solutions of protein-building acids come into a magical place where living protein structures can be reassembled from dead things, and then we get this magical substance, humus. It comes out of the life and the sensitivity of the clay. Clay itself is not alive, but it is sensitive to life. It has a quality of a metal. It is plastic and extremely sensitive to water, because it takes from the mica the shape of the little layers of the clay; these appear as little shingles when you look at them under a microscope. They are resting on one another, and when the clay dries out they stick to each other. However, as soon as it comes near water, the water flows in just as it did in mica between the layers of the clay; the particles can then slide over one another. This is the quality that makes pottery clay so plastic. It actually draws water in to lubricate these little shingles. The water between the clay particles creates a kind of condition that is very close to being alive. All it needs is some humic acids from dead things and the presence of light, and great chemistry that starts to make amino acid chains and, suddenly, there is a sense that something else is happening in the humus.

Therefore, out of the central space, following all the way down, we have awareness, forces, and archetypal and elemental spiritual beings, giving rise to amphoterics among metals active in mica that weather into clay and absorb amino acids and proteins, which create conditions for new life. We are starting to see that in the center—clay—is an analog for the human incarnation process, going from archetype to manifestation. It seems the Quran gives us a pretty good picture. The human was made of clay.

Let us look at the Celtic, or pagan, calendar. It includes a festival called Imbolc, which corresponds to Candlemas in the Christian calendar. In the Chinese calendar, the festival at this time of year is called the "setting of spring." It occurs when the forces of spring are beginning to stir, but the earth has not yet revealed it. In the Celtic calendar, this is the traditional feast of St. Brigid—or rather the Goddess Brigid. She is both the saint and the goddess of fertility and has the quality of resurrecting the light, rising

In the Celtic year, February 2 is Imbolc, or St. Brigid's Day, in the Christian calendar. In both calendars, this day brings a gesture of recovery from a sustained illness and celebrates the crafts. The cosmic significance is that at this time the Sun is beginning to accelerate out of the winter Solstice darkness and quicken the Earth's womb. Clay is the womb of the minerals. Adam is made of red dust in the Bible, but made of clay in the Quran.

from the ashes into a new form. This might also be a kind of Mother Kali—the destroyer so that new life can be created. St. Brigid resurrects life from what has been destroyed. At Candlemas, what has been in the darkness now becomes suddenly, strangely, magically fertile. Something is stirring, even though we cannot see it down in the womb. Given the picture that I tried to present earlier, we could say that clay is like the placenta of the mineral world. It is the sheaths in which humus is born so that fertile ground and the possibility of life can be created. At Candlemas, St. Brigid is celebrated by lighting candles and lights to say the darkness has turned and something new is happening. The winter solstice, when the Sun stops moving in declination, is the rest period that precedes Candlemas.[2] At the equinox, when the Sun is at the equinoctial point, if you look in an ephemeris you see that the Sun is moving rapidly, day by day, from north to south or from south to north in declination. You can feel this if you are paying attention and are outside daily around the equinox.

Once the Sun reaches the winter solstice, there is a staying place for its daily motion. As mentioned earlier, *solstice* means "Sun stays." It stays in the same degree of latitude for about a month and a half or two months. At this darkest time of the year, everything in the previous growth cycle is now being assimilated and pulled down into the ground. Everything is rotted. The crops are out of the fields; the leaves are off the trees; the birds have left;

2 The Sun's declination varies with the seasons. As seen from arctic or antarctic latitudes, the Sun is circumpolar near the local summer solstice, leading to the phenomenon of it being above the horizon at midnight, which is called midnight sun. Likewise, near the local winter solstice, the Sun remains below the horizon all day, which is called polar night.

bees have gone into their hives; and insect larvae have crawled up under the bark of trees to pupate. Life goes into hiding at the winter solstice. Then, at Candlemas, the Sun begins to move more quickly in declination again, this time coming up. It takes about two weeks after Candlemas to see it, but after those two weeks, everyone is aware.

With Candlemas, there is an intimation that something new is coming—that there is already something in the womb and it is going through a process. Now we just need to make the sheaths. The sheaths are now growing the new embryo of the new natural year that will emerge from the trees and the land. Now, in the "setting of spring," we are beginning the formation of sheaths for the new birth. This is the mood of Candlemas. Steiner suggests that we take clay and do things with it at Candlemas and spread it out over the fields, so that there is a kind of connection between the old and the new that what is down below gets lifted up by the amphoteric nature of the clay.

I would like to share with you now a quote from a book. It is in an old edition of a book called *Living Water: Viktor Schauberger and the Secrets of Natural Energy,* by Olaf Alexanderson.[3] In it is a particular episode in which Schauberger is visiting a farm and meets a man. I always had the feeling the man might have been Hugo Erbe. In any case, here is a quote from the book: "The old man was stirring rhythmically with a wooden stick in a big barrel with water. All the while he was singing tunes and throwing some earth in the water." Actually he was throwing clay into the mix. "Singing upward, he stirred against the clock [counterclockwise]. Singing downward he stirred clockwise this virgin hyme." *Hyme* is a shortened word for hymen, a membrane.

A virgin membrane, as the farmer called the mixture, he scattered over his land. It functioned as a thin layer of "skin" that lies as a violet-colored filter on the earth. It enabled the earth to "breathe" just in the right way. From past traditions, this method of making the earth breathe was called "clay singing." Schauberger concluded that the stirring of the clay in the water

3 See also Alick Bartholomew, *Hidden Nature: The Startling Insights of Victor Schauberger.*

creates a neutral "voltage." When this water is put over the land, after the water has evaporated, a very fine crystal-like layer remains. The old farmer called his fine layer of crystals "the virgin hymen of the earth." Clay singing builds up a fine membrane between the surface of the earth and the atmosphere. This insulating membrane or skin of fine crystals is necessary for maintaining a charge in the earth that is the source of the forces of growth. Without it, the sun constantly discharges the forces of growth and dissipates them into the atmosphere, creating a short circuit.

So when my sons were young, around the time of Candlemas we would get a five-gallon bucket and the kids would help in the clay singing. One kid would make a little clay model of a carrot, and we would throw the clay carrot into the water and stir it around; then someone would make a lettuce leaf out of clay and throw it in; and someone else would make a clay fennel or a squash. It was our little variation on the idea of clay singing, but its purpose was to stimulate the imaginations of the crops to come.

It is my understanding that the greatest force of the human being in the natural world is the human ability to imagine, to form inner pictures, and to work with the inner pictures so that they become precise as though alchemical sigils that one can dance as eurythmy forms. The forms and movements, the sequence of a plant's leaves that lead to a particular kind of flowering process, the form in which a cow or a crystal of mica is organized are all good imaginations for clay singing. Imagine the water flowing into and out of the crystal. Imagine the potassium ion channels being built within the crystal. Imagine the weathering process. Imagine the clouds bringing the rain, passing through the corpses of animals and manure down into the soil and creating an acid, going into the minerals and pulling it out and then coming up into the plants. These are fertile imaginations for Candlemas clay singing.

Thus clay can be a vehicle through which the imaginations of nature and the imaginations of human beings can come together. It can be any imagination you like. You can make little faces, little cows, or other things and it will support our imaginations. On a molecular level, that is its role in the great kingdom of nature and in the elemental worlds. Nature receives imaginations from the cosmos to support the entire natural world, but it also can receive imaginations from human beings.

I will end this chapter with this image. Human consciousness is the clay of the gods. The human soul is the clay of the gods. It is the raw material through which the spirit and matter can once again enter a new form. Our imagination is that clay, and we need to keep it well watered. We have to keep it active, and when we do, our human consciousness enters the natural world and creates imaginations of what could be becomes part of the fertilizing. I truly believe in the old saying, "The best fertilizer is the footsteps of the farmer." It is in the contact that you make with your fingers in the dirt. I would venture to say the clay in the dirt that your fingers touch is the interface between the way the natural world is *hoping* to unfold in the future and what you are bringing with you to meet it and help it along. Those imaginations tell you what fertilizer will be best, where you will have a problem with fungus, where you will have a dry spot, a plow soul, or something like that. Such imaginations are a kind of clay in the natural world.

CHAPTER 6

SILICA: THE GENIE IN THE BOTTLE

Substance • Process • Being • Magma Formation • Fraunhofer Spectrum • Molecular Silica • Colloids • Law of Minimal Surface • Silicate Weathering • Semi-Conductors • Computer Chip Formation • Occult Chips • Rudolf Steiner • Micro-bubbles

Silica is generally considered a substance. To alchemists, it is a process, a silica process. To esotericists, silica is a being. And these three levels—substance, process, and being—come together in what we could call the moral relationship that human beings have toward this remarkable entity silica. I say "moral" because beings have consciousness; substances do not. Processes resemble consciousness, but beings possess consciousness. When we look at substances without consciousness they become resources, and the Earth is the loser. It is my understanding as an esotericist and alchemist that the Earth requires human consciousness for its evolution. Without our human consciousness, there is no evolution of the Earth, and it must evolve as we are doing—into a spiritual existence. Before this can happen, it is necessary for us as human beings to understand that, when we manipulate a substance, we are interacting with the consciousness of the being, no matter what the substance is. Therefore, I would like to trace a path of silica as a being, a process, and a substance as a kind of evolutionary cycle of silica. In the end, we shall see that silica has a memory, which is the basis of our information technology.

As a substance, silica constitutes ninety percent of the Earth's surface, the crust of the Earth. This is not just silica as quartz, but also in the compound forms known as silicates. Silicates are silica as a material that has evolved into many different mineral forms. Silica as a silicate means that the element silicon (Si) is at the root of the original substance. Silicon is part of the periodic table, but pure silicon rarely shows up physically on the Earth. It has to be manufactured into a crystal to become manifest. Usually, silicon appears in combination with something else, usually with oxygen, to form silicon dioxide (SiO_2), or silica as we know it. Thus, the element silicon appears in combination and in untold variations with other minerals and metals, making up ninety percent of the substances in the crust of the Earth. This gives us a picture of the ubiquitous being of silica, which is a sacrificial being. It is a being that "comes to Earth to die." At face value, this sounds a bit anthropomorphic. However, scientists have analyzed this remarkable substance in so many ways that we could add another aspect; esoterically, we could say that silica has a biography; it has a destiny. In being a substance, it has an evolution. To describe how that is, I will give you a picture of how the substance silicon dioxide as quartz is formed in the body of the Earth.

The raw material of Earth is in the viscous form of magma, which is molten, or fluidic, silica in which minerals and metals are embedded. There is magnesium, iron, and feldspar, which is calcium, sodium, or potassium. Metals and minerals combine with silica to form the large mass of the Earth. Nevertheless, there is a peculiar relationship between silica and these metals that has to do with how silica comes into being as a crystal.

Under the sea floor, in the core of the Earth, silica is heavily impregnated with magnesium and iron. The magnesium and iron have compounded in the silicate form as the mantle of the Earth. The core of the Earth is metal, but the mantle of the earth is siliceous material rich in iron and magnesium. It is significant that iron and magnesium are the most abundant metals in the inner Earth, because iron and magnesium are the two metals that form a polarity in the respiration cycles of plants and animals. The blood of most animals is based on iron and the breathing properties that iron has with oxygen. The plant's sap contains magnesium in place of iron. Magnesium allows plants to engage in photosynthesis because of its relationship to light. Thus iron and magnesium in the Earth body are analogs for the basis for the reciprocal respiratory processes of plants and animals. In this polarity, animal life consumes oxygen (O_2) and gives off carbon dioxide (CO_2), whereas plants take in carbon dioxide to produce oxygen. Thus, the core of the Earth is made of silica married to the two metals that allow life forms to breathe reciprocally. This breathing polarity really has to do with the reciprocal relationship between blood and the sap.

This allows us to say that the Earth core is a kind of primal blood–sap formation embedded in a silica matrix. Geologists call this blood–sap polarity *mafic* (from magnesium, Ma, and iron, Fe). Magnesium and iron, in a siliceous matrix, constitute mafic rock and form the Earth's mantle. Mafic mantle is fluid and extremely hot; cracks in the sea floor are sources of new magma coming up continually through the sea floor. As magma rises through those cracks, it solidifies in the water and spreads out from the vents in two directions. As the new material comes to the surface, the sea floor widens. So the sea floor is made from a siliceous material, rich in iron and magnesium, but because it solidifies almost instantly the rock has very little time to form crystals. Before a crystal can form, the rock must remain molten for a much longer time. The solidifying sea floor spreads out from the undersea vent and eventually comes in contact with a continental plate. The continental plate, in a sense, is floating in an ocean of mafic rock. As it spreads from vents in the sea floor, the mafic rock hits and is subducted under the continental plate, where it becomes pressurized. It mixes with water and solutions, becomes liquid, and forms great, huge globules of liquid rock far beneath the continental plate. Those large globules, called plutons, could be as much as four hundred by a hundred miles and thirty miles deep. A pluton is a molten mass of what used to be the island arc of a continent in the ocean. This mass is subducted and pushed under the continent and turns into a liquid mass of magma.

At such a deep level there is a huge amount of what is called *overburden*. It prevents the hot liquid globule—a huge globule four hundred miles long—from reaching the surface. The overburden creates a thick insulation over the pluton. Consequently, it can remain fluid for an extremely long time. The longer it remains liquified, the larger the crystals that can form in that liquid mass. Geologists call this sort of liquid rock *milk*. It resembles a pudding. The mafic materials that were pushed under the plate stay still and, through a gradual separation of temperatures in the molten mass, the various constituents are separated out by a process similar to making a pudding as the raisins fall to the bottom. That is a useful picture. The "cream" of the rock milk rises to the top; the rock milk goes to the middle, and the raisins fall to the bottom. The raisins are the mafic parts because magnesium and iron are heavier metals—especially the iron. These metals pull the darker, heavier parts out of the solution to be deposited in the bottom of the pluton. The finer parts drift to the top like cream, and this becomes the siliceous-rich materials that become gems. This rock is very fine and very rich in silica. The pressure of the pluton pushing up on the overburden from below cracks the overburden and injects the very fine siliceous material into the cracks. Those injections, called pegmatites, are the source of masses of crystals.[1]

Pegmatites occur where you find gems. In the center, the rock milk is not too heavy and not too light. You have gems at the top with a very rich, clear siliceous material. Crystals form on the periphery at the top. The silica magma is still mixed with some iron, potassium, and whatever constituted the native rock that was subducted. That area of the pluton forms crystals like little grains of sand; this is called the *granodiorite* area, the grainy area we call granite. The same liquid material forms heavy dark masses, porphyry, and heavy gabbro rocks at the bottom, granite in the middle, and fine-grained crystals at the top. This settling out is a picture of the relationship of the whole Earth body, because the heavier magnesium and iron parts form the core and mantle of the Earth, but the siliceous parts of the crust of the Earth, are most abundant at the periphery.

The iron, the magnesium "blood gesture" in the center, and the siliceous, and membranous crystalline part on the outside all form a pattern analogous to the human organism. Most of the forms of your sensory organs resemble the patterning of silica formations. Skin and the membranes of the nervous system resemble the sheets of molecular structures in siliceous minerals. The inner side of your organism is where blood pools form, based on

1 A pegmatite is a very crystalline, intrusive igneous rock composed of interlocking crystals, usually larger than 2.5 cm.

the iron and magnesium polarity. Therefore, the divisions of the silicates in their evolution give us a picture of the human organism.

That is the physical side of the being of silica. Now we could investigate the source of silica. Science has some answers for that. They say it comes from plasma. In scientific terms, *plasma* is a very interesting phenomenon. Alchemy has a term for something similar to today's concept of plasma. It recognizes a state of aeroform metal. The alchemical concept of aeroform metal is similar to what science today calls plasma. If I have three molecules of silica spread out over a hundred square miles, this would reflect the density of silica plasma. Science says that plasmas in this level of density are the forms of minerals and metals in a gaseous state. These plasma metals are spread throughout the cosmos in vast clouds.

Imagine molecules of silica spread out at that distance all over England, the whole of Europe, or even all over the Earth. Now imagine that you are a beam of light traveling through that cloud of diffuse silica. You would experience a minute bit of thermal resistance. Now imagine that this plasma area is as large as the distance from here to the Sun. Moving at the speed of light, you would interact with what is known as a plasma cloud; compared to the ambient temperature of space, that plasma cloud would reach a very high temperature as you passed through it. The plasma cloud at that speed would seem very dense and hot. However, science gives us another picture of plasma, one that involves metals in a gaseous form. In that picture, metals in the cosmos exist in plasma clouds of metals in the form of glowing gasses spread through the whole universe. Not only are there plasmas, but plasmas form in the auras, or coronas, of stars. Stars emit plasmas, and the plasmas emit light.

So if I were to try to understand the source of silica, I would have to begin by looking at something called a Fraunhofer line. It is a scientific tool used to study the light of stars. In Germany, there was a scientist, Joseph von Fraunhofer (1787–1826), who was a lens maker. This was around 1814, the heyday of optical systems. A few years earlier, in the time of Kepler, telescopes and microscopes were very rudimentary, and lens grinders were mostly amateurs. However, by 1814 optical systems had become rather sophisticated, and Fraunhofer was a trained lens maker in a town with a glass lens manufacturer. One day there was a fire in the factory where the optical glass was made. After the fire, Fraunhofer went into the burned-out factory and found a crucible. In the crucible, there was a slug of glass that had become molten because it had contained the proper flux materials. The intense heat of the burning building coupled with the way the crucible was constructed rendered the glass into a perfect form for making lenses.

Being a lens maker, this was a great windfall for Fraunhofer. So he took the crucible, opened it, removed the glass, and made it into the best prism he could. He was interested in the way prisms operate with light, because he was a telescope maker. Therefore, he began Newton's spectral convergence prism experiment by directing light through the prism and creating a spectrum on the wall. This was the sort of cutting-edge experiment that people engaged in with optical systems in those days. At the time, scientists were discovering that, if they had a spectrum of visible light on a wall, they could put a thermometer in the spectrum and move it through the visible light and there would be no change. However, when they moved the thermometer out of the visible red into the non-visible red, the thermometer would rise. There was no visible light, but they had just discovered infrared light, which has a tremendous thermal capacity. They did not have the whole wavelength theory down yet, but scientists were beginning to perceive that something big was happening in the study of light. The visible spectrum was just a small part of something much larger.

In this kind of mood, Fraunhofer decided on a whim one day that he would take some of his perfect glass and make telescope lenses with the absolute minimum of distortion. One day he had the idea of looking at the spectrum that was cast through his new distortion free telescope. When he looked at the spectrum, he saw not just a continuous band of color, but also a band of color that had many dark lines in it. This amazed him. When he looked at it without the telescope, it was just a continuous spectrum, but when he looked at it through the telescope the colors of the spectrum were organized into discrete units separated by dark bands. Those bands today are called "Fraunhofer lines" in his honor.

Nevertheless, he did not know what to make of this. He performed many experiments trying to determine exactly where one line began and where the color began. He did this by measuring the distance between the colors. Then he discovered that, if he used this apparatus and observed the spectrum created by burning metal, a single color appeared. He burned sodium and observed that the light produced had one bar of a yellow. He burned calcium and saw a different colored bar. If he burned iron, it was another color. Over time, others began to work with his idea and found that each of the bars of color represents the light from some form of metal when burned. An alchemist would say that they had discovered a way to make the activity of aeroform metals visible.

Still, there was the mystery of why the colors are separated by dark lines, and it took a long while for people to figure out that one. They found that some metals and minerals produce lines that are hidden or forbidden by other substances. Today, there is great controversy over so-called forbidden silica. If you Google "forbidden silica," you will find yourself immersed in a controversy among those who work with the spectrographic analysis of starlight. It was found that when sulfur is burned, a certain colored bar would show up. However, as measuring technology became increasingly sophisticated, it was discovered that the lines of silica are hidden within the wavelengths of the ground state of sulfur. The silica is "forbidden" by the sulfur to show itself. This shows an intimate relationship between sulfur, which has a deep affinity for iron and magnesium in the core metals, and silica, the mineral of peripheral forces. Once again, we get this polarity.

It is very interesting that the wavelengths of silica are hidden within the wavelengths of sulfur. Silica is hidden in sulfur because the lines are so fine. Today, astrophysics says that, as light comes from a star, it passes through various plasma clouds of metals already in space. It is theorized that the light in those clouds of metals absorbs frequencies from the light passing through it. Moreover, the light passing through a plasma cloud when the light of that metal is taken out will show up as a dark line in the Fraunhofer spectrum. In other words, the dark lines are actually a result of the interaction of the rays of light from stars canceling out each other as they move through interstellar space. This is the Fraunhofer spectrum, the colors of the light from the incandescent substances that make up stars. We can call it starlight, metal, or silica. The choice is yours.

From this perspective, we could say that silica is starlight that came to Earth to die, and that sulfur is starlight that came to Earth to die. These substances came to Earth to manifest so that human beings could have bodies and sense organs. The act of sacrifice that created silica as a corpse needs to be resurrected and lifted back into light through human consciousness. This is a motif I would like to bring here because these substances are the building blocks of life. The light of the stars brings to us oxygen, sodium, helium, mercury, iron, magnesium, hydrogen, calcium, silica, and sulfur—the primary substances in the Fraunhofer spectrum. These are all of the metals and minerals of life. Within the light that bathes our planet exists the drama of a spiritual being coming from the stars to end up here as the very thing we live on, the very thing we walk on. In very rare conditions it forms a crystal, and as a crystal form this being is a foundation for the elemental process of silicon becoming silica.

Silicon combines with oxygen in a pattern to create what we call SiO_2. That substance has the form of a crystal lattice, known in crystallography as a space lattice. The crystal lattice is its shape according to science, because the molecules of silica, SiO_2, are tetrahedral. The tetrahedra in silica are organized so that one tetrahedron molecule is attached to another in a spiral form in which a plane exists on the surface of each tetrahedron that creates what are known to physicists as energy holes in the molecular structure of the silica crystal. They are energetic holes in the molecular structure, but they are not actual physical holes but energetic holes where the particular structure of this incredible material is capable of having spaces in which it is possible to "store things." Here we begin our eventual turn toward energetic storage devices.

To put this in a context, I will describe various forms of silica, because silica is a great chameleon. It can become a myriad of different things. It is kind of a sacrificial being destined to combine with all these other metals to make all kinds of substances happen. A while back, a friend and I were at Avebury, a Neolithic henge monument of three stone circles in Wiltshire, England. We were looking at a rock wall made of chalk, and he said, "Oh, there are flints in there." I asked, "What are flints?" He said, "Nobody really knows what a flint is." That piqued my interest, and I discovered that the structure of flint is known as cryptocrystalline—*crystalline,* meaning crystal, and *crypto,* meaning we do not know. There are species of cryptocrystalline flints, but there are also species of flint that begin to betray a whole other origin. Those are known as biosilicate flints, silicates that arise from life forms in prehistoric seas. There was so much chalk then that any silica in solution was picked up by particular organisms called radiolarians. They made little skeletons of any available silica in the ancient sea water. They formed elegant silica skeletons around their very soft bodies and built large colonies within a chalk matrix made of countless organisms that had chalk as their external skeletons. So we have a kind of salt, sulfur thing going on in the life forms of the ancient seas. This drama is cast in the polarities of calcium and silica in the ancient seas and in the colonies of the organisms that picked up those minerals.

Now when a biosilicate forms a flint, the flint changes its name. The biosilicate flint is known as *chert,* in which you can actually see the little remnants of these microorganisms. This means that silica must somehow be soluble in water if an organism can absorb it from the sea. Science tells us that, at 25 degrees Celsius (77° F.), silica will be soluble at the rate of 140 parts per million. At 340 degrees Celsius, however, silica is soluble at 1,660 milligrams per kilogram. This means that pressure and temperature affect the ability of silica to go into solution. In science, there are degrees of solubility. The word *solubility* comes from the word *sol,* meaning "Sun." In alchemical language, *Sun* is an acronym for levity. Levity is light to an alchemist. To become soluble means that silica has to start returning to the form from which it came. Solubility for silica means going back to its origin in starlight. To move it into a solution, where it becomes soluble, I have to make it interface with a solute. A solution of silica at 140 parts per million will never settle out, but neither will a solution at 1,660 milligrams per kilogram settle out at 340 degrees. If it starts to cool, however, crystals begin to fall out and you get the process I described in the formation of crystals from magma. Therefore, there is this kind of balance between the liquid state for a crystal, whereby it becomes a solution, and then as it starts to fall out it goes through a middle phase known as a colloid. If it falls out completely, that condition is called a suspension. However, in between a solution and a suspension is this curious form called a colloid.

Depending on the temperature and the pressure that it exists in, a colloid can either go into solution and remain in a liquid state or enter a solid state by falling out of the suspension. Colloids are said to have a remarkable form; minerals in a colloid are at such a size that the surface creates charges with its environment. The size for particles in a colloid can vary between just one nanometer (one billionth of a meter) and 1,000 nanometers. The particles in a colloid must have charges that keep all the other particles at a very specific distance. They can never settle out because they are all charged to keep the space between them at a very narrow range. In a colloid, the tiny particles all repel and attract one another equally. A colloid is a kind of liquid crystal. The liquid crystal display (LCD) in your cell phone contains a colloid. Most of the connective tissue in your body is a liquid crystal. Many of the proteins in your body are liquid crystals, or colloids. A colloid has an interesting relationship to liquid and solid states, because the particles are exactly the right size to be affected by the law of minimal surface.

The law of minimal surface guides the action of colloids in nature, which always appears when a solid has to remain elastic and yet can be solid or elastic. This is fundamentally mucous, but it occurs anywhere there is lubrication or a need to transfer materials from a solid form that can become liquid, and a liquid form that can become solid—gels, for example. The flux between solid and liquid happens every time you move your arm and the tendons flex with your bones in fluids. Your tendons solve and gel to allow movement. Colloids allow life to happen because of the charges in the particles that have a specific relationship to one another. They cannot fall out because every particle repels and attracts the others in a balanced way. They cannot fall out, because something curious happens in nature. This may seem a little strange, but I will describe what science understands in this area. I will give you an imagination.

Imagine that I have a sphere; it has a surface and a mass. The mass has a height, width, and depth; it is in three dimensions. The surface, however, has only height and width. When I look at a wall, the surface of the wall has height and width, but no depth. The surface of the wall has zero altitude if I travel from one spot to another spot on the wall, I do not change my altitude at all. Mathematically, that is a plane. I do not change my altitude moving from one location on the plane to another. The curious thing about a sphere is that, when I go from one place to another place on a sphere, I do not change my altitude. I am always the same distance from the center, meaning that the surface of a flat plane and the surface of a sphere are mathematically and topographically the same. They are planar.

This has interesting affects on the way the world operates. Imagine a sphere; we know that the surface has only two dimensions, like a flat plane, but the mass of the sphere has three dimensions. Imagine that I increase the size of the sphere. Its mass increases by one dimension (the distance from center to periphery) more quickly than does the surface dimensions. At a certain point, the surface will become compromised by the greater expansion rate of the mass. Think of a balloon. When I blow up the balloon, I expand the mass to a point where the surface cannot quite keep up with it—the original mass and surface. Something has got to give. In nature, if your cells exploded every time they became larger than a certain size, you would be out of business very quickly, because your cells are growing all the time. However, nature has a way of preventing ruptures as things grow; it simply forms a plane between the two sides of the growing cell. The cell breaks the nucleus in half, and you get a daughter cell. This prevents a lot of ugly things. The governing principle here is the law of minimal surface. In other words, in nature, there is a relationship in surfaces that must be maintained for optimal interaction with the environment. This governs the size of cells in organisms. This is why cells in a body of a whale are the same size as cells in the body of a mouse. The Law of minimal surface is an incredibly powerful natural law.

Now do a thought experiment in a different direction. Imagine a sphere whose surface has two dimensions and whose mass has three dimensions. Now imagine making this sphere smaller and smaller. It will become smaller in three dimensions more quickly than it does in two dimensions. At a certain point, therefore, its mass will become smaller than the surface. Because cells interact through their surfaces with their environment, if the surface is larger than the mass, the surface of a very small cell is very interactive. Now imagine grinding particles of silica on a sheet of glass. As we grind the particles smaller and smaller, the law of minimal surface comes into effect as the particles reach the nano range. Such a small particle is radically interactive with its environment because it has such a large surface in relation to its mass, because energy is exchanged on surfaces in nature. Energy is not exchanged in the mass of a thing, but on the surface. Surfaces are interfaces where predator meets prey. Surfaces are the sites of gaseous exchange and serve as active agents in molecular attraction. The enhanced surface action of very small particles becomes so reactive with the environment that a huge charge will jump onto each particle, hence, the unique nature of a colloid. Activity within a colloid is reminiscent of the sound of boxcars when the engine stops, and you hear boom, boom, boom, boom, boom. Such an action in a colloid is remarkable. Most proteins are colloidal because they need to be sensitive to changes; they give our body elasticity.

Now we are in the realm in which nanoparticles rule energy exchanges. Colloids are defined by the size of their particles. Particles larger than nano size settle out; those smaller than that go into solution. Colloids are extremely interesting, especially in the realm of silica. Natural colloids of silica occur in the form of substances such as opals. An opal is a colloidal silicate. The small particle size of colloids, of silica, also gives plasticity to minerals such as clays. A clay is a colloid that can absorb and give off water. The tiny particles in the finest clays, such as bentonite, are extremely plastic as a silicate. Bentonite can hold and give off huge amounts of water. Clay in the normal order of nature is a colloid that can take on and give off water. When this happens, the clay, being a silicate, takes on water, but very often the water has a pH charge in it, perhaps because something died upstream and made the water acidic. The acidic water acts on the inner surfaces of the colloid and creates strong reactions whereby molecules embedded in the silicate migrate out of that silicate into the surrounding water.

I have just described the humus and cation exchange capacity, which forms the basis for agriculture. Humus arises where organic acids interact with colloidal-sized silicates. Organic acids leach into the spaces of the silicates and pull out potassium, calcium, and sodium, allowing the minerals locked in the silicate to become mobile and interact with plant roots, which have fine hairs that need to reach and attach themselves to surfaces. The colloidal silicates in clays and humus have an infinity of inner surfaces because of their nano size. The root hairs find a tremendous inner surface in a tiny bit of humus. The tiny bit of humus, being hydrophilic, holds the water that the root hair can grab a hold of, and then you have protection during drought among plants, because you have humus that comes from the clay–humus complex in the soil. This is "cation exchange capacity" (CEC) in soil science, and it has to do with the ability of the soil and the plant to communicate. CEC arises because silica can become colloidal in relation to its environment.

I am describing, in the organic realm, this remarkable substance that allows us to move, breathe, form protein, and grow plants. When I said that silica is a sacrificial being, this is what I meant. It is continually giving itself to become all these other things through interaction. Now the ability of silica to form a colloid, a kind of liquid solid, requires that silica can become a liquid mineral that still has a crystalline interface to it. Today, you search Google for "nanoparticle silica colloid," and you will be led to websites related to the construction of computer chips. This is because Silicon Valley has found that there are certain ways in which the silica can absorb very fine dilutions of metals into its molecular structure, and because silica has a very refined crystalline structure, the molecules of silica and metal can be made to be extremely well organized. The metals disperse within the spaces and holes in the crystal matrix in a remarkably orderly way. It is possible to insert other materials such as boron or phosphorus into the structure of a silica crystal. To insert such metalloids into a liquid matrix of silica is called *doping*. Scientists are interested in doping silica because silica resists electric currents, while phosphorus conducts electrical currents. If we try to pass a current through a piece of silica, it will resist the current because its structure is already organized to such a degree that there may be energy that goes into it but it will not conduct out. If we make a batch of silica and find a way to intersperse very fine particles of phosphorus in it, the phosphorus molecules will conduct electricity, and the silica will resist electricity. Some parts of the mixture will conduct, and some parts conduct only part of the time. This is a semiconductor.

Semiconductors are the heart of IT, or information technology, because a semiconductor can be made to control current. If I make a switch out of doped silica by slicing it into very thin pieces and attaching wires to it, I can determine exactly the voltage that will allow conduction. As soon as I go above that level, the silica will resist the current and stop the flow. This is called a gate. If I attach one gate to another and the two have a certain relationship, and if I have a current that goes through and then stops and then goes through, I have a circuit through which I can predict how the current will flow or stop at will. When the current goes through, I call it a *1,* on. When it stops, I call it a *0,* off. I can arrange gates so that the 1s and 0s are in sequences, on and off, on and off. I can create bunches of gates and circuits, put them together, and get a "motherboard" that regulates minute currents through flow and resistance patterns of the doped silica. This is undoubtedly a very rough picture, but it illustrates the general idea. In addition, I can store patterns of gates information that is coded to regulate how the gates will open and close. Then, when a particular kind of current comes through, this section will hold this bit of information. When the current goes another way, that section holds the bit of information. Then I make bytes of bits, kilobytes of bytes, megabytes, gigabytes, terabytes, and so on. The gates are all opening and closing, and silica has stored memory. This is where we are today with this amazing being who has the capacity to remember.

This doping leads to recognizing a need to make highly accurate circuits from extremely pure silica. Our technology needs synthetic crystals formed in such a way that the crystal matrix in them is mathematically perfect all the way through. In a natural crystal, the way the molecules line up has to do with pressures in the pluton, the temperature, the dissolved minerals in the magma, and many other vectors. The crystals that form in nature have an inward rotation, because the molecular tetrahedra of the silicates act as little vortices. However, because of the way nature acts, this vortical formation gets tweaked a little. If you shine a light into a quartz crystal, it will rotate, because everything in that quartz crystal is based on tetrahedral planes that grow by rotating into one another. If I heat a crystal and melt it, that beautiful crystalline structure disappears and the silica becomes amorphous. It becomes more like a colloid serving life. This transformation from mathematical perfection to amorphousness is the strange thing.

Nevertheless, human beings are very clever. Because Silicon Valley said we need crystals that are more perfect than natural crystals, what can we do? We will take a vat of pure silica, as pure as we can form it, with no doping in it whatsoever. Into the center of that vat we will place a little model of a mathematically perfect crystal on a little spindle. We will slowly rotate that little crystal in one direction in the liquid silica and rotate the vat in the opposite direction very slowly. As one would wind paper towels around a core, we wind the liquid silica, molecule by molecule, around our perfect crystal to the point where we get a *boule*. A boule has a crystalline silicate structure in which every molecule is perfectly spaced from every other molecule, with absolutely no distortions. A boule is super-high-tech piece of glass. It resembles what water might look like if we put it in a plastic bottle, place it in a freezer, and then removed the bottle. The boule is a long cylinder of absolutely perfectly formed glass. Within it there is no variation whatsoever in the crystal matrix of the silica. This can then be sliced into extremely fine pieces, etched, and resisted with sulfur spray or arsenic to create even more complex small gates of super-microcircuits on a single chip. Instead of having big gates of doped silica, we can make ever-smaller chips with whole motherboards etched onto a single tiny chip.

I read about an experiment in which a man who was working with this kind of research put the number of bits in an *Encyclopedia Britannica* into an electron and pulled it out without distortion. The problem is, however, if you do this with one electron, and the one next to it has some type of affinity, you will no longer be able to get your encyclopedia out. When we get into the minute nature of this world, we find a class of phenomena that is very intriguing and profitable to many people.

Two similar experiences in my own life are related to this issue. Years ago I was on an airplane sitting next to a flight attendant who was off duty and traveling. We talked a little, and of course the subject of occupations came up. When I told her I am a teacher, she asked, "What do you teach?" I replied, "Uh...inner work, and I work with some things alchemically." And she said, "Oh, alchemy; could I talk to you about something that happened to me?"

She told me that she had worked in Silicon Valley before becoming a flight attendant. In Silicon Valley, she had worked for a man who was part of a group that had to do with making synthetic crystals. They would sit around the vat as it turned, and they would meditate into the vat in an attempt to capture a certain quality of consciousness in the layers of molecules and the formation of the boule. She told me that one person was chosen to be the one whose consciousness would be moved into the boule, while the others would act as a kind of community will-force to help them focus their consciousness into the boule of silica. That was their esoteric practice. I told her, "I actually teach meditation and follow the work of Rudolf Steiner." She did not know of Steiner, but one thing led to another and she said to me, "Could I contact you? Could you take care of my boule and keep it safe?" I told her she should probably go throw it into the ocean to keep it safe. I thought a long while after that about where we had come to in human technological development. Meanwhile, I thought she was a bit paranoid and forgot the issue.

Then, a year or two later, a man came to my course and continued for a whole year. He was a very unusual man. We talked, and I got to know him a little bit. At the end of the year, he said, "Can I have an office hour?" and I said, "Sure." He came into the office hour with a little blue velvet bag. I looked at the bag and recalled my encounter with the flight attendant. Sure enough, he was part of a cabal in Silicon Valley and they had fixed his soul into this silica boule in order to make chips that could interface with biological structures. He asked me the same question about keeping it safe, and I gave him the same answer: "I think you should give it back to the ocean for safekeeping." This man was living with the idea that his soul was in that bag of silica, and he was very nervous about it. As an esotericist, I think that these experiences point to something of great importance.

There is research today related to interfacing silica chips at the nano level with protein structures. There have been experiments in which the crystalline form of living protein can be trained to solve logic problems faster than silica chips. This man had a dish of protein that could solve logic problems faster than integrated circuits could. There are people working today to make nano-sized silica chips interact in protein structures. The concept of "electronic flesh" exists in the world today. Steiner talks about human beings who will begin to do this kind of work, because they are angels who have been given the task of inspiring people to make technological developments that go beyond their moral capacity to deal with the power that comes from the devices they design. Silica crystals and memory devices—technological genies that have been let out of the bottle.

Rudolf Steiner's use of silica is to take a silica crystal and render it down to a small particle size, probably somewhere in the nano level, so that the surface becomes larger than the mass. It is then put in water. Water is two gasses that decided to come to Earth to sacrifice. Basically, it has a cryptocrystalline form. As a crystal, it has a highly flexible inner form in the creation of inner planes. The inner planes arise as energetic portals within water when it is moved. When water moves, the energetic surfaces within the water begin to be organized in a crystalline form; moving water is a fluid crystal. Then, into that fluid crystal, we place silica that has been brought down to the nano level and has achieved a huge surface area; the highly reactive surfaces of the silica particles interact with the inner surfaces of the water to transfer a high level of order into the liquid crystal. This is the process of stirring biodynamic preparation 501.

Then we break the vortex and then stir in the other direction. As discussed earlier, when we break a vortex, incredibly small nano-sized spheres called micro-bubbles are created. People have been able to take photographs of micro-bubbles bursting. When a micro-bubble bursts, out of the bottom of it, briefly forms a vortex. In the realm of the micro-bubble, if a nanoparticle of silica with a huge surface gets involved in a micro-vortex, the result is a tremendous action between that substance and the water. Not only is a little vortex formed, but also a measurable sound is created. A relatively huge sound is generated, and the local temperature around a micro-bubble spikes up in the hundreds of degrees. This is a great deal of activity generated by a tiny bubble, and untold numbers of micro-bubbles are created when the flow is broken. In the fluid that is flowing with these crystals, a huge interaction of highly reactive surfaces is created. In this new ritual, the nature of the silica as a being of light and sacrifice is transferred into the water—that is, a being of sacrifice in addition to the water itself—and there is a marriage. That solution is then sprayed in the finest way possible on the surfaces of leaves (also planar). It is designed to receive light; the nano surfaces of the silica work to draw in the light. I now have surfaces and surfaces and surfaces and surfaces. We call this remarkable ritual "preparation 501"; it renders plants more capable of receiving the light—the light that comes from the stars and incarnates among us as the great sacrificial being, silica.

CHAPTER 7

VALERIAN

I. Medicinal Plant Organization • Growth and Flowering Polarity • Saturn Gesture in Valerian • Peripheral Forces • Chaos • Phosphorus • Count Keyserlingk • Amethyst Geode • Sheath Synergy

I begin with a picture of a "lawful" plant, one that listens to the fundamental plant archetype of a root with a crown, a stalk above it with leaves and nodes, and a node where the calyx then unfolds, and corolla, stamen, and pistil go into the seed. We can call that a lawful pattern for a plant. Many vegetables have that as a kind of ground plan and archetype. However, when we look at an herb or a plant with properties that go beyond simply nutrition, often there is a displacement of that normal pattern.

Somehow, within an herb, an organ moves either up or down from its place in the archetypical plant. That displacement creates a kind of tension within the plant, resulting in its pharmacological action. This can then be seen in its life gesture and its structure. For instance, members of the labiate family have a displacement of the flowering process down into the body of the plant. In the labiate there is an oil process that is normally associated with flowering and seeding in the archetypal plant brought down into the leaf. So the displacement of a particular organ, or organ principal gives the gesture of this culinary family. The actual flowers might not be displaced because labiates have flowers, but they have in the body of the plant the essential oil process that would normally be part of the flowering process; they have that flowering gesture dispersed all the way through the plant.

There is a kind of displacement of the archetype into a different position, and then that displacement produces the pharmacological qualities, and physiological effects of the substances within the plant. Sometimes the displacement is a result of a particular ecological niche or climatological niche that the plant has to overcome, that it will displace an organ in a particular way. The plant forms an active principle or organ in order to fit into a particular niche.

I want to give that principle as a background to build a picture of the interaction of the two great parts of the plant archetype, the flowering process, where we have the action of the cosmos acting on the plant from the periphery, and the growth process, where we have the action of the earth and the water lifting the plant away from the earth. Growth is terminated by the flowering process, and the flowering process is inhibited by growth.

The more fertilizer available to the plant, the more its growth develops. This growth represses the flowering process. Through fertilizing, I am supplying too much of the earthy side of the equation. When I restrict nutrition and even water, I am enhancing the flowering and fruiting process at the expense of the growing process. The cosmic side—flowering, fruiting, oil-producing, protein-producing, seed-producing, and fragrance-producing—has to do with limiting growth. The opposite pole—the development of strong leaves, nodes, and roots, especially at the crown—is the earthly side of the archetypal plant. The crown is where the root and the stalk meet; it is a very critical part in a plant. The organs around the crown represent an earthy gesture, because the plant is working out its relationship to lower soil sap and so on to make the growth substances that it needs. Then, gradually, in the lower saps of the body of the plant, the coarser substances are lifted into finer and finer forms—alcohols, esters, alkaloids, phenols, and so on.

Thus, when growing, the plant is related more to the earth; when flowering, it is related more to the cosmos. The particular relationships among the organs of the plant can be displaced owing to the forces that the plant is balancing. The displacing of a cosmic principle down or the displacing of an earthly principle up results in what we would call the life gesture of the plant. Therefore, in general, displacing a cosmic principal down creates more refined qualities further down in the plant. The displacement of the earthly gesture up gives a coarser quality in the flower and fruit. Life gesture is a tool for assessing the way a particular plant balances earth and cosmos.

Whether it is a vegetable or even a perennial, by looking at the particular families of plants and the way they organize these forces, we can learn to see a coarse quality that moves up or a fine quality that moves down. These properties create the life gesture of the plant.

Now let us consider the life gesture of the valerian by asking why we could relate valerian to the principal of Saturn. What does Saturn mean, anyway? If we could understand what Saturn as a plant would be, where would we find evidence of this? What kind of life gesture would we see in a saturnine plant, and what effect do the forces of Saturn have on the action of a compost heap? It is interesting to me that the valerians are also connected to Venus; it has a five-petaled corolla. In the assessment of plant organs, the number five often refers to the influence of Venus. Venus and Saturn are known as the virgin and the old man and are linked to each other as a polarity. Valerian has a strong life gesture of extreme polarity, so this Venus–Saturn polarity is just part of its particular niche.

Consider Saturn. It seems there are some really interesting pictures of what we could call a Saturn principle or a Saturn gesture that have a great deal to do with substances becoming more refined. Saturn, in the old world, was considered to be the gate through which the cosmos incarnated. Saturn is the gate. It represents the activity of peripheral forces. Those forces are warmth and refinement, since matter is going into spirit at the periphery. The periphery is where the unmanifest archetype behind the manifest plant lives as an active being. The periphery is the gate that the plant being comes through to become a manifest plant. Saturn is the code word for that process of incarnation from spirit into matter. The force of the periphery is of the unmanifest world of chaos. Yet that chaos is ordered so that particular forms arise and eventually manifest. Chaos denotes *potential,* not randomness. Chaos is the process whereby the unmanifest becomes manifest. We could also call it warmth. Moreover, when we call it warmth, we are using an alchemical code word. Warmth is the enthusiasm for incarnating. In chaos, you are not quite incarnated yet, but you are thinking about it pretty hard. You are just coming near the edge to put your toe in the water of incarnation, to see if you actually

Figure 6

will manifest. Steiner calls this Saturn gesture *chaos*.

The chaoticizing of the lower plant is a kind of volatizing of the organization. The plant, when forming seed and flower, is going to its own periphery, its own Saturn periphery. The plant when forming seed is volatizing everything from below and lifting what has manifest into a higher level of organization. This higher level is a more refined, less incarnated form of organization. This is what a fragrance is. It is a substance that is returning to heaven. The flowering pole and the substances found there have a lot of do with consciousness.

So Saturn, as a life gesture, represents the action of the periphery. Those forces bring the qualities of what is beyond the horizon—the unseen, the unknown, and yet still somehow ordered. This pole represents the dissolution of the plant organs to produce a seed. Peripheral forces chaoticize the plant to induce the archetype of a new generation into the seed germ. All that I am describing would be called Saturn. Saturn consciousness is the wisdom and warmth that holds everything that will eventually become manifest. Alchemically, all of this could be called Saturn, the gate, the door, and the entryway for the light to come in. That entryway for the activity of the light that manifests in a warmth process is another alchemical code word. When warmth activity becomes a substance, we call it phosphorus. Phosphorus means the one who carries the light. Therefore, phosphorus is the ultimate manifestation of the forces of warm enthusiasm and light. We can barely hold phosphorus in an incarnation because it wants to escape back to the periphery. That urge to return to the periphery is Saturn, the action of the periphery on what has become.

Whatever has become represents the action of the centric, or earthly, forces. They represent the forces of substance and matter, earth and water. These are the earthly forces of manifestation. What has become manifest has to have a way to go back out into the infinite, so it goes back out through the gate through which it entered, and that gate is then Saturn as a substance in plants that leads the more fixed elements of calcium and even sulfur as a kind of fixed fire back to the periphery. What leads the calcium out again to the periphery is phosphorus as an activity that is opposite the salt process of incarnation. The gesture of dispersing and going to the periphery is phosphorus as a substance, or Saturn as a cosmic life gesture. There is a particular life gesture when Saturn becomes dominant in a plant. There is a particular something we can look for as a structural key.

According to Adalbert Count Keyserlingk:

> Saturnine powers are evident in the unusual feature in which a plant holds its stamens up high, offering them to the heavens in a wonderful sea of flowers, while the pistils, drawn to the earth, develop low down on the stem. A space is created in the horizontal as stamens and pistil are held apart in a gesture that can also be seen in the ringed planet.[1]

This a strong polarity of the farthest out and the farthest in. It is a kind of inner space. Therefore, the creation of an inner space, according to Count Keyserlingk, is Saturn. It is exactly what we see in the growth pattern of valerian. In *figure 2*, we see teosinte, the original form of corn, or maize, in contrast to contemporary corn.

So the original corn cob was two inches tall. It was held on a grass plant and had two kernels. But in *figure 2* (next page), we see that there is a separation of pistil and the stamen. In most other grasses the pistil and the stamen are embedded one inside the other. In oats and wheat, the stamens hang out of the pistil for fructification and then fall off. However, in teosinte the pollen former and the seed former are separated by a stalk. Then over time, and forming through selection, the pistil has moved farther down the stalk, and the stamens have moved farther up, until now there is a foot and a half to two feet of stalk between the pollen former and the pistil. That separation has been developed over time.[2]

---

1 Keyserlingk, *Developing Biodynamic Agriculture*, p. 17.

2 The teosinte origin theory was proposed in 1931 by the Russian botanist Nikolai Ivanovich Vavilov and in 1932 by the American Nobel Prize winner, George

Figure 6

The irony of this is that researchers say this happened within a couple of generations, as though by magic—not through centuries of selective gathering, but something else. One would think it takes centuries for such a development, but the scholars say no, there were teosinte remnants found in pottery, and then suddenly there is an ear of corn, within a couple of generations. They have no idea how or why, but perhaps there was some other way of working with plants that has been lost.

Now look at *figure 3*, an image of the valerian. Below on the left we see a very fleshy crown and root, or rhizome, with a long stem with internodes going way up to an inflorescence, way up in the air. Again, there is a separation, with a kind of earthy gesture down below, a heavenly gesture up above, and a large space in between. Look at what a rhizome is and compare that to the archetypal plant. A rhizome is a stem that is normally above the ground, but it has been drawn down below. Therefore, in one part of the

Beadle. It is supported experimentally and by recent studies of the plants' genomes. Teosinte and maize are able to crossbreed and produce fertile offspring. However, a number of questions remain, including how the tiny archaeological specimens of 3500 to 2700 BC could have been selected from a teosinte and how domestication could have proceeded without leaving remains of teosinte or maize with teosintoid traits earlier than the earliest known until recently, dating from ca. 1100 BC. See http://en.wikipedia.org/wiki/Maize for more.

valerian we get a gesture of pulling the cosmos down; the stem is under the ground. That is the real nature of a rhizome. We have a downward pull of the cosmos. There is even the gesture of a strong aroma permeating all of the vegetative parts. Strong fragrances down in the vegetative parts of a plant point to the pulling of the cosmic down into the earthly. Then there is a long separation in between the crown and the inflorescence until we get the sea of flowers held high up above, which Count Keyserlingk described. The empty, or hollow, space between is the mark of the Saturn gesture.

Now look at the flowering principal up above in *figure 4*. We see a single flower, and a little fruit or seed at the bottom. Looking at the tube corolla, we see the same type of gesture in the tube corolla going way up above the seed and petals, and the stigma of the style goes up even above that. The tube corolla encloses the stamens, and that little organ sticking out of the top is the stigma. In the flowering, there is a pappus that is a modified calyx. It eventually becomes a little umbrella similar to that of a dandelion, which helps the seed to float away. Even in the flower, the seed gets lifted up and disseminated to the periphery. There are two gestures in the valerian plant. One is pulling the cosmos down into the root; the other is sending all parts of the flowering process out to the periphery. The gestural motif is the separation of the two principles in a Saturn gesture. Thus the Saturn gesture is the creating of an inner space where something new can come, a space where the earth is pulled down and the cosmos is pushed up.

Figure 3 (above)
Figure 4 (below)

A deeper insight leads to the perception that the cosmos is not pushed up from below but pulled up from above. We could ask whether the flower is moving away from a center, or if it going toward a center. How can the periphery be a center? This question is an example of the kind of thinking needed to understand the activity of Saturn. The conundrum arises from Steiner's insight that the forces of the periphery are not explosive, but are suctional. This means that they have a center in the infinitely distant. The imagination is that earthly beings die into the periphery of the cosmic forces. The spirit leaves the body of flesh and goes to the periphery of the cosmos. That is the one pole. However, spiritual forces and beings die in the center of the mass of an entity that is manifest here on earth. Cosmic forces die here. Entities that manifest on Earth have a center here and grow toward, and ultimately die in, the infinitely distant. Growth dies in the infinitely distant. Spiritual entities dwelling as archetypes with infinite potential in the infinitely distant grow toward their death within entities here that are manifest and have exhausted their potential in what they have become.

The plant archetype dies in the plant that is here on Earth as a manifestation. These two paths weave across each other in the growth patterns of plants. The plant archetype dies in that particular plant, but the growth of the plant moving it toward the periphery creates a condition whereby the growth of the plant gives back to the archetype—what we could call a sensory experience of everything it went through during its life process. That is then contained within the archetype of the plant as an energetic template that moves back toward the forces of the periphery as the plant goes to flower and seed. This drama becomes encoded in the seed process when it is driven to chaos. This is all an image of the work of Saturn, the gate of the cosmic forces at the periphery.

When I am not incarnated, I live in another universe, somewhere in the infinitely distant. When I am incarnated I am right here on this piece of earth in my own space. When those two processes come together in a form of life, we call that form a plant. Every plant has a different way of organizing those two sets of forces.

Is the plant we see the true plant? No, the true plant is an energetic being that leaves a wake when it passes, an image of the archetypal being that holds the template of its existence. What we see as a plant is the wake, or corpse, of the passing of the true plant. The true plant occupies everything between the seed and the next germination. Once germination occurs, the true plant is on the way to excarnation and leaves a matter-permeated wake for us to use. The true plant is a rhythmic, sensitive, invisible activity unfolding in space. It occupies a space between the unmanifest and the manifest. The plant on Earth occupies the space between the manifest and the unmanifest.

Some plants fill the space rapidly; others, such as valerian, stretch that space out so that something new can enter. Something different can happen when Saturn forces separate the manifest and the unmanifest. There is a kind of new enthusiasm for manifesting that can take place. Valerian is a specialist at bridging the poles of the manifest and the unmanifest, because it lives in the wet places yet loves the warmth and the light. All herbs are specialists at working out these kind of polarities in some way. However, valerian is a *special* specialist; it unites many paradoxes. Is it sweet or is it rank? The answer is yes.

Is it light and warmth? Does it love the water? The answer is yes. Does it put us to sleep, or does it wake us up? The answer is yes. In this Saturn activity, in the space in the center where there is actually nothing, strong forces can be developed. Looking at the picture of the valerian in *figure 3*, we will see there is a lot of internode space along the stem. There

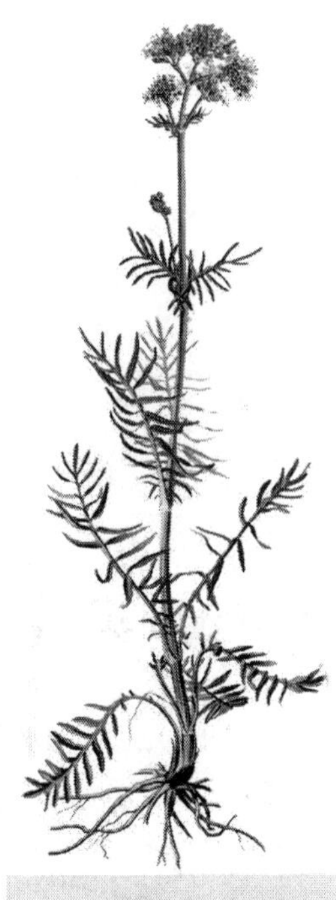

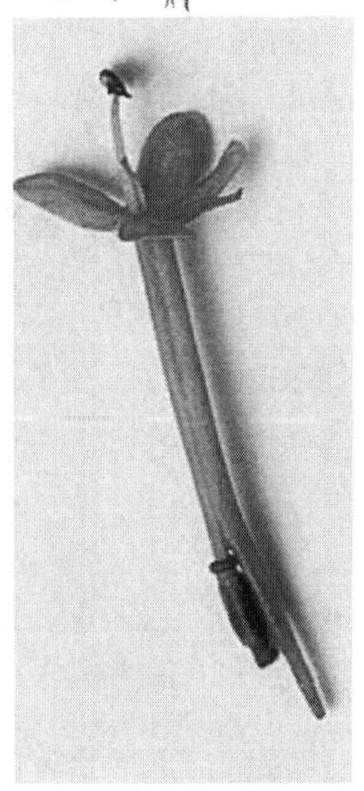

Figure 3 (above)
Figure 4 (below)

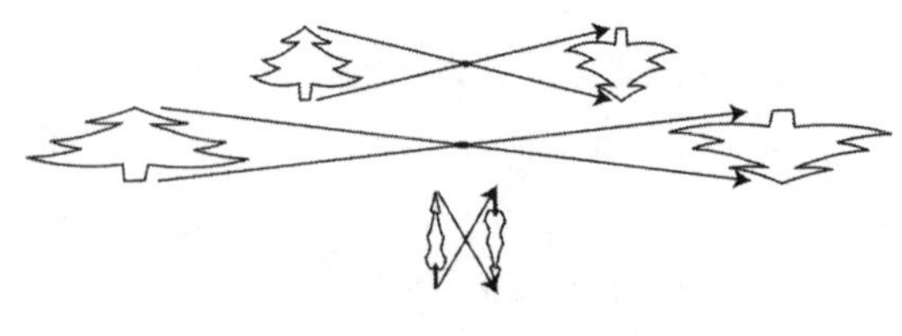

Figure 5

is a lot more internode space than there are nodes. In addition, look at that one little floret in *figure 4*. That tube corolla takes off like a rocket.

The gesture of creating an inner space, the Saturn gesture, then permeates the organs of the valerian. In *figure 5,* we see a curious diagram. It shows two upside-down pine trees, reminding us of a projector or a pinhole camera. The "x" in the center, where the lines cross, is the pinhole. If you look into pinhole camera lore, you will find the story that there was a yurt with shepherds sitting inside. One day on the wall of the yurt they saw the animals outside walking by upside-down. When they traced the source of the image, it turned out to be a pinhole in the side of the yurt through which light was coming. The yurt was dark and the pinhole in the yurt was projecting the image on the opposite wall of what was going on outside. Legend says that this is how the pinhole camera was discovered.

A pinhole camera is a camera that has no lens, just a hole. The hole gathers the light from whatever is on one side of the hole and focuses it on the other side as a "projector point." The projector point, or projector, then transmits the light to the other side of the hole completely intact, except that the projected image appears upside-down on the other. Saturn is all about the projection of light from one side of the cosmos onto Earth. As the cosmic gate, Saturn brings what is coming from the outside and projects it to the inside. In the middle, however, where the projector is, there is a whole lot of nothing.

In this cosmic dance between the unmanifest and the manifest, the more *nothing* we have, the more something gets interested in the nothing. "Nature abhors a vacuum." Where we have a space, there is nothing. It acts as an irresistible space for something to manifest. In the projector diagram in *figure 5*, we see a very curious phenomenon. We will call the pine trees in the top diagram normal. In the middle diagram, the pine tree on the left is pushed out to the left. In response, the image of the pine tree on the right will get pushed out to the right. If I push the pine tree on the left to the horizon it will appear smaller. At the same time the image of the pine tree on the right will recede and become smaller. If I move the pine tree on the left so far that it disappears into the horizon, the image of the pine tree on the right disappears at the same time that the pine tree on the left disappears. Where do they disappear to? Into the horizon. The horizon is the periphery surrounding the place where I am in the world; it is the place where everything disappears. My whole world disappears about seven miles from the center of the circle of my visible world. At the horizon is a vanishing point. The vanishing point is a kind of projector out there on the periphery.

Figure 6

In the middle section of *figure 5*, the projector moves things out to the periphery, to the horizontal. The real pine tree and the image of pine tree are linked by the action of the projector. Below that, we see that the real pine tree on the left has come much closer to the projector. The projected image on the right follows suit and also appears to come closer to the projector. Scaling happens just by letting the projector in the middle do the work. As I pull the real pine tree in toward the projector, it appears to get taller, and the projected image does the same. If I pull the pine trees all the way into the center, they get so tall and narrow that they become a vertical line. Actually they meet and cross in the projector. If we push the pine trees out in the middle and all the way to the periphery, they become a horizontal line. Therefore, the action of the projector in the middle, the space of the nothing, is a projector for the forces on either side of the projector. The projector is the space where the vertical and the horizontal meet in a point.

What do we see looking at *figure 6?* It is a planet with a strong vertical axis and a set of rings in the horizontal plane. This is a strong polarity between the horizontal and the vertical. In the plant world, every plant is an image of the two polarities of horizontal and vertical, but looking at valerian we see the horizontal force dominating down at the crown. In the flowering pole, the vertical force dominates. However, in the center, we have a strong separation of the two poles. In the valerian these polarities of horizontal and vertical appear extremely polarized. The archetypes of horizontal and vertical forces compose the field of activity that plants use to unfold their life gestures. The younger shoots of a plant, such as the flowering parts, appear to be more vertical. The older portions of the plant, such as the crown and root, appear to be more horizontal. The young leaves of a growing tip point upward. The older leaves of a growing tip point to the sides. Thus the vertical forces represent the new forces related to the cosmos, and the horizontal forces are the older forces in relation to the Earth. Therefore, in the plant there is a constant alternation between the vertical and the horizontal forces. The life gesture of a plant is an image of the way the plant is bringing these two forces into balance.

In the valerian, even in one flower, we see that the vertical and the horizontal are quite polarized. The stronger those two polarities are in the plant the stronger the plant needs to be to hold them in balance. The stronger the polarities, the more dynamic the action in the plant is. The more they are neutralized, the more they move toward a vegetable. The more they get separated, the more they show that a kind of tension is active in the plant; this is where we find the pharmacology in herbs and medicinal plants.

The bulk of the book *Developing Biodynamic Agriculture* is about experiments performed between the time Steiner passed away and just before World War II. Researchers in Europe were trying to develop different kinds of wheat and grain, and they did experiments with hollow trees as vessels for making special composts. *Figure 7* shows the hollow trunk of a plum tree; it is a Saturn gesture. People would go into the woods or find old fruit trees in orchards in search of hollow trees, and they would make composts in them. Then they would use the compost in the trunk of the hollow trees to spray and soak seed for particular grains. The goal was to change the phenotype of grain. They were trying to get a rye that had many more berries on it but with very short awns. They were trying to see if they could actually use the forces of Saturn found in the hollow trunks to create special composts to make sprays and seed soaks that would affect the way plants receive forces from the periphery.

Figure 7

Unfortunately, during World War II, these people lived in a war zone and many of their fields were destroyed by passing armies, and a lot of their records were also destroyed. Nevertheless, enough was saved to form that book. It is interesting that they would use hollow stems to build their special compost because they were trying to create a vessel in which the hollow space was active. The hollow space in a tree is a sheath, like a horn or an organ. They were using a different kind of organ to bring the forces into focus. Here we see a picture of the forces we can call the Saturn gesture. Saturn is action on the periphery with a kind of hollow space in the middle.

Now consider the action of valerian in the human being, particularly the valerian root because it is also a picture of these Saturn forces. As we start to look into the pharmacology of valerian, some interesting pictures come up. When doctors or chemists analyze what is in a plant, they work with three groups. First we have alkaloids, which are found in the group of substances that an alchemist terms "salt." Among the salts; you have acids and bases. Both of those poles would be called *sal*. When you mix an acid and a base together, a salt drops out of the solution as a precipitate. Alkaloids are in the sal pole alchemically.

Then chemists will look for *terpenes,* which are substances associated with the formation of alcohol. To alchemists, alcohol has always been *mercury.* All sorts of oils, fats, waxes, and so on come from various kinds of what could be called the Mercury principle. Mercury is the process in the middle, between sal and sulf.

Then the third set of qualities that chemists look for when doing a plant analysis are the phenolics. These are substances that bear similarities to both acids and alcohols, but are like neither. These substances are the source of aromatic compounds such as oils, fatty acids, and the like—what alchemists would call sulfur, or *sulf.*

So we have three qualities, a *sal*, which is precipitate, an alkaloid, and a terpene, or derivatives of alcohols. Enzymes and all kinds of things can come out of that, and the phenols are pretty aromatic oils. Those are the three kinds of substances that a chemist would look for in a plant. Now it turns out that chemists have discovered over the years that these three general categories of substances are highly active in Valerian. Valerian as a medicine has been known for centuries, and each new wave of chemists comes up with a new angle on what makes valerian work therapeutically.

Each new wave of chemists has a different way of analyzing substances, so each new discovery points to a different substance as the active principle. Then the next wave says, "Well, actually, it is not that, it is actually this." Over time, going back to the seventeenth or eighteenth centuries, when people first starting looking at such substances analytically, everyone thought that the active ingredient in valerian was something different, and there was good evidence to show that the active ingredient was something different but no one could ever pin it down. They still cannot really pin down the active principle, because it is all of them together acting synergistically. First the acids were active, then it was the oils, and then it was something else. Today no one can say for sure what the active component is, so the valerian is a chameleon. Nevertheless, valeric acid is the thing today that people are interest in and studying. Valeric acid has a lot of essential oils in it, and it affects the central nervous system in a particular way. Boron is one of the metals that is very active in valerian. Most active ingredients (terpenes and phenols)

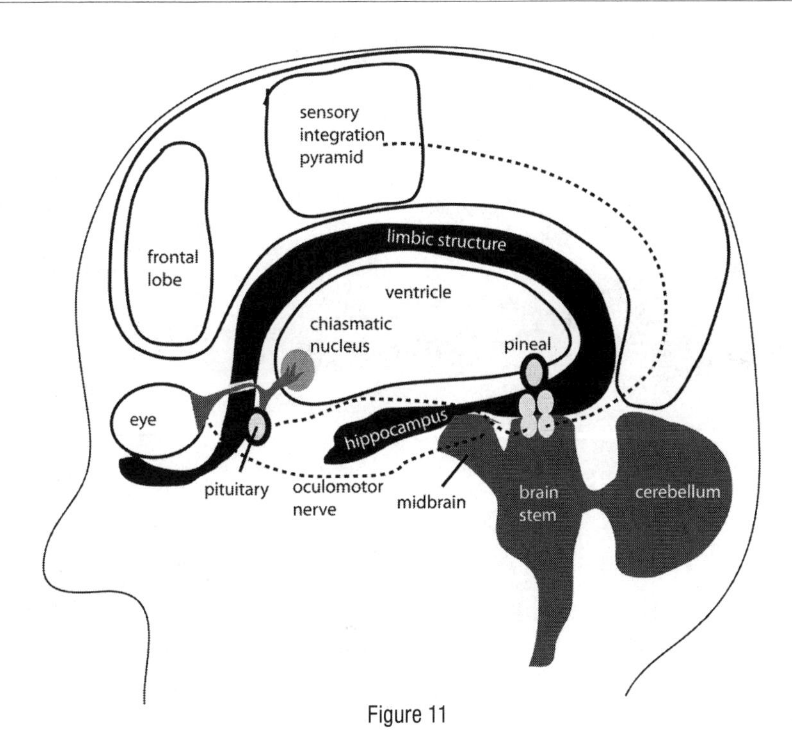

Figure 11

are antispasmodic and sedative. The object of much research and interest, valerianic acid is thought to be composed mainly of borneol esters (a component of many warming essential oils), bringing balance to the relationship between the acid and alcohols. The significant warming of the borneol is thought to be the source of the sedative and anti-spasmodic properties of the oil. It is considered to be a "stimulating sedative."

What is a "stimulating sedative"? This seems to fit the polarization picture of valerian very well. Valerian inhibits the destruction of GABA (inhibitory neurotransmitter in the central nervous system synapses). This releases GABA, which dampens the nervous transmissions in the synapses and depresses the central nervous system. When we get excited, a chemical is released in the synapses that creates an excitement that makes the synapses much more active, so there is much more transmission among the neurons. GABA inhibits too much activity, so the valeric acid supports the calming of the synapses. It is most effective in the suprachiasmatic nucleus.

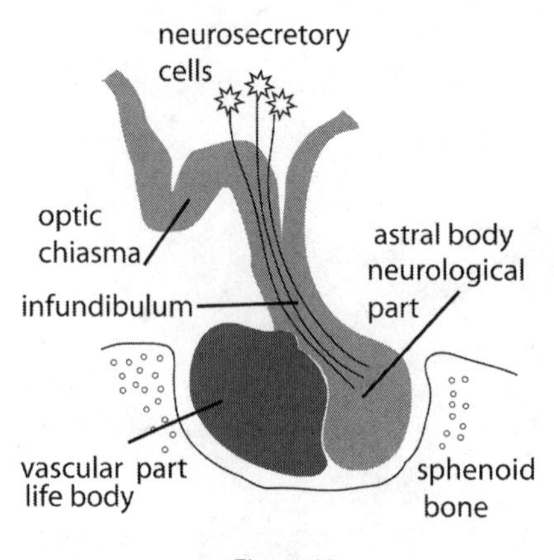

Figure 12

It acts especially on the sleeping and waking rhythms by affecting the suprachiasmatic nucleus (controls endogenous circadian rhythms). It is situated in the optic chiasma of the anterior hypothalamus where the optic nerves cross on the way back to the vision center of the brain. This organ acts chemically in response to light passing across the retina of the eye. The suprachiasmatic nucleus sends impulses to the pineal gland to regulate body temperature and the production of melatonin for sleep induction. Many neurological studies point to the action of the suprachiasmatic nucleus and endogenous diurnal patterns that are influenced by temperature.

Now we will ask, *what is the suprachiasmatic nucleus?* Of course, it is an important organ for us. In *figure 11,* we see the human eyeball. The optic nerve goes back to the suprachiasmatic nucleus in the very middle of the brain. If we enlarge that little pea-like thing labeled "pituitary" and look at it, we get the picture in *figure 12*.

The suprachiasmatic nucleus is where the transmission of light impulses that hit the retina in the eye is localized. With every sense impression in the eye, a nerve impulse is created that moves back through the optic nerves toward the optic centers in the back of the brain. On the way, the optic nerves cross, and the light transmission creates an impulse in the suprachiasmatic nucleus where they cross. It is like a relay.

In the diagram of the suprachiasmatic nucleus, we see little star-like, or flower-like, organs. They are nuclei that are stimulated by impulses of light from the optic nerve. Every time you look at something, those organs are activated. Every differentiation between light and dark and every movement and nuance is registering in those little organs. Nerve impulses are then transmitted down the nerve into the pituitary gland, which contains a vascular structure that creates a metabolic response to the light impulse. The pituitary gland secretes a hormone that stimulates the thyroid, which creates hormones to regulate reactions throughout your whole body. Those reactions stimulate further reactions that in turn influence the action of the pineal gland.

In *figure 11*, we see the pineal gland just to the right of the pituitary gland. The relationship between the pineal gland and the pituitary gland is that the pineal gland helps you sleep, while the pituitary gland helps you wake up. Valeric acid has a particular effect on the centers that help determine when we sleep and awake.

The pineal gland is actually a vestigial form of what used to be an eye, which is still present in rattlesnakes. Their pineal gland is hooked up to a heat receptor, with which they locate prey. We have a similar gland in our head, but instead of locating prey we use it to find our way back to heaven at night. We do so by looking for centers of warmth in the darkness—in other words, going to sleep. However, once we have spent enough time in darkness, we have to come back to other side to find light. Valeric acid has a particular action on the way light and warmth alternate in our consciousness.

The space between the pineal and pituitary is the bottom of the third ventricle of the brain. That space contains nothing but fluid. It is where, according to Steiner, we form the inner pictures that support our day consciousness. Through the activity of our inner pictures, we can assess the relationship between things we see with our senses and their archetypes in the spiritual world. This is how we recognize what we sense. We look at the flower and see that it is yellow. Internally, we recognize it as a dandelion. This is the bodily area where our experiences are processed. Those glands and organs allow me to unite the farthest reaches with my life here on Earth. They allow me to connect the periphery with the center, the cosmos and Earth. Therefore, in the realm of consciousness, the action of valeric acid on human consciousness is a picture of valerian's gesture as an herb; this is Saturn activity. Saturn is the door we go out through into sleep when melatonin is produced in the pineal gland, and it is the door through which we return in the morning when light hits the suprachiasmatic nucleus and moves us into a state of wakefulness as light impulses enter through the

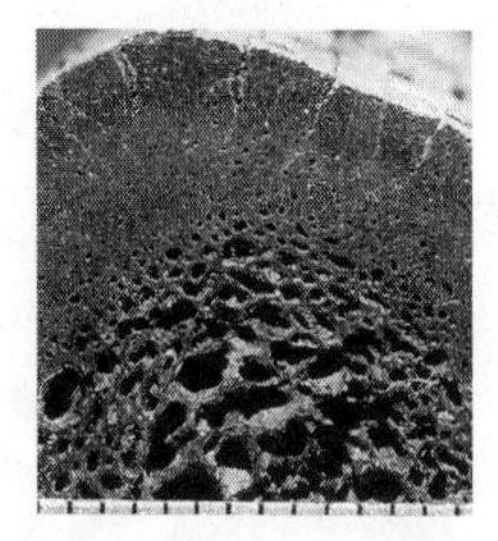
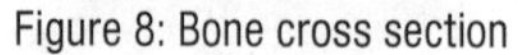

Figure 8: Bone cross section

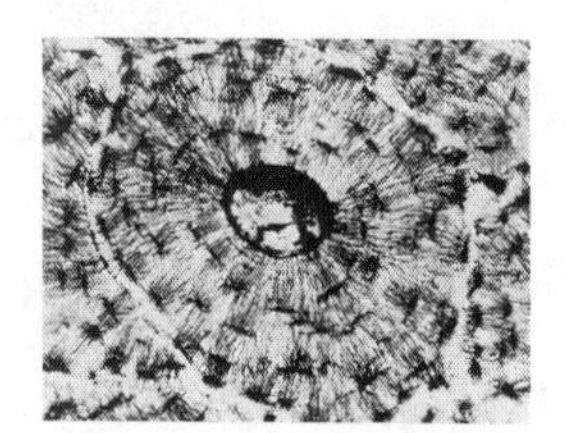

Figure 9: Bone cell

optic nerve. The alternation of these two important states of consciousness is specific to the action of valeric acid. This is a perfect analog for the gesture of the valerian plant.

Figure 8 and *figure 9* show the inner structure of a bone. The result of the Saturn influence is bone formation. If the warmth of Saturn were to move all the way out to the periphery, we would end up with ash. Ash is deposited in bone. The form that bone takes shows a very strong activity on the periphery of the bone; in the center is openness. *Figure 8* is a cross section of bone. It opens up as we go toward the center, and at the periphery we find a greater concentration. Compare this to the picture of the hollow log.

Figure 9 shows a single bone cell, where we see the same gesture. We get concentration on the periphery and open space in the center. The bones are Saturn. Bone is a saturnine organ. For example, we might take a cow femur, mix valerian juice with the marrow, and place that paste in a femur as a kind

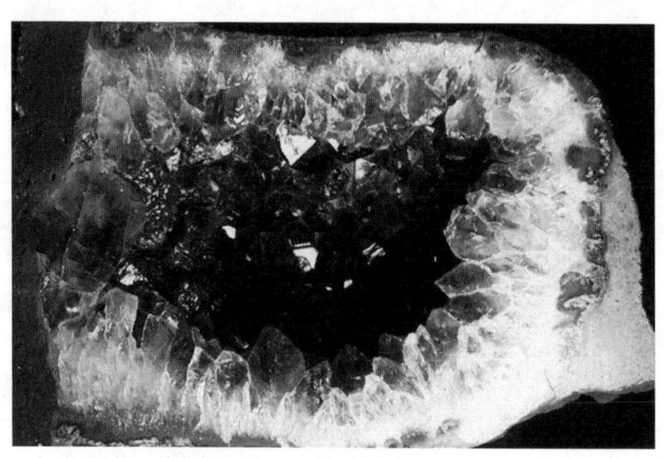

Figure 10: Amethyst geode

of an organ sheath for a biodynamic preparation. This is something to think about.

Figure 13 Cabbage

Figure 10 is a photo of an amethyst geode. We see here a gesture similar to the bone cross-section and the hollow tree. It shows a concentrating and crystallizing activity on the periphery of the geode, producing a bone-like process, and a form in the center where there appears to be nothing. In a typical quartz crystal, there is a molecular kind of growth from the periphery. Molecularly, this is how all crystals grow, adding peripheral layers of the crystal solution to a seed forming in the center. In quartz, the molecular layers lie around a central axis and rotate in one direction around that axis. With amethyst, the molecular growth of the crystal grows in two directions simultaneously. One molecular layer rotates right, the next left, alternating this way through the whole crystal. The result is that the finished crystal seldom shoots along the long axis, but tends to form along the horizontal axis and then is often present with adjacent crystals in the form of a hollow geode.

Years ago I became interested in amethyst as a possible alternative silica spray. In *figure 13* we see that Polish-looking Klocek guy holding a large cabbage. That was taken about twelve or thirteen years ago when I first started working with amethyst to make a spray for growing cabbages. I tried this because I have very heavy clay soil in the oven-like heat of the Central Valley in California. I cannot grow cabbages in the spring but only in the fall, because the clay soil heats up so slowly. However, in the Central Valley, it is often extremely hot in September and October. This is when I need to get them into the ground and off to a good start so that, by November when the light is shutting them down, they have grown to a good size. Because my soil is so slow in the spring, by the time the light shuts down in the fall, a cabbage would be only a one pounder, even in very good soil. Therefore, I looked for something that would give a gesture of wrapping the leaves around the center for late summer to give the cabbage an early boost of cabbage archetype.

I did not want to spray preparation 501 in September, because silica would turn my cabbage into beanstalks. I wanted something to bring a silica boost for light assimilation while keeping the light on the moist side. At the time, I did not know about the molecular aspects of amethyst, but I looked at a geode and something there said this is what you want. The gesture of the geode with the crystals on the periphery reminded me of a lettuce or cabbage with its leaves forming on the periphery. I ground up the amethyst and put it in the cow horn and sprayed it. The result was the cabbage in *figure 13*. That year the temperature hit 105 degrees in late September. The cabbage was in about its eighth leaf, and I thought, "Oh my god, this is the end of my cabbage." However, the cabbage never took a hit. When I cut into them, they were moist and sweet even though they never received any overhead watering. They were savoys, and the crinkled spaces inside the leaves were filled with beautiful rainbow-shining droplets of water. It was good soil, but to enhance the water I sprayed them with the amethyst spray when the moon moved through water every ten days. Over time, I have made that spray a regular part of my gardening practice.

Later, I was down in San Jose and read a book from the Rosicrucian library there. In it was an analysis of "gem ferments." I knew that Steiner, late in his life was very interested in gems. That fact made me decide to look into gems. The book from the Rosicrucian library said that the ferment for amethyst is red wine. So, I got some red wine and took my ground amethyst out of the cow horn, mixed it with the red wine, and starting spraying that. It seemed to have a little kick to it. I place the pulverized gem into the horn, and when I take it out, I put it into red wine and let it stew a bit, and then take the juice and use it very diluted in water as a spray.

I have been working on my garden soil for about eighteen years. I add all kinds of alfalfa and compost to it ever year. It was pottery clay when I moved in, but now it is really beautiful, dark loam and I know what to expect when I put in carrots or squash. I know that squashes grow one way, carrots another. I feel this helps me to assess my experiments because noticing irregularities is easier.

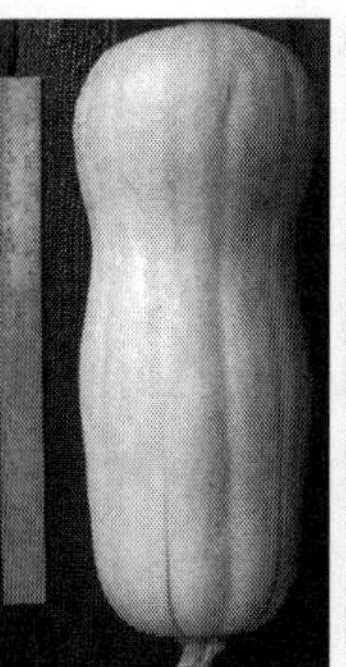
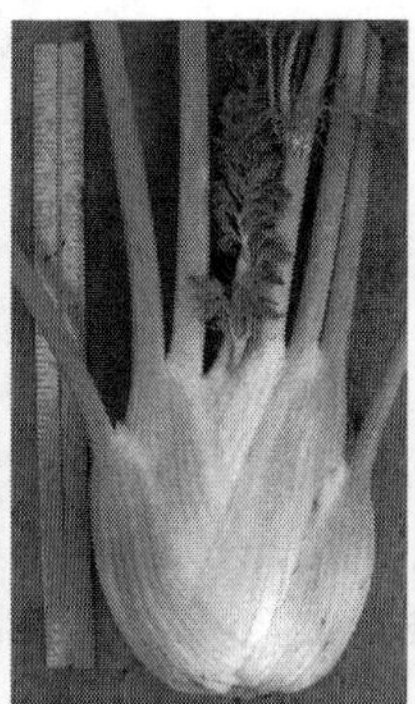

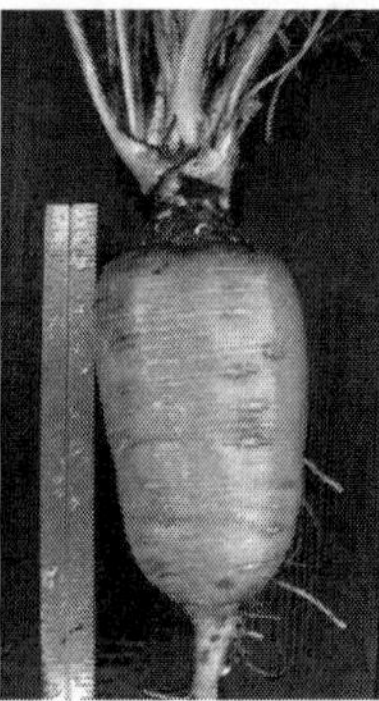
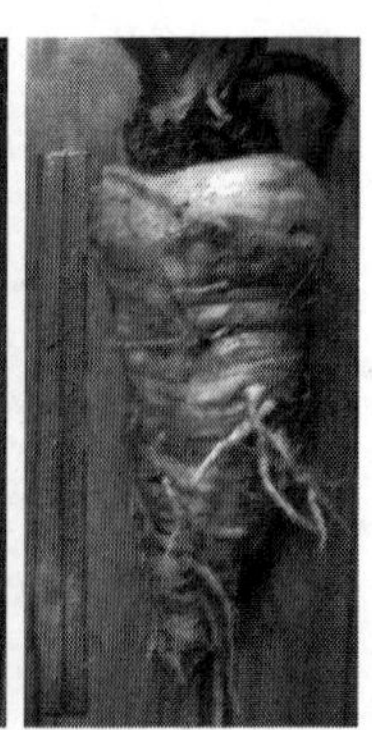

Left to right: Figure 14: Butternut squash; Figure 15: Fennel; Figure 16: Fennel section; Figure 17: Foot-long carrot; Figure 18: Foot-long parsnip

Figures 14 through *18* show produce that I planted and treated with the amethyst spray. Not all of the butternuts look like the one in figure 14, but there was a characteristic form in them. Instead of having a fat belly containing the seeds and a tapering fleshy part, the fleshy parts began to elongate in the plants sprayed with amethyst. Compare the picture of the squash fruit with the valerian flower. A similar elongation and separation of the two poles is present as a gesture. Not all of them were like that, but I typically select squash seeds for that quality. It is very nice to have a big fleshy neck on a butternut rather than a tapering small fleshy portion with a big seed cavity.

I sprayed the butternuts as an experiment and started to see that elongated type of fruit arising with more frequency each year. Then I thought, I will also spray my fennel, because a long fennel bulb would be great instead of a squat bulb with too many leaves. The little ruler next to the fennel is a foot long. *Figure 16* shows a cross section of a pretty nice fennel. If you look into the core, you can see very harmonious growth patterns. The hollow space is getting filled in with leaf after leaf of fennel, which is what I was looking for.

The carrot in *figure 17* has the same elongated center filled in with harmoniously structured tissue. That Danvers carrot is a foot long, but the shoulder and the tip of the root come out looking like a block. Danvers carrots typically taper sharply toward the tip. This is a carrot grown in clay soil.

Those who are gardeners know that you cannot fertilize a carrot to make it grow big. Put fertilizer into a carrot bed and you get something other than your carrot. Likewise for the parsnip in *figure 18;* it is a foot-long parsnip with three and a half- to four-inch crowns that have an elongated middle section of harmoniously formed tissue. These vegetables are sweet all the way through; there is no hypertrophy. The growth is balanced throughout.

I have included this because I think there are other things that can be done in the realm of preps, especially regarding the use of gems. I think tourmaline could be really a great prep somewhere down the road. In addition, we are working with wine diamonds—wine tartar—to see if we can get that to do something eventually. As a result of the amethyst experiments, I am adding valerian drops to the amethyst wine solution. It seems that amethyst as a gem has a mineral gesture of valerian. I use a half a teaspoon of the amethyst spray in a quart of water, and then add four drops or so of valerian into that. Then I agitate and stir as I would with a typical biodynamic preparation and spray that. I spray it in the evenings when there is frost in winter, and I spray it on the brassicas in the evenings in the fall.

This research is ongoing. It seems to be moving beyond the valerian, but there is a consciousness in valerian that I feel is part of how periphery and center can be united in a common space where new things can happen.

II. Questions and Answers

Question: Can I ask about wine and about what you think the wine is doing? Is it working somewhat like blood as a carrier?

I think the wine is very enzymatic. For the preps that do not have a sheath, substances such as wine or honey might be an interesting alternative. In the other preps, the sheaths provide a kind of organ-forming activity. There is a forming principle having to do with the action of an organism. I think the whole question of appropriate sheaths and what they contribute is a good subject for some really good research. There has been a lot of work with the substances in the preparations themselves, but the thing that intrigued me about the experiments with the hollow plum trees was that those people were looking for the sheath to contribute a specific something to the formation of grains.

If you talk to winemakers, it is easy to see them as alchemists. They understand how substance is an analog of what the plant went through that year. This is what wine appellations are about—the concept of terroir and the vintage properties. People can tell you what the climate was doing in the grape that year just by tasting it. There is something in perceived qualities that brings up the whole question of the sheath for quality enhancement, but these are largely unanswered questions. I pictured the wine in the spray as a kind of enzymatic sheath, or a kind of stomach of the grapevine. The stomach actually comes from the yeast that forms the bloom on the periphery of the grape.

Question: Do you think that there are other sheathes?

The quick answer is yes, but the issue is finding the right analog process related to the activity of the sheath. The important point is that, when the sheath was part of an organism it had a particular function, and that function is carried over into the substance as a synergy between the substance and the sheath. All of the phenolics that are valuable in a wine grape come from the skin. The winemaker's bank account is based on the skin of a grape—except for the white wines, then it may be the oak barrel. The skin is the sheath and the source of the tannins. Everything has its proper sheath that can be used for enhancement purposes.

Years ago, there was a book called *Common-sense Compost Making.*[3] The author was interested in the methods of biodynamics and was a vegan, so she did not like the idea of killing the cow. Therefore, she worked with honey and beeswax as sheaths. I got the idea that we could make bisque pottery, dip it in hot beeswax as a sheath, and mix minerals with honey. It seemed this kind of sheath might be good for barrel compost, for example. It is hard to get the little wooden nail kegs these days, so I try to use a sheath of clay. I get the most porous bisque pot I can find, paint it with beeswax, and

3 Maye E. Bruce, *Common-sense Compost Making by the Quick Return Method.*

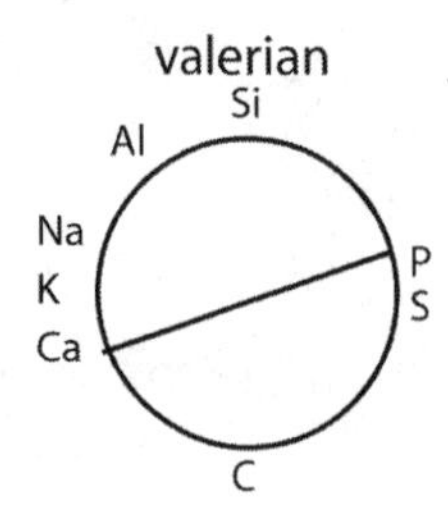

Figure 1: Valerian

then mix honey in with the manure and rock powders when I mix my barrel compost. I think it makes a much better mix. The mix that comes out resembles a piece of leavened bread. It has little air pockets and looks like a loaf.

We can experiment and play with sheaths using Maria Thun's recipe, but I think we can find more active sheaths. There are different ways of bringing those forces to bear in a slightly different way, so I look at the wine as a kind of sheath. Maybe we could get a goatskin, goat stomach, or something like the people of the past used—a goatskin as a ferment sheath to see if it would do something.

Comment: Since wine is a ferment, we could experiment with other ferments.

I think what we are doing in making preparations *is* a kind of fermentation. It enhances a substance in a particular way so that it morphs into something more potent than it was originally. This method of enhancement uses the principle of synergy. The skull of an animal has a particular relationship to calcium, silica, and phosphorus. A skull is calcium phosphate. This is exactly what the valerian chemistry is about. Look at figure 1. We see that the action is from calcium to phosphorus. This relationship links the most immobile calcium to the most mobile phosphorus.

Consider an herb such as comfrey. We see something very similar to valerian, though not quite as articulated. Comfrey is also known as *knitbone.* Bone is calcium phosphate, and the forces in comfrey knit calcium and phosphorus. The radiation in valerian is the dynamic in comfrey.

Question: How much honey are you putting into it?

Kind of like a chocolate chip cookie.

Question: How long are you fermenting the wine?

I make it in the spring, put in there, and I let it sit out until I use it; nothing special.

Question: I was wondering if you ever worked with or considered amber?

Yes, but amber is difficult to render because it is been fossilized. It requires pretty noxious substances to get it into solution—pretty strong hydrocarbons. Technically, the amethyst ferment is not actually a ferment. Chemists say, "Well, you are not actually fermenting the gem." This is true, technically, but grinding this gem to very fine parts produces a particle size that is highly interactive with a solution. Essentially, it is not a true ferment, but particles at this size are very reactive with solutions that are enzymatic and have an altered pH. Tourmaline, especially, is incredibly reactive with extremely small pH changes in the solution, which is why I think it is a very interesting gem to look at and to play with.

Question: Have you ever used an amethyst geode itself as a sheath?

I might if I could afford it. I get mine from a little store in Brazil when I visit there. If you have the patience and know what you are looking for, you can get some nice pieces for grinding.

Question: You mentioned the original type of corn plant and that you thought it might have been a couple of centuries for the plant to go through such changes. Then you commented that you had heard that the change took only a few generations. Ancient seed breeders were looking at the plant—the inner growth of the plant and specific times in that gesture where there is that space growing from one stage to the next. I wonder if there wasn't a certain sound that they could utter that created or caused that change. What do you think?

Possibly, but I think we could now add a dimension to that idea. That sound was given to them from the movement of the planets, from the periphery. Today, I would take this a step further by watching events in the planetary realms on the day when you grind the amethyst; watch what is happening on the day when you shake it; watch what is happening on the day you spray it; and try to work that way in time. The sounds that served the sacred works in the past really were the music of the spheres. If you pay attention to how you work in time, you will actually create a kind of sound potential in time. Your rhythm of working in time is actually a kind of melody. I think this is where Steiner would go. Maybe it could also be a kind of toning, but that would take a lot of research to corroborate.

Question: The role of the planets moving the time makes the tone. Is that what you are saying?

The movement of the planets in time *is* a tone, or a kind of singing. What we call "time" is actually a word like the *unfolding* of what could be called a melody. We can learn what that melody is by learning to read an ephemeris. If we start to replicate our actions rhythmically in time, the will that we use to do that would be the same kind of will that the people of ancient times used when they sang the seed. I think there is a way to do this, except we can do it in a more cognitive way. Personally, I think that would be a kind of contemporary way to avoid the pitfall of witchcraft, which is "I am going to sing you bad." That is the danger of this work. Even during Steiner's time, some people were a little nervous about the fact that he was burning things like animal skins at particular times of the year. It was nearing the edge of some things that were eventually used in a wrong way by some who were not nice people.

We know these things work. I think there is a cleaner way to do these kinds of things today, using our intellect instead of allowing the sound to come down in an unconscious or magical way. Later on, after a couple of evolutionary cycles, when we all meet again and have cleaned up our act a bit, we will probably be doing that. We will be singing the world again or doing eurythmy to make our gardens grow. For now, however, there is another interface that we have to learn, which challenges us about the moral aspect of what we are doing. We have to discover the nature of the harmony already being played by the cosmos. If we sing that, maybe we can sing some new things into being.

CITED WORKS AND SUGGESTED READING

Alexandersson, Olof. *Living Water: Viktor Schauberger and the Secrets of Natural Energy*. New York: MacMillan, 2002.

Barfield, Owen. *Poetic Diction: A Study in Meaning*. Oxford, UK: Barfield Press, 2010.

Bartholomew, Alick. *Hidden Nature: The Startling Insights of Viktor Schauberger*. Edinburgh: Floris Books, 2012.

Berrevoets, Erik. *Wisdom of the Bees: Principles for Biodynamic Beekeeping*. Great Barrington, MA: SteinerBooks, 2010.

Bruce, Maye E. *Common-sense Compost Making by the Quick Return Method*. London: Faber and Faber, 1946.

Cobbald, Jane. *Viktor Schauberger: A Life of Learning from Nature*. Edinburgh: Floris Books, 2007.

Cook, Wendy E. *The Biodynamic Food and Cookbook: Real Nutrition that Doesn't Cost the Earth*. London: Clairview Books, 2006.

———. *Foodwise: Understanding What We Eat and How It Affects Us: The Story of Human Nutrition*. London: Clairview Books, 2003.

Graves, Julia. *The Language of Plants: A Guide to the Doctrine of Signatures*. Great Barrington, MA: Lindisfarne Books, 2012.

Jenny, Hans. *Cymatics: A Study of Wave Phenomena and Vibration*. Newmarket, NH: Macromedia Press, 2001.

Joly, Nicholas. *Biodynamic Wine Demystified* (forwds. by M. Benziger & J. Greene). South San Francisco: Wine Appreciation Guild, 2008.

———. *Wine from Sky to Earth: Growing and Appreciating Biodynamic Wine*. Austin, TX: Acres U.S.A., 2005.

Keyserlingk, Count Adalbert. *The Birth of a New Agriculture: Koberwitz 1924 and the Introduction of Biodynamics*. London: Temple Lodge, 2008.

———. *Developing Biodynamic Agriculture: Reflections on Early Research*. London: Temple Lodge, 1999.

Klett, Manfred. *Principles of Biodynamic Spray and Compost Preparations*. Edinburgh: Floris Books, 2006.

Koepf, Herbert, *The Biodynamic Farm: Agriculture in Service of the Earth and Humanity*. Great Barrington, MA: SteinerBooks, 2006.

———. *Koepf's Practical Biodynamics: Soil, Compost, Sprays, and Food Quality*. Edinburgh: Floris Books, 2012.

Masson, Pierre. *A Biodynamic Manual: Practical Instructions for Farmers and Gardeners*. Edinburgh: Floris Books, 2011.

Osthaus, Karl-Ernst. *The Biodynamic Farm: Developing a Holistic Organism*. Edinburgh: Floris Books, 2011.

Pfeiffer, Ehrenfried. *Pfeiffer's Introduction to Biodynamics*. Edinburgh: Floris Books, 2011.

———. *Weeds and What They Tell Us*. Edinburgh: Floris Books, 2012.

Schwenk, Theodor. *Sensitive Chaos: The Creation of Flowing Forms in Water and Air*. London: Rudolf Steiner Press, 1965.

Schwenk, Theodor, and Wolfram Schwenk. *Water: The Element of Life*. Hudson, NY: Anthroposophic Press, 1989.

Steiner, Rudolf. *Agriculture Course: The Birth of the Biodynamic Method* (tr. G. Adams). London, Rudolf Steiner Press, 2005.

———. *Bees* (intro. by G. Hauk). Hudson, NY: Anthroposophic Press, 1998.

———. *Cosmic Memory: The Story of Atlantis, Lemuria, and the Division of the Sexes* (tr. K. E. Zimmer). Great Barrington, MA: SteinerBooks, 2006.

———. *Inner Reading and Inner Hearing: And How to Achieve Existence in the World of Ideas* (tr. M. Miller). Great Barrington, MA: SteinerBooks, 2008.

———. *Mystics after Modernism: Discovering the Seeds of a New Science in the Renaissance* (tr. K. E. Zimmer). Hudson, NY: Anthroposophic Press, 2000.

———. *An Outline of Esoteric Science* (tr. C. E. Creeger). Hudson, NY: Anthroposophic Press, 1997.

———. *Spiritual Foundations for the Renewal of Agriculture* (tr. C. E. Creeger and M. Gardner). Kimberton, PA: Bio-Dynamic Farming and Gardening Association, 1993.

———. *What Is Biodynamics? A Way to Heal and Revitalize the Earth* (7 selected lectures; intro. by H. Courtney). Great Barrington, MA: SteinerBooks, 2005.

Steiner, Rudolf, and Ita Wegman. *Extending Practical Medicine: Fundamental Principles Based on the Science of the Spirit*. London: Rudolf Steiner Press, 1996.

Thornton Smith, Richard. *Cosmos, Earth, and Nutrition: The Biodynamic Approach to Agriculture*. London: Rudolf Steiner Press, 2009.

Thun, Maria and Matthias K. Thun. *The North American Biodynamic Sowing and Planting Calendar*. Edinburgh: Floris Books, annual.

Wiencek, Henry. *Master of the Mountain: Thomas Jefferson and His Slaves*. New York: Farrar, Straus and Giroux, 2012.

Wildfeuer, Sherry (ed.). *Stella Natura: Biodynamic Planting Calendar: Planting Charts and Thought-Provoking Essays*. Kimberton, PA: Biodynamic Farming & Gardening Association, annual.

Wilkes, John. *Flowforms: The Rhythmic Power of Water*. Edinburgh: Floris Books, 2003.

Wright, Hilary. *Biodynamic Gardening: For Health and Taste*. Edinburgh: Floris Books, 2009.